Scattering and Diffraction in Physical Optics

Wiley Series in Pure and Applied Optics

The Wiley Series in Pure and Applied Optics publishes outstanding books in the field of optics. The nature of these books may be basic ("pure" optics) or practical ("applied" optics). The books are directed towards one or more of the following audiences: researchers in universities, government, or industrial laboratories; practitioners of optics in industry; or graduate-level courses in universities. The emphasis is on the quality of the book and its importance to the discipline of optics.

Scattering and Diffraction in Physical Optics

MANUEL NIETO-VESPERINAS
Instituto de Optica
National Research Council (C.S.I.C.)
Madrid, Spain

A Wiley-Interscience Publication

John Wiley & Sons, Inc.

New York / Chichester / Brisbane / Toronto / Singapore

Library of Congress Cataloging in Publication Data:

Nieto-Vesperinas, M., 1950–

Scattering and diffraction in physical optics / M. Nieto-Vesperinas.

p. cm.—(Wiley series in pure and applied optics, ISSN 0277-2493)

"A Wiley–Interscience publication."

Includes bibliographical references.

1. Light—Scattering. 2. Diffraction. 3. Optics, Physical. I. Title. II. Series.

QC427.4.N54 1991

535.2—dc20

ISBN 0-471-61529-3 91-8261

CIP

Printed in the United States of America

10 9 8 7 6 5 4 3 2 1

To Rosamary, Alberto, and Mario

Contents

Preface

In recent years there have been major advances in the theory of physical optics. This is particularly true for the models proposed for the propagation, scattering, and diffraction of wavefields and the analysis of their information content in connection with inverse optical problems.

This book is intended for graduate students and researchers in physical optics. The discussion progresses from the basic elements of the theory and their conceptual applications to topics of current research. End of chapter problems are intended to either reinforce the concepts presented or to illustrate them by introducing the reader to questions that arise in actual basic investigations. References are provided to aid those interested in a deeper study of the field or a particular topic. Although the primary emphasis is on optical theory, the basic elements of the theory often have more general application as, for example, to radio science or acoustics.

Chapter 1 introduces the extinction theorem and related integral formulas, which are widely applied in actual problems dealing with the scattering and propagation of wavefields, and will be used in much of the remaining text. A brief discussion of these concepts is followed by a classical tutorial paper of E. Wolf, which contains the details of the extinction theorem. This book considers only coherent fields. Those wishing information on partially coherent fields will find recent literature cited in the references.

Chapter 2 deals with the decomposition of wavefields into plane waves by means of the angular spectrum representation and also discusses the role of propagating and evanescent components in this formalism. The importance of the angular spectrum representation is manifested by the characterization of special wavefields, such as source-free fields or diffraction-free beams, and by the use made of this concept in the other chapters, both in the derivation of the resulting complex amplitudes after scattering or diffraction (direct optical problems) and in the interpretation of the structural information of object reconstructions by inversion procedures (inverse optical problems).

Chapter 3 addresses the radiation and scattering from volumes. The discussion here mainly concerns the representation of the radiated or scattered angular spectra in terms of their sources, either primary or induced by an incident wave, and the resulting expressions for the fields both inside and outside the medium. Also discussed are these results in the analysis of currently used models: the first Born, Rytov, and eikonal approximations. Multiple scattering methods in dense media are not considered here, as good reviews can be found in the references cited. The mathematical properties of fields due to localized sources and the consequences for their information content are examined in Chapter 4. Chapter 5 then discusses universal properties of the scattering process that do not depend on the details of the interaction. Among these are the relations of the generalized reflection and transmission coefficients which follow from the reciprocity and unitarity properties of the scattering matrix. They have important consequences in thin film theory and optical phase conjugation, among others.

In Chapter 6, the angular spectrum representation technique is applied to classical diffraction problems, and the connection of this decomposition with the Fresnel, Fraunhofer, and Debye approximations is examined; the attempts to obtain a consistent mathematical interpretation of the Kirchhoff diffraction integral by means of the concept of boundary diffraction wave are addressed and the relationship between the Kirchhoff approximation and the scalar and vector formulations is discussed.

The results and range of validity of both the Kirchhoff approximation and the method of small perturbations, scalar and vector, are further analyzed in Chapter 7, which considers scattering from rough surfaces and diffraction from reflection gratings. In addition, multiple scattering is discussed here, since recent progress on this topic has not been widely reported. Here the reader will find Monte Carlo procedures to address such effects as enhanced backscattering.

Chapters 8, 9, 10 consider further useful applications of the angular decomposition of wavefields into plane waves. In Chapter 8, reciprocity and unitarity concepts and relations are used for the understanding of the propagation and distortion correction of phase-conjugate fields. Chapter 9 discusses inverse diffraction of waves, the stability of the inversion procedures, and the information content of the results. Finally in Chapter 10, the questions of uniqueness of the inverse source and scattering problems, characterization of nonradiating sources, and tomographic procedures of data inversion and object reconstruction are reviewed.

It is a pleasure to express my appreciation to Professor E. Wolf from whom I have learned much about many of the topics covered here. I could also always count on his help, comments, and suggestions on preliminary drafts of this book. Further I wish to thank Professor J. C. Dainty with whom I have maintained a long scientific cooperation that has been reflected in several parts of this book, and who has made many important remarks on the manuscript. Professor R. P. Porter deserves acknowledgment for the numerous comments that he made, and, particularly, for the very important points he raised on a previous version of Chapter 10. I also express my recognition to Professor A. A. Maradudin for several things that I learned from him concerning rough surface scattering in penetrable media, which prevailed in the content of Chapter 7. Dr. G. Ross and Dr. M. A. Fiddy played active roles in my learning when I first studied physical optics. A collaboration with Professor N. García started my interest in rough surface scattering. The interaction with my former students has also influenced several parts of this volume; my association with them through scholarships from the Ministry of Education and research grants from the CICYT made it possible to obtain some original results that are now included; among them are Drs. J. M. Soto-Crespo and J. A. Sanchez-Gil (Chapter 7), Dr. G. Lera, who also made me familiar with the LATEX word-processor (Section 3.7), and Dr. M. J. Perez-Ilzarbe (Section 9.10). My thanks also to those individuals and institutions that granted permission to reproduce figures. Dr. W. H. Carter kindly sent me original photographs for Figs. 2.7–2.10.

I would also like to thank the staff of Keyword Publishing Services for the many editing improvements that they introduced to an imperfect manuscript; to Miss B. Shube, former editor of Wiley, who encouraged me so much in the early phases of this task, and to the current editor, Mr. G. J. Telecki, who has always provided me with guidance and suggestions.

Finally, I want to express my gratitude to my wife, Rosamary, and children, Alberto and Mario, for their resignation and patience during the many extra hours that this undertaking kept me away from them.

MANUEL NIETO-VESPERINAS

Madrid, Spain
May 1991

1

Basic Equations for Electromagnetic Fields

Maxwell's equations, and the integral theorems derived from them, with appropriate boundary conditions, constitute the basic laws underlining propagation and scattering of light and other electromagnetic waves.

In this chapter we briefly review Maxwell's equations, as well as the wave equations that govern the propagation, scattering, diffraction and radiation of electromagnetic fields in optics, and in general in electromagnetic theory. We shall also quote the conservation laws and discuss the arbitrariness in the definition of density of flow of energy. Further, we shall discuss certain fundamental theorems that constitute the basis of modern theories of wave propagation, such as the integral theorem of Helmholtz and Kirchhoff and the extinction theorem, together with the integro-differential equations obtained by means of the Hertz vectors.

1.1. MAXWELL'S EQUATIONS

Electromagnetic fields are characterized by their *electric vector* **E** and *magnetic vector* **H**. In material media, the response to the excitation produced by these fields is described by the *electric displacement* **D** and the *magnetic induction* **B**. These quantities satisfy Maxwell's equations, which in Gaussian units are:

$$\nabla \times \mathbf{H} - \frac{1}{c}\frac{\partial \mathbf{D}}{\partial t} = \frac{4\pi}{c}\mathbf{j}, \tag{1.1a}$$

$$\nabla \times \mathbf{E} + \frac{1}{c}\frac{\partial \mathbf{B}}{\partial t} = 0, \tag{1.1b}$$

$$\nabla \cdot \mathbf{D} = 4\pi\rho, \tag{1.1c}$$

$$\nabla \cdot \mathbf{B} = 0. \tag{1.1d}$$

where c is the velocity of light in vacuum, and $\mathbf{j}$ and ρ denote the *electric current density* and the *charge density*, respectively. Equations (1.1) imply the conservation of charge by means of the equation of continuity:

$$\nabla \cdot \mathbf{j} = -\frac{\partial \rho}{\partial t}.$$

The vectors **D**, **E**, **B**, and **H** are related by the expressions:

$$\mathbf{D} = \mathbf{E} + 4\pi \mathbf{P}, \tag{1.2a}$$

$$\mathbf{B} = \mathbf{H} + 4\pi \mathbf{M}. \tag{1.2b}$$

Where **P** and **M** are the *polarization* and *magnetization* vectors, respectively. They characterize the effect of the interaction of the electromagnetic field with the material. The connection of these quantities with the electric and magnetic vectors is given by the so-called *constitutive relations*, which in general are nonlinear equations; however, for most media, and when the fields are not too intense, they reduce to the linear relationships:

$$\mathbf{P} = \chi \mathbf{E}, \tag{1.3a}$$

$$\mathbf{M} = \eta \mathbf{H} \tag{1.3b}$$

χ and η being scalar quantities that are called *dielectric* and *magnetic susceptibility*, respectively.

From Eqs. (1.2) and (1.3) a direct relation between **D** and **E**, and between **B** and **H** is obtained, namely:

$$\mathbf{D} = \varepsilon \mathbf{E}, \tag{1.4a}$$

$$\mathbf{B} = \mu \mathbf{H}; \tag{1.4b}$$

where the quantity ε, called *dielectric permittivity*, is defined by:

$$\varepsilon = 1 + 4\pi\chi;$$

and the *magnetic permeability* μ is given by:

$$\mu = 1 + 4\pi\eta.$$

In addition to Eqs. (1.4), the electric current density $\mathbf{j}$ is related to the electric field by:

$$\mathbf{j} = \sigma_c \mathbf{E}, \tag{1.5}$$

σ_c being the *specific conductivity* of the material. In many problems encountered in optics and electromagnetic theory one has to solve Maxwell's equations in regions in which there exist interfaces separating different media. Across an interface the electromagnetic field satisfies certain *continuity* or *saltus conditions*; namely, let $\mathbf{n}$ be the local unit normal to the surface separating two media 1 and 2, and pointing from region 1 into region 2, these conditions are:

$$\mathbf{n} \times (\mathbf{E}_2 - \mathbf{E}_1) = 0, \tag{1.6a}$$

$$\mathbf{n} \times (\mathbf{H}_2 - \mathbf{H}_1) = \frac{4\pi}{c} \mathbf{J}, \tag{1.6b}$$

$$\mathbf{n} \cdot (\mathbf{D}_2 - \mathbf{D}_1) = 4\pi\sigma, \tag{1.6c}$$

$$\mathbf{n} \cdot (\mathbf{B}_2 - \mathbf{B}_1) = 0. \tag{1.6d}$$

Equations (1.6) express the continuity of the tangential component of **E** and of the normal component of **B** across the interface, as well as the discontinuity of both the tangential component of **H** and the normal component of **D**. In these equations the subscripts 1 and 2 are to be understood as the limiting values of the fields as the interface is approached from the medium 1 and from the medium 2, respectively. **J** and σ represent the electric current density and the charge density at the interface, respectively; they are always zero, except when one of the media is a perfect conductor.

1.2. WAVE EQUATIONS

By taking the curl in Eqs. (1.1a) and (1.1b) and eliminating **D** and **B** by means of Eqs. (1.2), one easily obtains the following wave equations that satisfy the time dependent fields $\mathbf{E}(\mathbf{r}, t)$ and $\mathbf{H}(\mathbf{r}, t)$:

$$\nabla \times \nabla \times \mathbf{E} + \frac{1}{c^2}\frac{\partial^2 \mathbf{E}}{\partial t^2} = -\frac{4\pi}{c}\frac{\partial}{\partial t}\left[\frac{1}{c}\mathbf{j} + \frac{1}{c}\frac{\partial \mathbf{P}}{\partial t} + \nabla \times \mathbf{M}\right], \tag{1.7a}$$

$$\nabla \times \nabla \times \mathbf{H} + \frac{1}{c^2}\frac{\partial^2 \mathbf{H}}{\partial t^2} = \frac{4\pi}{c}\left[\nabla \times \mathbf{j} + \nabla \times \frac{\partial \mathbf{P}}{\partial t} - \frac{1}{c}\frac{\partial^2 \mathbf{M}}{\partial t^2}\right]. \tag{1.7b}$$

For monochromatic fields all vector quantities appearing in Eqs. (1.7) have a time dependence factor $\exp(-i\omega t)$, ω being the *frequency*. It is straightforward to obtain from Eqs. (1.7) the following wave equations for the time independent parts in this case:

$$\nabla \times \nabla \times \mathbf{E}(\mathbf{r}) - k^2\mathbf{E}(\mathbf{r}) = \mathbf{F}_e(\mathbf{r}), \tag{1.8a}$$

$$\nabla \times \nabla \times \mathbf{H}(\mathbf{r}) - k^2\mathbf{H}(\mathbf{r}) = \mathbf{F}_m(\mathbf{r}); \tag{1.8b}$$

where we have defined the wavenumber k:

$$k = \frac{\omega}{c}. \tag{1.9}$$

Equations (1.8) are vector inhomogeneous Helmholtz equations. The right-hand sides $\mathbf{F}_e$ and $\mathbf{F}_m$ play the role of source terms characterizing the generation of electromagnetic waves. These terms are:

$$\mathbf{F}_e(\mathbf{r}) = 4\pi\left[\frac{ik}{c}\mathbf{j}(\mathbf{r}) + k^2\mathbf{P}(\mathbf{r}) + ik\nabla \times \mathbf{M}(\mathbf{r})\right], \tag{1.10a}$$

$$\mathbf{F}_m(\mathbf{r}) = 4\pi\left[\frac{1}{c}\nabla \times \mathbf{j}(\mathbf{r}) - ik\nabla \times \mathbf{P}(\mathbf{r}) + k^2\mathbf{M}(\mathbf{r})\right]. \tag{1.10b}$$

It should be emphasized that, although it is convenient to represent the fields by complex quantities for calculation purposes, only their real parts constitute the actual physical fields.

Often the time average of the product of two quantities is needed. It is straightforward to see that if they have harmonic time dependence, namely, if two quantities $A(\mathbf{r}, t)$ and $B(\mathbf{r}, t)$ can be expressed as:

$$A(\mathbf{r}, t) = A(\mathbf{r}) \exp(-i\omega t), \tag{1.11a}$$

$$B(\mathbf{r}, t) = B(\mathbf{r}) \exp(-i\omega t), \tag{1.11b}$$

then the time average $\overline{(\text{Re}\, A(\mathbf{r}, t))(\text{Re}\, B(\mathbf{r}, t))}$ is:

$$\overline{(\text{Re}\, A(\mathbf{r}, t))(\text{Re}\, B(\mathbf{r}, t))} = \tfrac{1}{2}\,\text{Re}[A(\mathbf{r})B^*(\mathbf{r})]. \tag{1.12}$$

(Re stands for "real part.")

In deriving the relation (1.12) we have used the following results: $\overline{\cos^2 \omega t} = \overline{\sin^2 \omega t} = \frac{1}{2}$; $\overline{\sin \omega t \cos \omega t} = 0$.

1.3. CONSERVATION LAWS

By recalling the vector identity:

$$\nabla\cdot(\mathbf{E}\times\mathbf{H}) = \mathbf{H}\cdot(\nabla\times\mathbf{E}) - \mathbf{E}\cdot(\nabla\times\mathbf{H}), \tag{1.13}$$

and making use of Eqs. (1.1a) and (1.1b), one easily obtains:

$$\nabla\cdot(\mathbf{E}\times\mathbf{H}) + \frac{1}{c}\mathbf{H}\cdot\frac{\partial\mathbf{B}}{\partial t} + \frac{1}{c}\mathbf{E}\cdot\frac{\partial\mathbf{D}}{\partial t} = -\frac{4\pi}{c}\mathbf{j}\cdot\mathbf{E}. \tag{1.14}$$

By integrating Eq. (1.14) in an arbitrary volume V and applying the Gauss theorem, we obtain:

$$\frac{c}{4\pi}\int_S (\mathbf{E}\times\mathbf{H})\cdot\mathbf{n}\,ds + \frac{1}{8\pi}\frac{\partial}{\partial t}\int_V (\mathbf{E}\cdot\mathbf{D} + \mathbf{B}\cdot\mathbf{H})\,dv = -\int_V \mathbf{j}\cdot\mathbf{E}\,dv, \tag{1.15}$$

where S is the surface enclosing the volume V and $\mathbf{n}$ is the unit outward normal.

Equation (1.15) is the relation that governs the balance of *electromagnetic energy*. The vector $\mathscr{S}$:

$$\mathscr{S} = \frac{c}{4\pi}(\mathbf{E}\times\mathbf{H}) \tag{1.16}$$

is the *Poynting vector* and represents the *density of flow of energy*. The quantity:

$$\mathscr{W} = \frac{1}{8\pi}(\mathbf{E}\cdot\mathbf{D} + \mathbf{B}\cdot\mathbf{H}) \tag{1.17}$$

is the *density of electromagnetic energy* inside the volume V. The *electric* and *magnetic energies* are given by the first and second terms of Eq. (1.17), respectively. On the other hand, the right-hand side of Eq. (1.15) constitutes the *dissipation of energy* due to Joule heating.

The density of flow of energy $\mathscr{S}$ in Eq. (1.15) can be, however, defined with certain freedom. It is evident from the first term of this equation that any other quantity obtained by adding to $\mathscr{S}$ the *curl* of an arbitrary vector, also satisfies Eq. (1.15) (this fact has been remarked in many textbooks, e.g. Stratton [1.1] or Born and Wolf [1.2]). This has given rise to alternative definitions of density of flow of energy by choosing appropriate Lagrangians (in this connection, see for instance Hines [1.3] or Wolf [1.4]). The observable quantity is however the *time average* flow of energy integrated over a finite region.

Due to the high frequency of optical signals, detection is always done by averaging the flow of energy over a time interval. According to what was said in Section 1.2, and to Eq. (1.12), the *time averaged* Poynting vector is:

$$\bar{\mathscr{S}}(\mathbf{r}) = \frac{c}{8\pi}\,\mathrm{Re}[\mathbf{E}(\mathbf{r}) \times \mathbf{H}^*(\mathbf{r})]. \tag{1.18}$$

It is customary to identify the modulus of the vector $\bar{\mathscr{S}}$ with the *optical intensity*.

Using the constitutive relations (1.4), the *time averaged* density of electromagnetic energy may be written as:

$$\bar{\mathscr{W}}(\mathbf{r}) = \frac{1}{16\pi}\,[\varepsilon\mathbf{E}(\mathbf{r})\cdot\mathbf{E}^*(\mathbf{r}) + \mu\mathbf{H}(\mathbf{r})\cdot\mathbf{H}^*(\mathbf{r})]. \tag{1.19}$$

1.4. SCALAR THEORY OF OPTICAL PROBLEMS

The ambiguity in the definition for the density of flow of energy is even deeper than the discussion of Section 1.3 reveals, and also exists for the density of electromagnetic energy. Green and Wolf [1.5] obtained alternative expressions for these quantities in terms of a single complex scalar function, leaving the vectors **E** and **H** unchanged.

Diffraction and scattering problems in optics and electromagnetics are often approached by means of a scalar wavefield, or *complex disturbance* $U(\mathbf{r}, t)$, that satisfies the scalar wave equation:

$$\nabla^2 U(\mathbf{r}, t) - \frac{1}{c^2}\frac{\partial^2}{\partial t^2}U(\mathbf{r}, t) = -4\pi\rho(\mathbf{r}, t), \tag{1.20}$$

$\rho(\mathbf{r}, t)$ representing a source distribution of the wavefield. For monochromatic fields, in a similar manner as discussed in Section 1.2, the spatial part of this wavefunction, $U(\mathbf{r})$, satisfies the inhomogeneous Helmholtz equation:

$$\nabla^2 U(\mathbf{r}) + k^2 U(\mathbf{r}) = -4\pi\rho(\mathbf{r}), \tag{1.21}$$

where: $\rho(\mathbf{r}, t) = \rho(\mathbf{r})\exp(-i\omega t)$.

However, the meaning and origin of the wavefunction U that approximates the actual electromagnetic field by a scalar quantity are not explained. In problems in which no depolarization takes place, like two-dimensional problems with the electric vector transversal to the plane in which propagation is considered, or when no multiple interactions occur in the scattering or diffraction process, it is possible to characterize the electromagnetic field by a scalar which corresponds to one transversal component of either the electric or the magnetic vector. However, this is not the only case in which a scalar theory has been employed; even then, it is not obvious why the usual definition of intensity:

$$I(\mathbf{r}) = |U(\mathbf{r})|^2$$

agrees with experiments, and hence, why it is in accordance with the time averaged Poynting vector defined in Eq. (1.18).

Green and Wolf [1.5] have pointed out the freedom in the definition of the density of flow of energy and density of electromagnetic energy, and the corresponding invariance of the observable quantities; namely, of the total flow and energy over finite regions. They introduced a definition for those densities in free space in terms of a single scalar complex wavefunction $U(\mathbf{r}, t)$ by expressions analogous to those for the probability current and the probability density in quantum mechanics:

$$\mathscr{S}(\mathbf{r}, t) = -\frac{1}{8\pi}\left(\frac{\partial U^*}{\partial t}\nabla U + \frac{\partial U}{\partial t}\nabla U^*\right), \tag{1.22a}$$

$$\mathscr{W}(\mathbf{r}, t) = \frac{1}{8\pi}\left(\frac{1}{c^2}\frac{\partial U^*}{\partial t}\frac{\partial U}{\partial t} + \nabla U \cdot \nabla U^*\right), \tag{1.22b}$$

where the function $U(\mathbf{r}, t)$ satisfies the homogeneous wave equation associated with Eq. (1.20) (namely, with $\rho(\mathbf{r}, t) = 0$). This theory was subsequently extended by Roman [1.6] to regions with sources. Other kinds of representations by scalar functions were also considered by Whittaker [1.7].

The definitions (1.22) were shown later by Wolf [1.8] to justify rigorously the use of a scalar theory under a wide variety of conditions encountered in optical systems, providing the wavefunction $U(\mathbf{r}, t)$ of Eqs. (1.22) was identified with the optical wavefield, and its spatial part $U(\mathbf{r})$ was used to define the optical intensity as $I(\mathbf{r}) = C|U(\mathbf{r})|^2$, with C being a constant, thus placing the scalar theory of optics in a formal basis. In this connection, an interesting discussion can be also found in Chapter 7 of the book by Marathay [1.9], where the use of the scalar theory is analyzed by illustrating the interpretation of the well-known Wiener's experiment [1.10]–[1.12] on standing waves formed by the interference between an incident and a reflected wave at a metallic mirror.

1.5. LORENTZ'S RECIPROCITY THEOREM

Let us consider monochromatic electromagnetic fields in an isotropic, time independent medium. From Eqs. (1.1a) and (1.1b) and the constitutive relations, one has for the spatial parts of the field vectors one solution that satisfies:

$$\nabla \times \mathbf{H}_1 + ik\varepsilon \mathbf{E}_1 = \frac{4\pi}{c} \sigma_c \mathbf{E}_1, \tag{1.23a}$$

$$\nabla \times \mathbf{E}_1 - ik\mu \mathbf{H}_1 = 0. \tag{1.23b}$$

Let us take the scalar product of (1.23a) and (1.23b) with the vectors $\mathbf{E}_2$ and $\mathbf{H}_2$ of another solution, respectively. Conversely, let us write equations like (1.23) for $\mathbf{E}_2$ and $\mathbf{H}_2$ and perform the scalar product with $\mathbf{E}_1$ and $\mathbf{H}_1$, respectively. Making use of the identity (1.13) one obtains in a straightforward manner:

$$\nabla \cdot (\mathbf{E}_1 \times \mathbf{H}_2 - \mathbf{E}_2 \times \mathbf{H}_1) = \mathbf{H}_2 \cdot (\nabla \times \mathbf{E}_1) + \mathbf{E}_2 \cdot (\nabla \times \mathbf{H}_1)$$

$$- [\mathbf{E}_1 \cdot (\nabla \times \mathbf{H}_2) + \mathbf{H}_1 \cdot (\nabla \times \mathbf{E}_2)], \tag{1.24}$$

which is easy to see, from the first pair of Maxwell's equations, becomes:

$$\nabla \cdot (\mathbf{E}_1 \times \mathbf{H}_2 - \mathbf{E}_2 \times \mathbf{H}_1) = 0. \tag{1.25}$$

On making use of the Gauss theorem, Eq. (1.25) becomes:

$$\int_S (\mathbf{E}_1 \times \mathbf{H}_2 - \mathbf{E}_2 \times \mathbf{H}_1) \cdot \mathbf{n} \, dS = 0, \tag{1.26}$$

S being a closed surface and $\mathbf{n}$ its unit outward local normal. Equation (1.26) constitutes *Lorentz's reciprocity theorem* and, as will be seen in Chapter 5, it forms the basis of interesting reciprocity relations in scattering theory.

1.6. INTEGRAL EQUATIONS FOR THE ELECTROMAGNETIC FIELD. THE EXTINCTION THEOREM

The solution to the wave equations with appropriate boundary conditions in optics and electromagnetics is, in many cases, successfully sought from integral theorems that are consequences of Maxwell's equations. These integral equations also contain physical insights about the phenomenon under study.

In this section we shall discuss the *extinction theorem* and other laws such as the *Helmholtz–Kirchhoff integral theorem.* The extinction theorem was originally investigated by Ewald [1.13] and Oseen [1.14] in the domain of molecular optics; an account in this context may be found in Chapter 2 of the book by Born and Wolf [1.2]. However, in recent years it has been shown that this theorem can be obtained in the framework of Maxwell's equations due to the studies of Sein [1.15], Pattanayak [1.16], Pattanayak and Wolf [1.16] and [1.18], de Goede and Mazur [1.19]. On the other hand, Waterman [1.20] derived independently a theorem from the wave equations, which is known as the *extended boundary condition.* Later, Agarwal [1.21] showed that both theorems are in fact two different forms of writing the same law.

The extinction theorem essentially describes the suppression inside a material medium of the field incident from vacuum, and its replacement by another field radiated by the dipoles induced in the medium. In addition, Pattanayak and Wolf [1.16]–[1.18] proved that this formula can be used as a nonlocal boundary condition in diffraction and scattering problems.

The Helmholtz–Kirchhoff integral theorem and related expressions form the basis of diffraction and scattering formulations. In this section we shall outline these integral theorems. We shall use the notation used by Wolf [1.17]. The details of the derivation can be found in Wolf's tutorial paper which is reproduced in Appendix 1.1 for convenience of the reader.

1.6.1. The Vector Form of Green's Theorem

We shall introduce the dyadic Green function $\mathscr{G}(\mathbf{r}, \mathbf{r}')$, which is defined by the expression:

$$\mathscr{G}(\mathbf{r}, \mathbf{r}') = \left(\mathscr{I} + \frac{1}{k^2}\nabla\nabla\right)G(\mathbf{r}, \mathbf{r}'), \tag{1.27}$$

$\mathscr{I}$ being the unit dyadic. $G(\mathbf{r}, \mathbf{r}')$ is the scalar Green function of the Helmholtz equation (1.21) represented by the outgoing spherical wave:

$$G(\mathbf{r}, \mathbf{r}') = \frac{\exp(ik|\mathbf{r} - \mathbf{r}'|)}{|\mathbf{r} - \mathbf{r}'|}, \tag{1.28}$$

which, as is well known (see for instance Chapter 7 of Ref. 1.22), satisfies the equation:

$$(\nabla^2 + k^2)G(\mathbf{r}, \mathbf{r}') = -4\pi\delta(\mathbf{r} - \mathbf{r}'), \tag{1.29}$$

δ being Dirac's delta function. $G(\mathbf{r}, \mathbf{r}')$ also fulfils *Sommerfeld's radiation condition*:

$$\lim_{r\to\infty} r\left(\frac{\partial G}{\partial r} - ikG\right) = 0, \tag{1.30}$$

where $r = |\mathbf{r}|$.

By means of Eqs. (1.27) and (1.29) it is straightforward to prove that $\mathscr{G}(\mathbf{r}, \mathbf{r}')$ satisfies:

$$\nabla \times \nabla \times \mathscr{G}(\mathbf{r}, \mathbf{r}') - k^2\mathscr{G}(\mathbf{r}, \mathbf{r}') = 4\pi\delta(\mathbf{r} - \mathbf{r}')\mathscr{I}; \tag{1.31}$$

$\mathscr{G}$ also holds the *vector form* of *Sommerfeld's radiation condition*:

$$\lim_{r\to\infty} r(\nabla \times \mathscr{G} - ik\hat{\mathbf{r}} \times \mathscr{G}) = 0, \tag{1.32}$$

$\hat{\mathbf{r}}$ being the unit vector in the direction of $\mathbf{r}$.

The vector form of *Green's theorem* reads (see e.g. Chapter 13 of Ref. 1.22):

Let $\mathbf{P}$ and $\mathbf{Q}$ be two vector functions of position that are well behaved, together with their first and second derivatives, then:

$$\int_V (\mathbf{Q}\cdot\nabla \times \nabla \times \mathbf{P} - \mathbf{P}\cdot\nabla \times \nabla \times \mathbf{Q})\, dv = \int_S (\mathbf{P} \times \nabla \times \mathbf{Q} - \mathbf{Q} \times \nabla \times \mathbf{P})\cdot\mathbf{n}\, dS \tag{1.33}$$

V being a certain volume, S denoting the surface enclosing it and $\mathbf{n}$ being the unit outward normal.

1.6.2. Integral Theorems

Let us consider a material medium occupying a volume V limited by a surface S, denoting by $\mathbf{r}_<$ and $\mathbf{r}_>$ the position vectors of a generic point inside and outside V, respectively.

Applying Eq. (1.33) to $\mathbf{P} = \mathscr{G}(\mathbf{r}, \mathbf{r}')\mathbf{C}$ ($\mathbf{C}$ being a constant vector) and to $\mathbf{Q} = \mathbf{E}(\mathbf{r})$, one obtains, depending on whether $\mathbf{r}$ and $\mathbf{r}'$ are inside or outside V:

$$\mathbf{E}(\mathbf{r}_<) = \frac{1}{4\pi}\int_V \mathbf{F}e(\mathbf{r}')\cdot\mathscr{G}(\mathbf{r}_<, \mathbf{r}')\, d^3r' - \frac{1}{4\pi}\sum\nolimits_e^{(-)}(\mathbf{r}_<), \tag{1.34}$$

when $\mathbf{r}$ and $\mathbf{r}'$ belong to V.

$$\mathbf{E}^{(i)}(\mathbf{r}_<) + \frac{1}{4\pi}\mathbf{S}_e = 0, \tag{1.35}$$

when $\mathbf{r}$ belongs to V and $\mathbf{r}'$ is outside V.

$$\mathbf{E}(\mathbf{r}_>) = \mathbf{E}^{(i)}(\mathbf{r}_>) + \frac{1}{4\pi}\mathbf{S}_e(\mathbf{r}_>), \tag{1.36}$$

when $\mathbf{r}$ and $\mathbf{r}'$ are outside V, and:

$$0 = \frac{1}{4\pi}\int_V \mathbf{F}_e(\mathbf{r}')\cdot\mathscr{G}(\mathbf{r}, \mathbf{r}')\, d^3r' - \frac{1}{4\pi}\sum\nolimits_e^{(-)}(\mathbf{r}_>), \tag{1.37}$$

when $\mathbf{r}$ is outside V and $\mathbf{r}'$ is inside V.

In Eqs. (1.35) and (1.36) $\mathbf{E}^{(i)}$ is the incident field upon the volume V, and $\mathbf{S}_e$ is given by:

$$\mathbf{S}_e(\mathbf{r}) = \int_{S^-}\left[\left(\mathbf{n}\times(\nabla\times\mathbf{E} - 4\pi ik\mathbf{M}) + \frac{4\pi ik}{c}\mathbf{J}\right)\cdot\mathscr{G}(\mathbf{r}, \mathbf{r}') + (\mathbf{n}\times\mathbf{E})\cdot\nabla\times\mathscr{G}(\mathbf{r}, \mathbf{r}')\right] dS; \tag{1.38}$$

and:

$$\sum\nolimits_e^{(-)}(\mathbf{r}) = \int_{S^-}[(\mathbf{n}\times\nabla\times\mathbf{E})\cdot\mathscr{G}(\mathbf{r}, \mathbf{r}') + (\mathbf{n}\times\mathbf{E})\cdot\nabla\times\mathscr{G}(\mathbf{r}, \mathbf{r}')]\, dS. \tag{1.39}$$

The symbol S^- in Eqs. (1.38) and (1.39) indicates that the limiting values of the integrands on S are taken by approaching the surface S from inside the volume V; both the surface element dS and the unit outward normal $\mathbf{n}$ depend on $\mathbf{r}'$.

Equation (1.34) expresses the field at an arbitrary point $\mathbf{r}_<$ of a volume V in terms of the limiting values of the tangential component of the electric and magnetic vectors at the boundary S, when sources are present in V and they are represented by the term $\mathbf{F}_e(\mathbf{r})$.

When $\mathbf{F}_e(\mathbf{r}) = 0$, Eq. (1.34) acquires one of the several equivalent forms of the vector formulation of the theorem of Helmholtz and Kirchhoff of diffraction theory, which constitutes the mathematical representation of Huygens' principle. We discuss further this theorem in Chapter 6. Detailed studies can be found in the works by Baker and Copson [1.23] and by Hönl *et al.* [1.24] (brief summaries are given in textbooks like Chapter 8 of Ref. 1.2).

Equation (1.35) is satisfied at every point $\mathbf{r}_<$ *inside* the volume V, and constitutes one of the forms of the *extinction theorem* describing the cancellation of the field $\mathbf{E}^{(i)}$, incident upon the volume V, by the field induced inside V which is given by the integral term $\mathbf{S}_e$. According to Pattanayak and Wolf [1.25], this theorem can be envisaged as a *nonlocal* boundary condition for the field at the surface S. The exterior field can then be calculated from Eq. (1.36).

Similar equations can be derived for the magnetic vector. They can be found in Appendix 1.1.

In addition, alternative forms of Eqs. (1.35) and (1.36) can be obtained (see Refs. 1.16 and 1.17), by means of the alternative expression for $\mathbf{S}_e(\mathbf{r})$ (see also Problem 1.2):

$$\mathbf{S}_e(\mathbf{r}) = \nabla \times \nabla \times \left[\frac{1}{k^2} \int_{S^-} \left(\mathbf{E} \frac{\partial G}{\partial \mathbf{n}} - G \frac{\partial \mathbf{E}}{\partial \mathbf{n}} \right) dS - \frac{4\pi}{k^2} \int_{S^-} (\mathbf{n}\nabla \cdot \mathbf{P}) G(\mathbf{r}, \mathbf{r}')\, dS - \frac{4\pi i}{k} \int_{S^-} (\mathbf{n} \times \mathbf{M}) G(\mathbf{r}, \mathbf{r}')\, dS \right]. \tag{1.40}$$

If the medium is nonmagnetic, nonconductive and spatially nondispersive, $\mathbf{S}_e$ reduces to the first integral of Eq. (1.40) and Eqs. (1.35) and (1.36) adopt simple familiar forms in terms of the boundary values of both $\mathbf{E}$ and G, and their normal derivatives, on the surface. Similarly, when the medium occupying the volume V is a perfect conductor, $\mathbf{S}_e$ becomes:

$$\mathbf{S}_e(\mathbf{r}) = \frac{4\pi i}{kc} \nabla \times \nabla \times \int_{S^-} \mathbf{J}(\mathbf{r}') G(\mathbf{r}, \mathbf{r}')\, dS; \tag{1.41}$$

and Eqs. (1.35) and (1.36) become equal to well known expressions used to solve scattering problems in electromagnetics; this will be discussed in Chapter 7.

Equations (1.34) and (1.37) can be expressed in terms of the *electric* and *magnetic Hertz vectors*, $\mathbf{\Pi}_e$ and $\mathbf{\Pi}_m$ [1.18], which are defined as:

$$\mathbf{\Pi}_e(\mathbf{r}) = \int_V \left[\mathbf{P}(\mathbf{r}') + \frac{i}{kc}\mathbf{j}(\mathbf{r}')\right] G(\mathbf{r}, \mathbf{r}')\, d^3r' + \frac{i}{kc}\int_{S^-} \mathbf{J}(\mathbf{r}')G(\mathbf{r}, \mathbf{r}')\, dS, \quad (1.42)$$

and:

$$\mathbf{\Pi}_m(\mathbf{r}) = \int_V \mathbf{M}(\mathbf{r}')G(\mathbf{r}, \mathbf{r}')\, d^3r'. \quad (1.43)$$

From Eq. (1.34) one obtains:

$$\mathbf{E}(\mathbf{r}_<) = \mathbf{E}^{(i)}(\mathbf{r}_<) + \nabla \times \nabla \times \mathbf{\Pi}_e(\mathbf{r}_<) + ik\nabla \times \mathbf{\Pi}_m(\mathbf{r}_<) - 4\pi\mathbf{P}(\mathbf{r}_<) - \frac{4\pi i}{kc}\mathbf{j}(\mathbf{r}_<), \quad (1.44)$$

and a similar equation for the magnetic field:

$$\mathbf{H}(\mathbf{r}_<) = \mathbf{H}^{(i)}(\mathbf{r}_<) + \nabla \times \nabla \times \mathbf{\Pi}_m(\mathbf{r}_<) - ik\nabla \times \mathbf{\Pi}_e(\mathbf{r}_<) - 4\pi\mathbf{M}(\mathbf{r}_<). \quad (1.45)$$

On the other hand, from Eq. (1.37) the following expression is derived:

$$\mathbf{E}(\mathbf{r}_>) = \mathbf{E}^{(i)}(\mathbf{r}_>) + \nabla \times \nabla \times \mathbf{\Pi}_e(\mathbf{r}_>) + ik\nabla \times \mathbf{\Pi}_m(\mathbf{r}_>), \quad (1.46)$$

with the corresponding expression for the magnetic vector:

$$\mathbf{H}(\mathbf{r}_>) = \mathbf{H}^{(i)}(\mathbf{r}_>) + \nabla \times \nabla \times \mathbf{\Pi}_m(\mathbf{r}_>) - ik\nabla \times \mathbf{\Pi}_m(\mathbf{r}_>). \quad (1.47)$$

Equations (1.44)–(1.47) are integro-differential equations for the electric and magnetic vectors of application in radiation, scattering, and diffraction problems.

1.6.3. Integral Theorems for Scalar Fields

The integral theorems discussed in Section 1.6.2 can be also established for scalar waves (Pattanayak and Wolf, [1.25]). As far as light fields are concerned, they correspond to the classical way in which a scalar description of light is employed, as discussed in Section 1.4, even if one imposes somewhat idealized continuity conditions for the wavefields across the surface separating two different media. This excludes the case of a perfect conductor. Such continuity conditions are considered to be:

$$U_2 = U_1; \tag{1.48a}$$

$$\frac{\partial U_2}{\partial \mathbf{n}} - \frac{\partial U_1}{\partial \mathbf{n}} = 0, \tag{1.48b}$$

Where the subindices 1 and 2 denote each of the two media separated by the surface.

More details on the scalar theory, to be discussed next, can be found in Appendix 1.1.

Let the spatial part of the field wavefunction satisfy the inhomogeneous Helmholtz equation, Eq. (1.21), with the source density $\rho(\mathbf{r})$ being given by:

$$\rho(\mathbf{r}) = -\frac{1}{4\pi} F(\mathbf{r})U(\mathbf{r}), \tag{1.49}$$

where the *scattering potential* $F(\mathbf{r})$ is defined by:

$$F(\mathbf{r}) = \begin{cases} -k^2[n^2(\mathbf{r}) - 1], & \text{if } \mathbf{r} \text{ belongs to } V, \\ 0, & \text{if } \mathbf{r} \text{ is outside } V, \end{cases} \tag{1.50}$$

$n(\mathbf{r})$ being the refractive index of the material occupying the volume V: $n = \sqrt{\varepsilon\mu}$. Equation (1.49) will be discussed later in Section 3.10.

Let a plane wave $U^{(i)}(\mathbf{r}) = \exp(ik\mathbf{n}_0 \cdot \mathbf{r})$, with propagation direction given by the unit vector $\mathbf{n}_0$, be incident from vacuum upon the medium occupying the volume V, then, by using *Green's theorem* applied to $U(\mathbf{r})$ and $G(\mathbf{r}, \mathbf{r}')$, one obtains, in a similar manner to that in Section 1.6.2, for points either inside and outside V:

$$U^{(i)}(\mathbf{r}_<) + \frac{1}{4\pi} \sum (\mathbf{r}_<) = 0, \tag{1.51}$$

for $\mathbf{r}$ in V and $\mathbf{r}'$ outside V.

$$U(\mathbf{r}_>) = U^{(i)}(\mathbf{r}_>) + \frac{1}{4\pi} \sum (\mathbf{r}_>), \tag{1.52}$$

for $\mathbf{r}$ and $\mathbf{r}'$ outside V.

$$U(\mathbf{r}_<) = U^{(i)}(\mathbf{r}_<) - \frac{1}{4\pi}\int_V U(\mathbf{r}')F(\mathbf{r}')G(\mathbf{r}_<, \mathbf{r}')\, d^3r', \tag{1.53}$$

for $\mathbf{r}$ and $\mathbf{r}'$ inside V.

$$U(\mathbf{r}_>) = U^{(i)}(\mathbf{r}_>) - \frac{1}{4\pi}\int_V U(\mathbf{r}')F(\mathbf{r}')G(\mathbf{r}_>, \mathbf{r}')\, d^3r', \tag{1.54}$$

for $\mathbf{r}$ outside V and $\mathbf{r}'$ in V.

$\sum(\mathbf{r})$ denotes the surface integral:

$$\sum(\mathbf{r}) = \int_S \left(U\frac{\partial G}{\partial \mathbf{n}} - G\frac{\partial U}{\partial \mathbf{n}} \right) dS. \tag{1.55}$$

Equation (1.51) constitutes the *scalar* form of the *extinction theorem*. It was derived in the context of quantum mechanics by Lax [1.26]. Once the field on the surface S is obtained from Eq. (1.51), the exterior field can be calculated from Eq. (1.52). On the other hand, Eqs. (1.53) and (1.54) are well known in potential scattering, giving the wavefield, both inside and outside V, in terms of the scattering potential.

Equations (1.51) and (1.53) also imply the equation:

$$U(\mathbf{r}_<) = -\frac{1}{4\pi}\int_V U(\mathbf{r}')F(\mathbf{r}')G(\mathbf{r}, \mathbf{r}')\, d^3r' - \frac{1}{4\pi}\sum(\mathbf{r}_<), \tag{1.56}$$

Equation (1.56) expresses the wavefield $U(\mathbf{r})$ inside a volume V in terms of its boundary value at the surface S that limits V.

When $F(\mathbf{r}) = 0$, namely, when the volume V is source-free, Eq. (1.56) becomes identical to the scalar form of the *Helmholtz–Kirchhoff integral theorem* of the scalar theory of diffraction.

1.6.4. Natural Modes

The natural modes of the wave equations are well behaved outgoing solutions that cannot be generated by any incident field. These modes may be either *radiative* (complex eigenfrequencies ω), or *nonradiative* (real eigenfrequencies ω), or *surface* or *bulk* modes. Wolf [1.18] wrote the equations for these modes in terms of the expressions discussed in Sections 1.6.2 and 1.6.3. According to this analysis, since these modes are not generated by any incident field, one has, from Eqs. (1.35) and (1.36):

$$\mathbf{S}_e(\mathbf{r}_<) = 0, \tag{1.57}$$

$$\mathbf{E}(\mathbf{r}_>) = \frac{1}{4\pi}\,\mathbf{S}_e(\mathbf{r}_>). \tag{1.58}$$

There are similar equations for the magnetic vector.

Likewise, in the scalar case, the equation for the modes obtained from Eq. (1.51), is:

$$\sum(\mathbf{r}_<) = 0. \tag{1.59}$$

Equations (1.57)–(1.59) define the set of eigenvalues k (or equivalently, of eigenfrequencies ω) of the modes that constitute solutions to these equations.

Equation (1.59) has been applied by Pattanayak and Wolf [1.25] to a central potential in order to characterize the nonradiative modes (*bound states*); this provided an alternative method to the one previously used by Humblet and Rosenfeld [1.27] for *resonance states* of the Schrödinger equation, of which the dispersion formula for nuclear reactions of Kapur and Peierls [1.28] is a particular case. Further references in connection with some problems of condensed matter physics can be found in Ref. 1.25. Also Agarwal [1.29] has derived the dispersion relation for *surface polaritons* and *Brewster modes* by applying Eqs. (1.57) and (1.58) to a plane interface separating two media.

1.6.5. Other Extensions and Uses of the Extinction Theorem

The extinction theorem has been generalized to spatially dispersive media by Agarwal *et al.* [1.30]; to bounded media by Birman and Sein [1.31], and to bounded gyrotropic media by Puri and Birman [1.32]. In Chapters 6 and 7 we discuss its applications to diffraction problems and scattering from rough surfaces. A mapping in the complex plane for two-dimensional scalar wavefields has been shown [1.33] to lead to the extinction theorem, as well as to the Helmholtz–Kirchhoff theorem, when the Cauchy integral is applied to a certain class of generalized analytic functions.

PROBLEMS

1.1. Prove that the dyadic Green function $\mathscr{G}(\mathbf{r}, \mathbf{r}')$ can be written as:

$$\mathscr{G}(\mathbf{r}, \mathbf{r}') = \frac{1}{k^2} \nabla \times \nabla \times G(\mathbf{r}, \mathbf{r}')\mathscr{I} - 4\pi\delta(\mathbf{r} - \mathbf{r}')\mathscr{I};$$

$G(\mathbf{r}, \mathbf{r}')$ being the scalar Green function and $\mathscr{I}$ denoting the unit dyadic.

1.2. Using the result of Problem 1.1, and the vector Green theorem for two vectors $\mathbf{P}$ and $\mathbf{Q}$ (cf. e.g. Ref. 1.22, Eq. (13.1.7)):

$$\int_V [\mathbf{P} \cdot \nabla^2 \mathbf{Q} - \mathbf{Q} \cdot \nabla^2 \mathbf{P}]\, d^3r = \int_S [(\mathbf{P} \nabla \cdot \mathbf{Q} - \mathbf{Q} \nabla \cdot \mathbf{P}) \cdot \mathbf{n}$$
$$- (\mathbf{P} \cdot (\mathbf{n} \times (\nabla \times \mathbf{Q})) + \nabla \times \mathbf{P} \cdot (\mathbf{n} \times \mathbf{Q}))]\, dS.$$

Prove that Eq. (1.38) can be expressed in the form given by Eq. (1.40).

1.3. By using the result of Problem 1.1 and the vector Green theorem prove that $\sum_e^- (\mathbf{r}_<)$ given by Eq. (1.39) may be written as:

$$\textstyle\sum_e^- (\mathbf{r}_<) = \displaystyle\int_S [(\mathbf{n} \times (\nabla \times \mathbf{E}(\mathbf{r}')))G(\mathbf{r}_<, \mathbf{r}') + (\mathbf{n} \times \mathbf{E}(\mathbf{r}')) \times \nabla G(\mathbf{r}_<, \mathbf{r}')$$
$$+ (\mathbf{n} \cdot \mathbf{E}(\mathbf{r}')) \nabla G(\mathbf{r}_<, \mathbf{r}')]\, dS.$$

This expression of $\sum_e^- (\mathbf{r})$, when substituted into Eq. (1.34), yields the field at a point $\mathbf{r}_<$, inside the volume V, in terms of the boundary values on a surface S. When $\mathbf{F}_e(\mathbf{r}) = 0$ the resulting equation is known as the Stratton–Chu formula (see Section 4.15 of Ref. 1.1 or Section 9.9 of Ref. 1.34).

1.4. Prove that Eq. (1.34) may be expressed in the form of Eq. (1.44) with the Hertz vectors $\mathbf{\Pi}_e(\mathbf{r})$ and $\mathbf{\Pi}_m(\mathbf{r})$ given by Eqs. (1.42) and (1.43), respectively.

1.5. Prove that the Hertz vectors $\mathbf{\Pi}_e(\mathbf{r})$ and $\mathbf{\Pi}_m(\mathbf{r})$ satisfy the Helmholtz equations:

$$(\nabla^2 + k^2)\mathbf{\Pi}_e(\mathbf{r}) = -4\pi \mathbf{P}(\mathbf{r}),$$
$$(\nabla^2 + k^2)\mathbf{\Pi}_m(\mathbf{r}) = -4\pi \mathbf{M}(\mathbf{r}).$$

REFERENCES

1.1. J. A. Stratton, *Electromagnetic Theory*, McGraw-Hill, New York, 1941.

1.2. M. Born and E. Wolf, *Principles of Optics*, Pergamon Press, Oxford, 1980, 6th edition.

1.3. C. O. Hines, *Can. J. Phys.* **30**, 123 (1952).

1.4. E. Wolf, *I.R.E. Trans.* **AP-3**, 228 (1955).

1.5. H. S. Green and E. Wolf, *Proc. Phys. Soc.* **A 66**, 1129 (1953).

1.6. P. Roman, *Acta Phys. Hung.* **4**, 209 (1955).

1.7. E. T. Whittaker, *Proc. Lond. Math. Soc.* **1**, 367 (1904).

1.8. E. Wolf, *Proc. Phys. Soc.* **74**, 269 (1959).

1.9. A. S. Marathay, *Elements of Optical Coherence Theory*, J. Wiley, New York, 1982.

1.10. O. Wiener, *Ann. Phys.* **40**, 203 (1890).

1.11. R. S. Longhurst, *Geometrical and Physical Optics*, J. Wiley, New York, 1967, 2nd edition.

1.12. F. A. Jenkins and H. E. White, *Fundamentals of Optics*, McGraw-Hill, New York, 1976.

1.13. P. P. Ewald, Dissertation, University of Munich (1912); *Ann Phys.* **49**, 1 (1916).

1.14. C. W. Oseen, *Ann. Phys.* **48**, 1 (1915).

1.15. J. J. Sein, Ph.D. Thesis, New York University (1969); *Opt. Comm.* **2**, 170 (1970); *Opt. Comm.* **14**, 157 (1975).

1.16. D. N. Pattanayak, Ph.D. Thesis, University of Rochester (1973); D. N. Pattanayak and E. Wolf, *Opt. Comm.* **6**, 217 (1972).

1.17. E. Wolf, in *Coherence and Quantum Optics*, L. Mandel and E. Wolf, eds., Plenum Press, New York, 1973, p. 339.

1.18. E. Wolf, *Symposia Mathematica* **18**, 33 (1976).

1.19. J. de Goede and P. Mazur, *Physica* **58**, 568 (1972).

1.20. P. C. Waterman, *Phys. Rev. D* **3**, 825 (1971); *Alta Frequenza* **38** (speciale), 348 (1969); *Proc. I.E.E.E.* **53**, 805 (1965).

1.21. G. S. Agarwal, *Phys. Rev. D* **14**, 1168 (1976).

1.22. P. M. Morse and H. Feshbach, *Methods of Theoretical Physics*, McGraw-Hill, New York, 1953.

1.23. B. B. Baker and E. T. Copson, *The Mathematical Theory of Huygens Principle*, Clarendon Press, Oxford, 1950, 2nd edition.

1.24. H. Hönl, A. W. Maue and K. Westpfahl, in *Handbuch der Physik*, S. Flügge, ed., Vol. XXVI/1, Springer-Verlag, Berlin, 1961, p. 218.

1.25. D. N. Pattanayak and E. Wolf, *Phys. Rev. D* **13**, 913 (1976); *Phys. Rev. D* **13**, 2287 (1976).

1.26. M. Lax, *Phys. Rev.* **85**, 646 (1952).

1.27. J. Humblet and L. Rosenfeld, *Nucl. Phys.* **26**, 529 (1961).

1.28. P. L. Kapur and R. Peierls, *Proc. Roy. Soc. A* **166**, 277 (1938).

1.29. G. S. Agarwal, *Phys. Rev. B* **8**, 4768 (1973).

1.30. G. S. Agarwal, D. N. Pattanayak and E. Wolf, *Opt. Comm.* **4**, 255 (1971); *Phys. Rev.* **10**, 1447 (1974); *Phys. Rev.* **11**, 1342 (1975).

1.31. J. L. Birman and J. J. Sein, *Phys. Rev. B* **6**, 2482 (1972).

1.32. A. Puri and J. L. Birman, *Opt. Comm.* **37**, 81 (1981).

1.33. M. Nieto-Vesperinas, *J. Math. Phys.* **25**, 1592 (1984).

1.34. J. D. Jackson, *Classical Electrodynamics*, J. Wiley, New York, 1965.

Appendix 1.1

In this appendix, the paper by E. Wolf entitled "A Generalized Extinction Theorem and its Role in Scattering Theory" is reproduced. This paper was published in: *Coherence and Quantum Optics*, pages 339–357, L. Mandel and E. Wolf, eds., Plenum Press, New York, 1973. Permission to reproduce the paper was granted by Prof. E. Wolf and Plenum Publishing Corporation.

A GENERALIZED EXTINCTION THEOREM AND ITS ROLE IN SCATTERING THEORY*

Emil Wolf

University of Rochester, Rochester, N.Y.

When an electromagnetic wave is incident on a homogeneous medium with a sharp boundary, it is extinguished inside the medium in the process of interaction and is replaced by a wave propagated in the medium with a velocity different from that of the incident wave. A classic theorem of molecular optics due to P.P. Ewald (1912) and C.W. Oseen (1915) expresses the extinction of the incident wave in terms of an integral relation, that involves the induced field on the boundary of the medium. Various generalizations of this theorem have recently been proposed and it was also shown that the customary physical interpretation of the theorem is incorrect.

In this paper results of a recent investigation carried out in collaboration with D.N. Pattanayak are presented, which provide a generalization of the extinction theorem to any medium. Like the recent generalization due to J.J. Sein our derivation is based entirely on Maxwell's theory and not on molecular optics. A hypothesis is put forward as to the true physical significance of the extinction theorem and it is shown how the theorem may be used to solve scattering problems in a novel way. An analogous extinction theorem for non-relativistic quantum mechanics is also presented.

***

One of the most poorly understood theorems of classical electrodynamics is undoubtedly the so-called *extinction theorem* first formulated by P.P. Ewald [1] in 1912 in his basic investigations on the foundations of crystal optics and later by C.W. Oseen [2] in 1915 in his studies of dispersion of light in material media.

* Research supported by the Air Force Office of Scientific Research and the Army Research Office (Durham).

Let me first say a few words about the usual formulation of the theorem.

Suppose that a plane electromagnetic wave is incident from vacuo on a material medium with a sharp boundary. The medium will for the moment be assumed to be of the simplest kind - a linear, homogeneous, isotropic, non-magnetic dielectric.

We know that under the influence of the incident electromagnetic field another field will be generated inside the dielectric, which will have a different wave number and hence a different phase velocity. We may, therefore, say that inside the medium, the incident wave, propagated with the vacuum velocity of light c, is somehow *extinguished* by the interaction with the medium and is replaced by a new wave propagated with the velocity c/n, where n is the refractive index of the medium. The question then arises: how does the extinguishing of the incident wave come about? The Ewald-Oseen theorem provides an answer to this question. In mathematical terms the theorem may be expressed in the form: [3]

$$\underline{E}^{(i)}(\underline{r}_<) + \frac{1}{4\pi k^2} \nabla\times\nabla\times \int_S \left\{\underline{E}(\underline{r}') \frac{\partial}{\partial n} G_0(\underline{r}_<, \underline{r}') - G_0(\underline{r}_<, \underline{r}') \frac{\partial}{\partial n} \underline{E}(\underline{r}')\right\} dS = 0, \qquad (1)$$

valid at every point $\underline{r}_<$ inside the volume V bounded by a surface S (see Fig. 1) occupied by the medium. Here $\underline{E}^{(i)}$ and $\underline{E}$ represent the Fourier transforms (for frequency components $\omega = kc$, c being the vacuum velocity of light) of the incident electric field and of the total electric field generated inside the medium respectively (and taken in the integral in (1) in the limit as the surface is approached from inside V),

$$G_0(\underline{r}, \underline{r}') = \exp\{ik|\underline{r} - \underline{r}'|\}/|\underline{r} - \underline{r}'| \qquad (2)$$

is the outgoing free-space Green's function of the Helmholtz equation and $\partial/\partial n$ denotes differentiation along the outward normal to the boundary surface.

The relation (1), which is essentially in the form as formulated by Oseen, was originally derived not from the macroscopic Maxwell theory, but rather from molecular optics, which is a microscopic theory. In this later theory the response of the medium to the incident field is expressed in terms of elementary dipole fields, generated by the interaction of the incident wave with the individual molecules of the medium. We note that in (1), the second term formally cancels the incident electric field at every point inside the medium. Since the second term involves the values of the total field $\underline{E}$ on the boundary surface S only it has been generally asserted that Eq. (1) implies that the incident field is extinguished entirely by those molecular dipoles that are situated on the boundary S of the dielectric. This is the original formulation of the Ewald-Oseen extinction theorem.

In the last few years the Ewald-Oseen extinction theorem has attracted a good deal of attention and various modifications and generalizations of it for more complicated media have been proposed and applied to numerous problems of current research interest. Here is a partial listing of the relevant publications, indicating the authors, year of publication and topics. The complete references are given in footnote 5.

(a)	A. Wierzbicki	(1961,1962)	Quadrupole radiation, reflection, refraction
(b)	N. Bloembergen and P.S. Pershan	(1962)	Non-linear optics
(c)	B.A. Sotskii	(1963)	Metals, optically active media
(d)	R.K. Bullough	(1968)	Many-body optics
(e)	J.J. Sein	(1969,1970)	Spatial dispersion, excitons
(f)	É. Lalor	(1969)	New formulation
(g)	É. Lalor and E.Wolf	(1971)	Interaction of charged particle with a dielectric
(h)	T. Suzuki (1971)		Diffraction
(i)	G.S. Agarwal, D.Pattanayak and E. Wolf	(1971)	Spatial Dispersion
(j)	É. Lalor and E. Wolf	(1972)	Refraction and reflection
(k)	J.R. Birman and J.J. Sein	(1972)	Polaritons in bounded media

It is thus clear that the extinction theorem is playing an increasingly greater role in widely different areas. Of the numerous investigations those of J.J. Sein [4,5e] are of particular relevance to the subject matter of this talk. Sein showed that the extinction theorem which, as I already noted, was derived originally from molecular optics may also be derived from Maxwell's theory and he also showed that the traditional interpretation of the theorem is incorrect [4].

In this talk I will present results of a recent investigation that I carried out in collaboration with D. Pattanayak [6], (see also [7]), in which we have attempted to answer the following two questions:

(1) Can the extinction theorem be generalized within the framework of Maxwell's theory to a medium of any kind, i.e. with arbitrary response?

(2) What is the true meaning of the theorem?

The first question has been partially answered already by Sein, but we will present quite a general and rigorous answer to it. Let me add that Sein's recognition that the extinction theorem follows also from Maxwell's theory represents an important contribution, since attempts to generalize it within the framework of molecular optics encounters formidable difficulties because of the local field corrections (associated with the Lorentz internal field).

As regards the second question - namely what is the true meaning of the theorem - we will put forward a hypothesis, supported by a few explicit solutions that we obtained with the help of the theorem.

We will also show that the extinction theorem has a strict analogue in potential scattering in non-relativistic quantum mechanics.

The full derivation of our main results is rather lengthy and I will only indicate the main steps.

Let us then consider the scattering of monochromatic electromagnetic wave incident from vacuo on a body with a sharp boundary. We assume the body to be of arbitrary kind; its response could be, for example, non-linear or non-local as in the case of spatial dispersion.

From Maxwell's equations for monochromatic fields, on eliminating, the electric displacement vector $\underline{D}$ and the magnetic induction vector $\underline{B}$ via the relations

$$\underline{D} = \underline{E} + 4\pi\,\underline{P}, \qquad \underline{B} = \underline{H} + 4\pi\,\underline{M}, \tag{3}$$

where $\underline{P}$ and $\underline{M}$ denote the polarization and magnetization vectors respectively, we obtain the four equations

$$\hat{L}\,\underline{E} = \underline{F}_e, \tag{4a}$$

$$\hat{L}\,\underline{H} = \underline{F}_h, \tag{4b}$$

$$\nabla.\underline{E} = 4\pi(\rho - \nabla.\underline{P}) \tag{4c}$$

$$\nabla.\underline{H} = -4\pi\ \nabla.\underline{M}, \tag{4d}$$

where $\hat{L}$ is the operator

$$\hat{L} = -k^2 + \nabla x \nabla x \tag{5}$$

and the source terms $\underline{F}_e$ and $\underline{F}_h$ are given by

$$\underline{F}_e = 4\pi[\frac{ik}{c}\underline{j} + k^2\underline{P} + ik\ \nabla x\ \underline{M}]\quad , \tag{6a}$$

$$\underline{F}_h = 4\pi[\frac{1}{c}\nabla x\underline{j} - ik\ \nabla x\ \underline{P} + k^2\underline{M}]\ . \tag{6b}$$

The vectors $\underline{E}$, $\underline{H}$, $\underline{j}$, $\underline{P}$ and $\underline{M}$ and the scalar ρ are, of course, functions of position ($\underline{r}$) and are taken at a fixed frequency ω.

The sources of the incident field are assumed to be in the domain $\tilde{V}$ exterior to V (see Fig. 1) and we take them to be at a finite distance [8] from V. It is clear then, since the exterior $\tilde{V}$ of V is vacuo, that

$$\underline{F}_e(\underline{r}) = 4\pi[\frac{ik}{c}\underline{j}_c + k^2\underline{P} + ik\ \nabla x\ \underline{M}\ ] \qquad \text{if } \underline{r}\ \varepsilon\ V \tag{7a}$$

$$= \frac{4\pi ik}{c}\underline{j}_{ext} \qquad \text{if } \underline{r}\ \varepsilon\ \tilde{V}\ , \tag{7b}$$

$$\underline{F}_h(\underline{r}) = 4\pi[\frac{1}{c}\nabla x\ \underline{j}_c - ik\ \nabla x\ \underline{P} + k^2\ \underline{M}\ ] \qquad \text{if } \underline{r}\ \varepsilon\ V \tag{8a}$$

$$= \frac{4\pi}{c}\ \nabla x\ \underline{j}_{ext} \qquad \text{if } \underline{r}\ \varepsilon\ \tilde{V}\ . \tag{8b}$$

In Eqs. (7) and (8), $\underline{j}_c$ denotes the conduction current density and $\underline{j}_{ext}$ denotes the external current density (representing the source). The corresponding charge densities are, of course, related to the current densities by the continuity equation. Further the polarization vector $\underline{P}$ and the magnetization vector $\underline{M}$ in (7a) and (8a) are assumed to be given functions of the electromagnetic field vectors $\underline{E}$ and $\underline{H}$, whose exact form depends on the nature of the medium. It is to be noted that because of this fact, the equations (4a) and (4b) are in general *coupled* to each other.

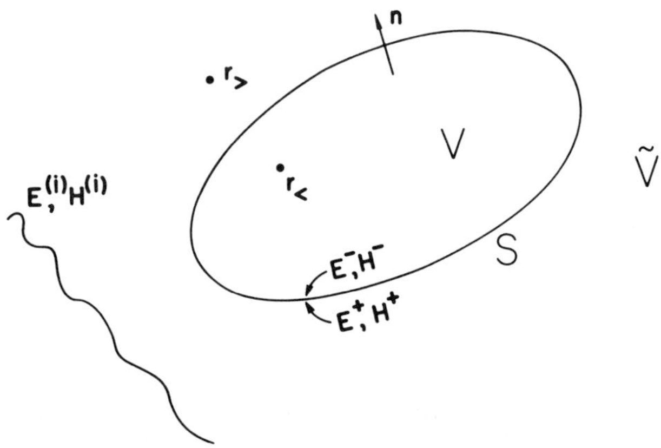

Fig. 1 Notation relating to scattering of an electromagnetic wave on a material medium.

To complete the formulation we must specify the behavior of the fields at the boundary. As is well known Maxwell's equations imply that across the boundary

$$\underline{n} \times (\underline{E}^{+} - \underline{E}^{-}) = 0, \qquad \underline{n} \times (\underline{H}^{+} - \underline{H}^{-}) = \frac{4\pi}{c} \underline{K} \quad , \tag{9}$$

where the superscripts plus and minus denote limiting values as the boundary surface S is approached from outside and inside of the medium respectively (see Figure 1), $\underline{n}$ is the unit outward normal to the boundary surface S and $\underline{K}$ represents the surface current density which will be non-zero only for a perfect conductor. Although these conditions are generally referred to as *boundary conditions*, they are, actually jump conditions - or *saltus* conditions as they are called in the older literature. A clear appreciation of the difference between true boundary conditions and the saltus conditions is, as we shall see later, at the heart of the proper interpretation of the extinction theorem.

For later purpose we also note that the corresponding problem in non-relativistic quantum mechanics, namely the scattering from a finite step potential is mathematically much simpler since in place of several coupled equations involving the vector fields

$\underline{E}$ and $\underline{H}$, the quantum mechanical problem involves only a single equation for the Schrödinger scalar wave function $\psi(\underline{r})$. Moreover, since ψ and its normal $\partial\psi/\partial n$ must, according to basic postulates of quantum mechanics be *continuous* across a finite potential step we now have in place of the conditions (9) the conditions

$$\psi^{+} - \psi^{-} = 0, \qquad \left(\frac{\partial\psi}{\partial n}\right)^{+} - \left(\frac{\partial\psi}{\partial n}\right)^{-} = 0. \tag{10}$$

These are *continuity conditions* and not true boundary conditions either.

Returning to the electromagnetic problem we introduce a dyadic Green's function associated with the $\hat{L}$-operator,

$$\hat{L}\,\underline{\underline{G}} = 4\pi\delta(\underline{r}-\underline{r}')\underline{\underline{U}}, \tag{11}$$

($\underline{\underline{U}}$ = unit dyadic), which obeys the vectorial form of the Sommerfeld radiation condition at infinity:

$$\lim_{r\to\infty} r\,[\nabla\times\underline{\underline{G}} - ik\,\hat{\underline{r}}\times\underline{\underline{G}}] = 0, \tag{12}$$

where $\hat{\underline{r}}$ is the unit vector in the direction of $\underline{r}$ and $r = |\underline{r}|$. It is known that [9]

$$\underline{\underline{G}}(\underline{r},\underline{r}') = (\underline{\underline{U}} + \frac{1}{k^2}\nabla\nabla)\,G_o(\underline{r},\underline{r}') , \tag{13a}$$

where G_o is the outgoing free-space Green's function of the Helmholtz equation, viz.

$$G_o(\underline{r},\underline{r}') = \frac{e^{ik|\underline{r}-\underline{r}'|}}{|\underline{r}-\underline{r}'|} . \tag{13b}$$

Now our equation involving the electric field is of the form

$$\hat{L}\,\underline{E} = \underline{F}_e , \tag{14}$$

with a similar equation involving $\underline{H}$. From (11) and (14) one obtains, if one also uses the vectorial form of Green's theorem, the following identity valid for integration through any domain V' bounded by a closed surface S':

$$\int_{V'} E(\underline{r}')\delta(\underline{r}-\underline{r}')d^3\underline{r}' = \frac{1}{4\pi}\int_{V'} \underline{F}_e(\underline{r}')\cdot \underline{\underline{G}}(\underline{r},\underline{r}')d^3\underline{r}' - \frac{1}{4\pi}\sum_e(\underline{r}), \tag{15}$$

where

$$\sum_e(\underline{r}) = \int_{S'} \{[\underline{n}\times \nabla\times\underline{E}(\underline{r}')]\cdot\underline{\underline{G}}(\underline{r},\underline{r}')+[\underline{n}\times \underline{E}(\underline{r}')]\cdot \nabla\times\underline{\underline{G}}\}dS' \tag{16}$$

and n is the unit normal to S' pointing outward from the volume V'.

Let us now take the volume V' to coincide either with the scattering volume V or with the exterior $\tilde{V}$ of it. Also we can take the field point $\underline{r}$ to be either in V or in $\tilde{V}$. Applying the theorem (15) separately to each of these four cases we obtain the following four relations:

(a) $\underline{r}\varepsilon V$, $\underline{r}'\varepsilon V$:

$$\underline{E}(\underline{r}_<) = \frac{1}{4\pi}\int_V \underline{F}_e\cdot\underline{\underline{G}}\, d^3\underline{r}' - \frac{1}{4\pi}\sum_e^{(-)}(\underline{r}_<), \tag{17a}$$

(b) $\underline{r}\varepsilon V$, $\underline{r}'\varepsilon\tilde{V}$:

$$0 = \frac{ik}{c}\int_{\tilde{V}} \underline{j}_{ext}\cdot \underline{\underline{G}}\, d^3\underline{r}' + \frac{1}{4\pi}\cdot\sum_e^{(+)}(\underline{r}_<), \tag{17b}$$

(c) $\underline{r}\varepsilon\tilde{V}$, $\underline{r}'\varepsilon\tilde{V}$:

$$\underline{E}(r_>) = \frac{ik}{c}\int_{\tilde{V}} \underline{j}_{ext}\cdot \underline{\underline{G}}\, d^3\underline{r}' + \frac{1}{4\pi}\sum_e^{(+)}(r_>), \tag{17c}$$

(d) $\underline{r}\varepsilon\tilde{V}$, $\underline{r}'\varepsilon V$:

$$0 = \frac{1}{4\pi}\int_V \underline{F}_e\cdot\underline{\underline{G}}\, d^3\underline{r}' - \frac{1}{4\pi}\sum_e^{(-)}(\underline{r}_>). \tag{17d}$$

Here

$$\sum_e^{(+)} = \int_{S\pm} \{[\mathrm{n}\times \nabla\times\underline{E}]\cdot\underline{\underline{G}} + [\underline{n}\times\underline{E}]\cdot\nabla\times\underline{\underline{G}}\}dS, \tag{18}$$

and the upper or lower signs are taken on $\Sigma_e^{\pm}$ and on $S^{\pm}$ according as the limiting values are taken from outside (S^+) or inside (S^-) of the scattering volume. In deriving (17b) and (17c) we also used the radiation condition (12) which ensures that there is no contribution from a sphere of infinitely large radius (the outer boundary of the volume V).

Now the integral containing the external current, taken over the exterior $\tilde{V}$ of our scattering volume has a clear physical meaning. From the significance of $\underline{\underline{G}}$ as the outgoing free-space Green's function of the $\tilde{L}$-operator, this integral must evidently represent the unperturbed incident field, i.e.

$$\frac{ik}{c}\int_{\tilde{V}} \underline{j}_{ext}(\underline{r}')\cdot\underline{\underline{G}}(\underline{r},\underline{r}')d^3\underline{r}' = \underline{E}^{(i)}(\underline{r}), \tag{19}$$

irrespective whether the point $\underline{r}$ is situated inside or outside the scattering volume. Hence the equations (17b) and (17c) may be expressed in the compact form

$$\underline{E}^{(i)}(\underline{r}_<) + \frac{1}{4\pi}\sum_e^{(+)}(\underline{r}_<) = 0, \tag{20a}$$

$$\underline{E}(\underline{r}_>) = \underline{E}^{(i)}(\underline{r}_>) + \frac{1}{4\pi}\sum_e^{(+)}(\underline{r}_>) . \tag{20b}$$

We note that (20a) has some resemblance to the Ewald-Oseen extinction theorem, since it expressed the cancellation of the incident field at every point $\underline{r}$ inside the scattering medium V in terms of an integral involving the field on the boundary of the medium only. However because $\Sigma_e^{(+)}$ rather than $\Sigma_e^{(-)}$ appears in this surface integral, the integral involves the limiting values of the field taken from the outside, rather than from the inside of the scattering volume; but one can easily transform (20a) and also(20b) so as to involve the limiting values from the inside, since from the definition (18) of $\Sigma_e^{(-)}$ and $\Sigma_e^{(+)}$ and from the saltus conditions (9) one easily finds that

$$\frac{1}{4\pi}\left[\sum_e^{(+)}(\underline{r}) - \sum_e^{(-)}(\underline{r})\right] = -ik\int_{S^-}\left(\underline{n}\times\underline{M} - \frac{1}{c}\underline{K}\right)\cdot\underline{\underline{G}}ds \tag{21}$$

Using this result in (20a) and (20b) and the expressions for $\Sigma_e^{(+)}$ one then obtains the following two relations:

$$\underline{E}^{(i)}(\underline{r}_<) + \frac{1}{4\pi}\underline{S}_e(\underline{r}_<) = 0, \tag{22}$$

$$\underline{E}(\underline{r}_>) = \underline{E}^{(i)}(\underline{r}_>) + \frac{1}{4\pi}\underline{S}_e(\underline{r}_>), \tag{23}$$

where

$$\underline{S}_e(\underline{r}) = \int_{S^-} \{[\underline{n}\times(\nabla\times\underline{E} - 4\pi ik\underline{M}) + \frac{4\pi ik}{c}\underline{K}].\underline{\underline{G}}(\underline{r}.\underline{r}') + [\underline{n}\times\underline{E}].\nabla\times\underline{\underline{G}}(\underline{r},\underline{r}')\}dS. \tag{24}$$

The relation (22) must be satisfied at each point $\underline{r}_<$ inside the scattering volume V. It is one form of our *generalized* Ewald-Oseen theorem, valid for scattering by *any* medium, irrespective of the nature of the constitutive relations. I will indicate shortly how it reduces to the usual form of the Ewald-Oseen theorem when the medium is of the simplest kind. But first I want to say a little about what I believe is the true meaning of the theorem and also discuss briefly the significance of the complementary relation (23). For this purpose we also must note there is an analogous set of relations to (22) and (23), involving the *magnetic* rather than the electric field. They can be derived in a similar way and are

$$\underline{H}^{(i)}(\underline{r}_<) + \frac{1}{4\pi}\underline{S}_h(\underline{r}_<) = 0, \tag{25}$$

$$\underline{H}(\underline{r}_>) = \underline{H}^{(i)}(\underline{r}_>) + \frac{1}{4\pi}\underline{S}_h(\underline{r}_>), \tag{26}$$

where

$$\underline{S}_h(r) = \int_{S^-} \{[\underline{n}\times(\nabla\times\underline{H} + 4\pi ik\underline{P} - (4\pi/c)\underline{j})].\underline{\underline{G}} + [\underline{n}\times\underline{H} + (4\pi/c)\underline{K}].\nabla\times\underline{\underline{G}}\}\, dS. \tag{27}$$

We know that inside the medium, the $\underline{E}$ and $\underline{H}$ fields obey the equations (4a) and (4b),

$$\hat{L}\,\underline{E}(\underline{r}_<) = \underline{F}_e(\underline{r}_<)\;, \qquad \hat{L}\,\underline{H}(\underline{r}_<) = \underline{F}_h(\underline{r}_<)\;.$$

However these are *general field equations* valid inside the medium. They do not completely specify the scattered field since they involve neither the incident field, nor any boundary conditions. Our hypothesis is that the *two extinction theorems (22) and (25) represent boundary conditions subject to which the (generally coupled) field equations (4) provide unique solution for the fields* $\underline{E}$, $\underline{H}$ *inside the scattering medium (i.e. inside the volume* V), *when an electromagnetic field* $\underline{E}^{(i)}$, $\underline{H}^{(i)}$ *is incident on the medium.* Thus, according to this hypothesis, the two extinction theorems allow us to replace the original saltus problem - involving the solution both inside and outside the medium - by a *boundary value problem* for determining the field inside the scattering medium. The boundary conditions for this later problem are of a somewhat unusual kind, having the form of *non-local* relations. Once the solution inside the scattering medium has been obtained, the solution outside it may be determined from Eqs. (23) and (26) by substituting the boundary values into the surface integrals occuring in these formulae. Our new interpretation of the extinction theorems is supported by explicit solutions that were obtained for several special cases [10].

Up to this point we have considered only two of the four relations (17), namely (17b) and (17c). If the other two relations, viz. (17a) and (17d) are also used, as well as the relations (19) and (21) and the corresponding formulae involving the magnetic fields one obtains an alternative set of equations in the form of *integro-differential equations* valid both inside and outside the medium - for the unknown electromagnetic fields $\underline{E}$ and $\underline{H}$. Lack of time prevents a discussion of this point here but we will later consider briefly the analogous situation for the case of quantum mechanical potential scattering.

With the help of various vector identities and Maxwell's equations, our general extinction theorems may be expressed in many alternative but equivalent forms. The extinction theorem for the electric field may, for example, be transformed into the form

$$\underline{E}^{(i)}(\underline{r}_<) + \frac{1}{k^2}\nabla\times\nabla\times\left[\,\underline{I}^{(E)}(\underline{r}_<) + \underline{I}^{(P)}(\underline{r}_<) + \underline{I}^{(M)}(\underline{r}_<) + \underline{I}^{(J)}(\underline{r}_<)\right] = 0\;, \tag{28}$$

where

$$\underline{I}^{(E)}(\underline{r}_<) = \frac{1}{4\pi}\int_{S^-}\{\underline{E}(\underline{r}')\frac{\partial G_0(\underline{r}_<,\underline{r}')}{\partial n} - G_0(\underline{r}_<,\underline{r}')\frac{\partial \underline{E}(\underline{r}')}{\partial n}\}\,dS, \tag{29}$$

$$\underline{I}^{(P)}(\underline{r}_<) = -\int_{S^-}[\underline{n}\,\nabla.\,\underline{P}(\underline{r}')]\,G_0(\underline{r}_<,\underline{r}')\,dS, \tag{30}$$

$$\underline{I}^{(M)}(\underline{r}_<) = -ik\int_{S^-}[\underline{n}\times\underline{M}(\underline{r}')]\,G_0(\underline{r}_<,\underline{r}')\,dS, \tag{31}$$

$$\underline{I}^{(J)}(\underline{r}_<) = -\frac{i}{kc}\int_{S^-}[\underline{n}\,\nabla.\,\underline{j}(\underline{r}')-\underline{K}(\underline{r}')]\,G_0(\underline{r}_<,\underline{r}')\,dS. \tag{32}$$

Clearly if the medium is non-magnetic, $\underline{I}^{(M)} \equiv 0$, if it is a non-conductor $\underline{I}^{(J)} \equiv 0$. For a linear, homogeneous ,spatially non-dispersive non-magnetic dielectric, not only do these two terms vanish, but so does also the term $\underline{I}^{(P)}$, unless the frequency of the incident field coincides with a frequency at which the dielectric constant vanishes; for except in this case the polarization field is necessarily transverse [11] (i.e. $\nabla.\underline{P} = 0$). If we exclude this exceptional case, (28) reduces to

$$\underline{E}^{(i)}(\underline{r}_<) + \frac{1}{4\pi k^2}\,\nabla\times\nabla\times\int_{S^-}\{\underline{E}(\underline{r}')\frac{\partial G_0(\underline{r}_<,\underline{r}')}{\partial n} - G_0(\underline{r}_<,\underline{r}')\frac{\partial \underline{E}(\underline{r}')}{\partial n}\}dS=0, \tag{33}$$

which is seen to be identical with the Oseen formulation (1) of the extinction theorem. In this case ($\underline{M} = \nabla.\underline{P} = \underline{j} = \underline{K} = 0$, $\underline{P} = \chi\underline{E}$, χ being a constant), the equation of motion (4a) is not coupled to (4b) (since no magnetic term now occurs on the r.h.s. of (6a)) and reduces to

$$\nabla^2\underline{E}(\underline{r}_<) + n^2k^2\underline{E}(\underline{r}_<) = 0, \tag{34}$$

where

$$n^2 = 1 + 4\pi\chi\,. \tag{35}$$

According to our hypothesis the electric field $\underline{E}$ inside the medium is that solution of (34), which obeys the condition (33) at every point $\underline{r}$ inside the medium. Once this solution has been found, the field outside the medium is obtained from the formula

$$\underline{E}(\underline{r}_>) = \underline{E}^{(i)}(\underline{r}_>) + \frac{1}{4\pi k^2} \nabla x \nabla x \int_{S^-} \{\underline{E}(\underline{r}')\frac{\partial G_o(\underline{r}_>,\underline{r}')}{\partial n} - G_o(\underline{r}_>,\underline{r}')\frac{\partial \underline{E}(\underline{r}')}{\partial n}\}dS, \tag{36}$$

to which Eq. (23) may be shown to reduce in the present case.

One can also show that in some other special cases, our general extinction theorem (22) for the electric field reduces to various extinction theorems derived in recent years by other authors. Moreover, one can readily show that in the special case when no scattering medium is present at all our general extinction theorem for the electric field reduces to

$$\underline{E}^{(i)}(\underline{r}_<) + \frac{1}{4\pi}\int_{S^-} \{\underline{E}^{(i)}(\underline{r}')\frac{\partial G_o(\underline{r}_<,\underline{r}')}{\partial n} - G_o(\underline{r}_<,\underline{r}')\frac{\partial \underline{E}^{(i)}(\underline{r}')}{\partial n}\}dS = 0, \tag{37}$$

which will be recognized as the classic *integral theorem of Helmholtz and Kirchhoff* (cf. for example, ref. 3a, p.377).

Returning to the general forms (22) and (25) of the extinction theorems it seems quite remarkable that for a completely arbitrary medium - e.g. an inhomogeneous, anisotropic, non-linear or spatially dispersive medium - the cancelation of the incident field inside the medium is expressible entirely by the values that the field takes at the boundary of the medium.

Finally I will show that the main results that we obtained for electromagnetic scattering have a strict analogue in non-relativistic quantum-mechanical potential scattering. Consider scattering of a free particle of momentum $\underline{p}$ on a three-dimensional potential barrier or potential well, characterized by a potential $\mathcal{V}(\underline{r})$ that vanishes outside a finite volume V, bounded by a surface S. (Fig. 2). For simplicity we assume that the potential $\mathcal{V}(\underline{r})$ has at most a finite discontinuity on S.

The Schrödinger equation for this problem may be written in the form

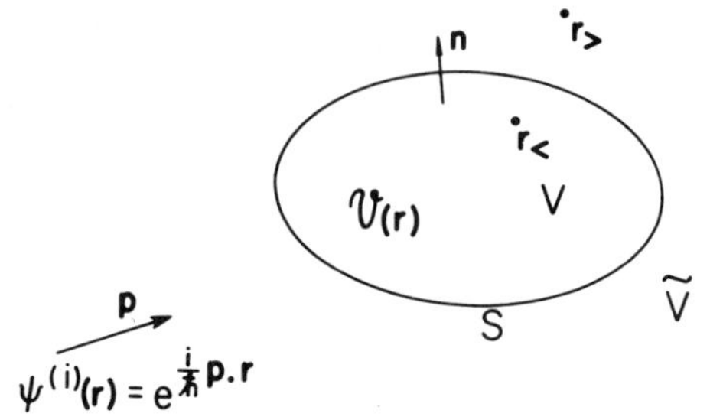

Fig. 2 Notation relating to quantum mechanical potential scattering.

$$(\nabla^2 + k^2)\, \psi(\underline{r}) = U(\underline{r})\psi(\underline{r}), \tag{38}$$

where

$$k^2 = \frac{2m}{\hbar^2} E, \qquad U(\underline{r}) = \frac{2m}{k^2} \mathcal{V}(\underline{r}) \quad \text{if } r \in V$$
$$= 0 \quad \text{if } r \in \tilde{V}. \tag{39}$$

Here m denotes the mass of the particle and $E = p^2/2m$ its energy, $\hbar$ is the Planck constant divided by 2π and $\tilde{V}$ denotes the (infinite) domain outside V.

The equation for the associated Green's function is

$$(\nabla^2 + k^2)\, G_0(\underline{r},\underline{r}') = -\, 4\pi\delta(\underline{r}-\underline{r}'), \tag{40}$$

the Green's function being, of course, the outgoing spherical wave [Eq. (13b) above].

From (38) and (39) we obtain, if we also use Green's theorem, the following identity valid for integration throughout any domain V' bounded by a closed surface S' :

$$\int_{V'} \psi(\underline{r}')\delta(\underline{r}-\underline{r}')d^3\underline{r}' = -\frac{1}{4\pi}\int_{V'} U(\underline{r}')\psi(\underline{r}')G_o(\underline{r},\underline{r}')d^3\underline{r}' - \frac{1}{4\pi}\sum(\underline{r}), \quad (41)$$

where

$$\sum(\underline{r}) = \int_{S'} \{\psi(r')\frac{\partial G_o(\underline{r},\underline{r}')}{\partial n} - G_o(\underline{r},\underline{r}')\frac{\partial\psi(\underline{r}')}{\partial n}\}dS \quad (43)$$

and $\partial/\partial n$ denotes differentiation along the outward normal to S'.

Let us now take the volume V' to coincide either with the scattering volume V or with the exterior $\tilde{V}$ of it. Again the field point $\underline{r}$ may be taken to be either in V or in $\tilde{V}$. Thus, in analogy with the electromagnetic case, we obtain four formulae:

(a) $\underline{r}\varepsilon V,\ \underline{r}'\varepsilon V$

$$\psi(\underline{r}_<) = -\frac{1}{4\pi}\int_V \psi(\underline{r}')U(\underline{r}')G_o(\underline{r}_<,\underline{r}')d^3\underline{r}' - \frac{1}{4\pi}\sum(\underline{r}_<), \quad (43a)$$

(b) $\underline{r}\varepsilon V,\ \underline{r}'\varepsilon\tilde{V}$

$$0 = -\frac{1}{4\pi}\int_{\tilde{V}} \psi(\underline{r}')U(\underline{r}')G_o(\underline{r}_<,\underline{r}')d^3r' + \frac{1}{4\pi}\sum(\underline{r}_<) - \frac{1}{4\pi}\sum^{(\infty)}(\underline{r}_<), \quad (43b)$$

(c) $\underline{r}\varepsilon\tilde{V},\ \underline{r}'\varepsilon\tilde{V}$

$$\psi(\underline{r}_>) = -\frac{1}{4\pi}\int_{\tilde{V}} \psi(\underline{r}')U(\underline{r}')G_o(\underline{r}_>,\underline{r}')d^3\underline{r}' + \frac{1}{4\pi}\sum(\underline{r}_>) - \frac{1}{4\pi}\sum^{(\infty)}(\underline{r}_>), \quad (43c)$$

(d) $\underline{r}\varepsilon\tilde{V},\ \underline{r}'\varepsilon V$

$$0 = -\frac{1}{4\pi}\int_V \psi(\underline{r}')U(\underline{r}')G_o(\underline{r}_>,\underline{r}')d^3\underline{r}' - \frac{1}{4\pi}\sum(\underline{r}_>), \quad (43d)$$

where

$$\sum(\underline{r}) = \int_S \{\psi(\underline{r}')\frac{\partial G_o(\underline{r},\underline{r}')}{\partial n} - G_o(\underline{r},\underline{r}')\frac{\partial\psi(\underline{r}')}{\partial n}\}dS, \quad (44a)$$

$$\sum^{(\infty)}(r) = \lim_{R\to\infty} \int_{S_R} \{\psi(\underline{r}') \frac{\partial G_o(\underline{r},\underline{r}')}{\partial n} - G_o(\underline{r},\underline{r}') \frac{\partial \psi(\underline{r}')}{\partial n}\} dS, \tag{44b}$$

$\Sigma^{(\infty)}$ being a contribution from a sphere of limitingly large radius $R\to\infty$ and $\partial/\partial n$ denotes differentiation along the outward normals to the respective volume regions. In the integral (44a) for $\Sigma(\underline{r})$ we need not distinguish between limits from inside and outside of V since ψ and $\partial\psi/\partial n$ must be continuous at the surface S [cf. Eq. (10) above].

The integrals over V in (43b) and (43c) vanish since $U(\underline{r}) = 0$ in $\tilde{V}$. Also the contribution from the surface at infinity must evidently represent the incident wave, i.e.

$$-\frac{1}{4\pi}\sum^{(\infty)}(\underline{r}) = \psi^{(i)}(\underline{r}). \tag{45}$$

Hence (40b) and (40c) reduce to

$$\psi^{(i)}(\underline{r}_<) + \frac{1}{4\pi}\sum(\underline{r}_<) = 0, \tag{46}$$

$$\psi(\underline{r}_>) = \psi^{(i)}(\underline{r}_>) + \frac{1}{4\pi}\sum(\underline{r}_>). \tag{47}$$

Further from (43a) we obtain if we also use (46)

$$\psi(\underline{r}_<) = \psi^{(i)}(\underline{r}_<) - \frac{1}{4\pi}\int_V \psi(\underline{r}')U(\underline{r}')G_o(\underline{r}_<,\underline{r}')d^3\underline{r}', \tag{48}$$

and from (43d) if we use (47)

$$\psi(\underline{r}_>) = \psi^{(i)}(\underline{r}_>) - \frac{1}{4\pi}\int_V \psi(\underline{r}')U(\underline{r}')G_o(\underline{r}_>,\underline{r}')d^3\underline{r}'. \tag{49}$$

The relations (46) and (47) are strictly analogous to those that we found for the electromagnetic case. In particular (46) represents *an extinction theorem* expressing the cancellation of the incident wave function at every point inside the potential barrier or potential well in terms of the values of the total wave function ψ and its normal derivative $\partial\psi/\partial n$ at all points on the boundary of the potential barrier or potential well. We assert that this theorem has the same kind of significance as we postulated for the electromagnetic extinction theorem: It is a (non-local) boundary condition subject to which the Schrödinger equation (38) has to be solved inside the scattering volume V. Once this solution is known the wave function outside V can be determined from (47) by substitution. Note that (47) involves only a surface integral $\Sigma(\underline{r}_>)$, not a volume integral as it does in the usual formulation. The other two equations (47) and (48), which are seen to be of the same form irrespective whether the field point is inside or outside V will be recognized as the *usual integral equations for potential scattering*. Mathematically they are equivalent to the Schrödinger equation (38) together with the two equations (46) and (47).

We made use of the quantum mechanical extinction theorem (46) and the associated formula (47) to solve simple scattering problems, in order to verify the correctness of this new formulation of scattering. The results agree with those obtained by conventional methods based on the integral equations (48) and (49). I might add that the extinction theorem also leads correctly to *bound states* under appropriate conditions. One finds in these cases that the Schrödinger equation can then be solved subject to our non-local boundary condition (expressed by the extinction theorem (46))only when $\psi^{(i)}(\underline{r}) \equiv 0$.

Let me remark here that a quantum-mechanical extinction theorem, was also derived, from a different approach by Melvin Lax [12] in 1952 in his treatment of multiple scattering.

I will end by summarizing our main conclusions:

(1) We obtained on the basis of Maxwell's theory generalization of the classic Ewald-Oseen extinction theorem, valid rigorously for scattering from medium of any prescribed macroscopic response.

(2) We have put forward a hypothesis as to the true meaning of the theorem: it is a non-local boundary condition for the solution of the equation of motion for the *interior* scattering problem.

(3) We have shown that once the boundary values have been determined the *exterior* scattering problem can be solved in a novel way, involving only surface integrations.

(4) We have shown that these results have strict analogues in non-relativistic quantum-mechanical potential scattering and provide a new approach to solving such problems.

Finally let me say that scattering problems involving sharp boundaries, such as we have considered here are, in general, hard to solve even approximately, since the Born approximation cannot be used in such cases. It is possible that our new formulation might provide a basis for the development of useful approximate techniques for solving problems of this kind. For this new formulation takes very explicitly into account the sharp boundary, the very presence of which makes the usual perturbation methods inapplicable. We plan to discuss this and related problems in other publications.

References

1. P.P. Ewald, (a) Dissertation, Univ. of Munich, 1912;
   (b) Ann. Phys. *49*, 1 (1915).
2. C.W. Oseen, Ann. Phys. *48*, 1 (1915).
3. For a detailed account of the theorem see
   (a) M. Born and E. Wolf, *Principles of Optics* (Pergamon Press, Oxford and New York, 1970) 4th ed., §2.4,

   or

   (b) L. Rosenfeld, *Theory of Electrons* (North-Holland Publishing Co., Amsterdam, 1951) Chapt. VI, §4.
4. J.J. Sein, *An Integral-Equation Formulation of the Optics of Spatially-Dispersive Media,* Ph.D. Dissertation, New York University, 1969, Appendix III.
5. (a) A. Wierzbicki, Bul. Acad. Polonaise des Sciences *9*, 833 (1961); Acta Phys. Pol. *21*, 557 (1962), ibid *21*, 575 (1962).
   (b) N. Bloembergen and P.S. Pershan, Phys. Rev. *127*, 206 (1962).
   (c) B.A. Sotskii, Opt. Spectro. *14*, 57 (1963).
   (d) R.K. Bullough, J. Phys. A. (Proc. Phys. Soc.) Ser. 2, *1*, 409 (1968).
   (e) J.J. Sein, ref. 4 above and Opt. Comm. *2*, 170 (1970).
   (f) É. Lalor, Opt. Comm. *1*, 50 (1969).
   (g) É. Lalor and E. Wolf, Phys. Rev. Lett. *26*, 1274 (1971).
   (h) T. Suzuki, J. Opt. Soc. Amer. *61*, 1029 (1971).
   (i) G.S. Agarwal, D.N. Pattanayak and E. Wolf, Opt. Comm. *4*, 260 (1971).
   (j) É. Lalor and E. Wolf, J. Opt. Soc. Amer. *62*, 1165 (1972).
   (k) J.L. Birman and J.J. Sein, Phys. Rev. *B6*, 2482 (1972).

6. Preliminary results of this investigation were presented in a lecture at the annual meeting of the Optical Society of America held in Ottawa in October 1971 (Abstr. WC16, J.Opt.Soc.Amer., *61*, 1560 (1971)) and in a note published in Optics Commun. *6*, 217 (1972).
7. While this manuscript was being prepared for publication a paper reporting some closely related results was published by J. de Goede and P. Mazur, Physica *58*, 568 (1972). This paper also contains some additional references to publications concerning the extinction theorem.
8. A slightly different argument to that given below is needed if the incident field is a plane wave (i.e. if the source is at infinity), but the final formulae remain the same. The case of plane wave incidence is discussed explicitly in connection with the quantum mechanical extinction theorem, in the last part of this paper.
9. See, for example, Chen-To Tai, *Dyadic Green's Functions in Electromagnetic Theory* (In-text Educational Publishers, Scranton and San Francisco, 1971).
10. One of them was presented in reference 5j .
11. That this is so follows at once from the Maxwell equation $\nabla.\underline{D} = 0$. For this implies that $0 = \nabla.(\varepsilon\ \underline{E}) = \varepsilon\ \nabla.\underline{E}$ (ε = dielectric constant), so that $\nabla.E = 0$ and hence also $\nabla.\underline{P} = 0$ (because of the linearity of the medium), unless $\varepsilon = 0$.
12. M. Lax, Phys. Rev. *85*, 646 (1952).

2

Angular Spectrum Representation of Wavefields

The usual integro-differential equations for scattering, diffraction, and propagation of wavefields represent a superposition of waves, usually spherical or cylindrical, which is often rather difficult to handle and interpret, and usually requires the use of several approximations.

The angular spectrum representation constitutes a method, based on plane wave expansions of the wavefields, which is useful in dealing with some of the above difficulties, and sometimes permits circumvention of the use of approximations. Also, it is of great interpretative value in connection with the information content of scattered and propagated fields.

In this chapter we discuss the representation of wavefields in terms of their angular spectrum, and some consequences are analyzed. We shall use the notation given by Shewell and Wolf [2.1], Wolf [2.2], and Mandel and Wolf [2.3]. (In these works, a discussion on the foundations of this method and some applications can be found.) We shall further present some illustrations.

2.1. INTRODUCTION

The angular spectrum representation is a mathematical tool for solving propagation, radiation, scattering, and diffraction problems. Among other applications, it has been used in connection with the scattering of radio waves and radiation patterns of antennas [2.4]–[2.8], models of energy transport of electromagnetic fields [2.9], the theory of multipole radiation [2.10] and Cerenkov radiation [2.11]; the theory of image formation [2.12], diffraction theory [2.13]–[2.17], and in many other recent topics such as inverse scattering [2.18], holography [2.19], rough surface scattering [2.20], diffraction by gratings [2.21], foundations of radiometry [2.22], radiative transfer [2.23] and phase conjugate optics [2.24]–[2.25]. This list is far from exhaustive.

An introduction to the angular spectrum can be found in several textbooks [2.3], [2.26], [2.27]. Here, we shall illustrate this technique for monochromatic scalar wavefields and then indicate its formulation for vector fields. Some consequences will be discussed in subsequent chapters.

2.2. EXPANSION OF SCALAR WAVEFIELDS INTO PLANE WAVES IN SOURCE-FREE REGIONS

Let us consider a monochromatic scalar wavefield traveling into the half-space $z \geqslant 0$ (see Fig. 2.1). This field is assumed to be generated by sources that will be considered located in the region $z < 0$. In the zone $z \geqslant 0$ the spatial part $U(\mathbf{r})$, $\mathbf{r} = (x, y, z)$, of the complex amplitude of this field satisfies the homogeneous Helmholtz equation:

$$\nabla^2 U + k^2 U = 0, \qquad z \geqslant 0. \tag{2.1}$$

A time dependence $\exp(-i\omega t)$ will be assumed from now on, and thus omitted.

We shall introduce a representation of $U(\mathbf{r})$ into plane waves such that $U(\mathbf{r})$ fulfils the boundary condition at the plane $z = 0$:

$$U(x, y, 0) = f(x, y), \tag{2.2}$$

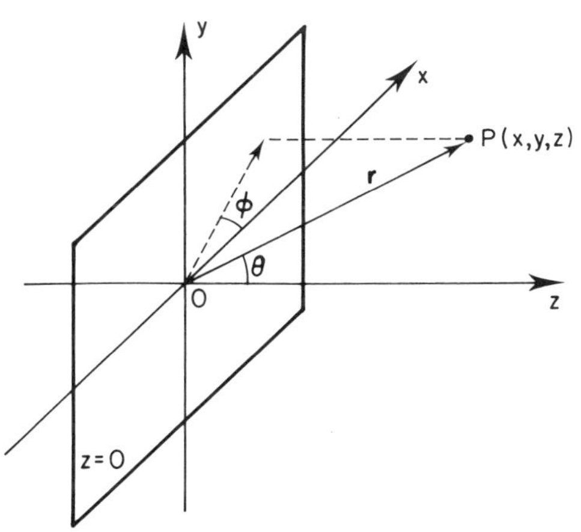

Figure 2.1. Notation for field propagation in a source-free half-space.

where $f(x, y)$ is a well behaved function with finite energy:

$$\iint_{-\infty}^{\infty} |f(x, y)|^2 \, dx \, dy < \infty; \tag{2.3}$$

i.e., $f(x, y)$ belongs to the functional space L^2.

The wavefield should also have finite energy at any distance $z \geqslant 0$:

$$\iint_{-\infty}^{\infty} |U(x, y, z)|^2 \, dx \, dy < \infty, \qquad z \geqslant 0. \tag{2.4}$$

In addition, it will be required that in $z \geqslant 0$, and at large distances from $\mathbf{r} = 0$ compared with the wavelength λ, $U(\mathbf{r})$ behaves as an outgoing spherical wave, namely:

$$U(\mathbf{r}) \sim F(\theta, \phi) \frac{\exp(ikr)}{r}, \tag{2.5}$$

for $kr \to \infty$, where $F(\theta, \phi)$ is a function that depends on the polar and azimuthal angles, θ and ϕ, characterizing the $\mathbf{r}$-direction (see Fig. 2.1). The asymptotic behavior given by Eq. (2.5) implies that $U(\mathbf{r})$ satisfies Sommerfeld's radiation condition in $z \geqslant 0$, namely:

$$\lim_{r \to \infty} r\left(\frac{\partial U}{\partial r} - ikU\right) = 0. \tag{2.6}$$

We shall assume that at any plane $z = constant$ the wavefield $U(\mathbf{r})$ has a two-dimensional Fourier transform ([2.1]–[2.3], [2.9] and [2.17]):

$$U(x, y, z) = \iint_{-\infty}^{\infty} \tilde{U}(p, q, z) \exp[ik(px + qy)] \, dp \, dq. \tag{2.7}$$

Substituting Eq. (2.7) into the Helmholtz equation (2.1) one obtains:

$$\frac{\partial^2}{\partial z^2} \tilde{U}(p, q, z) + k^2 m^2 \tilde{U}(p, q, z) = 0, \tag{2.8}$$

where:

$$m = \sqrt{1 - p^2 - q^2}, \qquad p^2 + q^2 \leqslant 1, \tag{2.9a}$$

$$m = i\sqrt{p^2 + q^2 - 1}, \qquad p^2 + q^2 > 1. \tag{2.9b}$$

There are two solutions to the differential equation (2.8):

$$\tilde{U}_1(p, q, z) = a(p, q) \exp(ikmz), \tag{2.10}$$

and:

$$\tilde{U}_2(p, q, z) = b(p, q) \exp(-ikmz). \tag{2.11}$$

The first solution, obtained when Eq. (2.10) is introduced into (2.7), leads to the following expression for $U(\mathbf{r})$:

$$U(x, y, z) = \iint_{-\infty}^{\infty} a(p, q) \exp[ik(px + qy + mz)]\, dp\, dq; \tag{2.12}$$

whereas, in a similar manner, the solution (2.11) leads to:

$$V(x, y, z) = \iint_{-\infty}^{\infty} b(p, q) \exp[ik(px + qy - mz)]\, dp\, dq. \tag{2.13}$$

Equations (2.12) and (2.13) are the angular spectrum representation of a wavefield in a source-free half space. Due to Eq. (2.9), the integral of Eq. (2.12) is convergent in $z \geqslant 0$, and corresponds to sources in $z < 0$, whereas the integral of Eq. (2.13) is convergent in $z \leqslant 0$, and valid for sources situated in $z > 0$.

It should be noticed that the plane wave components of Eq. (2.12) propagate into $z \geqslant 0$ with directional cosines p, q, and m. Conversely, the components of Eq. (2.13) propagate into $z \leqslant 0$ with directional cosines p, q, and $-m$. These components are called *homogeneous* when Eq. (2.9a) holds. On the other hand, when Eq. (2.9b) is satisfied they are called *inhomogeneous* or *evanescent* waves, as they propagate in directions parallel to the plane $z = 0$ with directional cosines p and q, and decay exponentially along the z-axis.

Fields in a source-free region that is not infinite, like e.g. the one limited by two planes at $z = L_1$ and $z = L_2$, with sources at the left of $z = L_1$ and at the right of $z = L_2$ (see Fig. 2.2), will require an angular spectrum representation given by the addition of Eqs. (2.12) and (2.13).

2.3. CONNECTION BETWEEN $U(x, y, z)$ AND ITS BOUNDARY VALUE AT $z = 0$

The field $U(\mathbf{r})$ propagating into $z \geqslant 0$ can be easily obtained in terms of its boundary value $f(x, y)$ at $z = 0$ (Shewell and Wolf [2.1]), as discussed next.

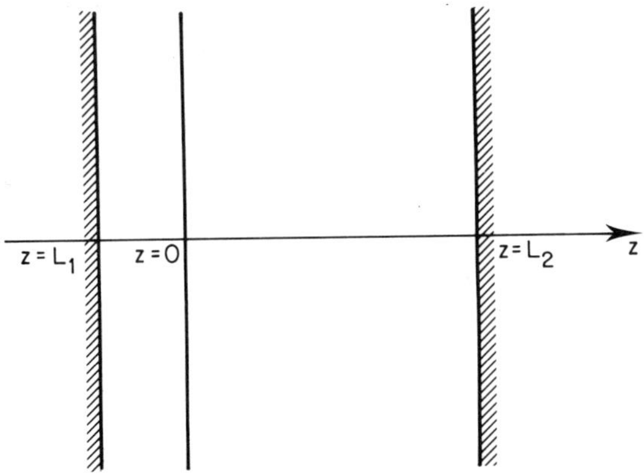

Figure 2.2. A source-free region limited by two planes $z = L_1$ and $z = L_2$. The sources are situated in the half-space $z < L_1$ and $z > L_2$.

Let $f(x, y)$ be:

$$f(x, y) = U(x, y, 0), \tag{2.14}$$

From Eqs. (2.14) and (2.12) one is led to:

$$f(x, y) = \iint_{-\infty}^{\infty} a(p, q) \exp[ik(px + qy)] \, dp \, dq. \tag{2.15}$$

The inversion of Eq. (2.15) yields:

$$a(p, q) = \left(\frac{k}{2\pi}\right)^2 \iint_{-\infty}^{\infty} f(x, y) \exp[-ik(px + qy)] \, dx \, dy. \tag{2.16}$$

By introducing $a(p, q)$ given by Eq. (2.16) into Eq. (2.12) we obtain:

$$U(x, y, z) = \left(\frac{k}{2\pi}\right)^2 \iint_{-\infty}^{\infty} f(x', y') \times \left[\iint_{-\infty}^{\infty} \exp[ik(p(x - x') + q(y - y') + mz)] \, dp \, dq\right] dx' \, dy'. \tag{2.17}$$

Equation (2.17) gives the relationship, under free space propagation, between the field $U(\mathbf{r})$, at any point $\mathbf{r} = (x, y, z)$ of the half space $z \geqslant 0$, and its boundary value $f(x, y)$ at $z = 0$. This is also the equation of an *invariant filter* expressing the *output* $U(x, y, z)$ at distance $z > 0$ from the input $f(x, y)$ at $z = 0$:

$$U(x, y, z) = \iint_{-\infty}^{\infty} f(x', y')\mathscr{K}(x - x', y - y', z)\, dx'\, dy', \tag{2.18}$$

where the *impulse response*, or wavefield *propagator* [2.1], from $z = 0$ to $z > 0$, is:

$$\mathscr{K}(x - x', y - y', z) = \left(\frac{k}{2\pi}\right)^2 \iint_{-\infty}^{\infty} \exp[ik(p(x - x') + q(y - y') + mz)]\, dp\, dq. \tag{2.19}$$

2.4. WAVELENGTH RESOLUTION LIMIT

Equation (2.12) may be inverted to obtain the angular spectrum from the field $U(x, y, z_0)$ propagated to $z = z_0$:

$$a(p, q) \exp(ikmz_0) = \left(\frac{k}{2\pi}\right)^2 \iint_{-\infty}^{\infty} U(x, y, z_0) \exp[-ik(px + qy)]\, dx\, dy. \tag{2.20}$$

However, Eq. (2.20) expresses the product of $a(p, q)$ with the function $\exp(ikmz_0)$ rather than $a(p, q)$ only. Considering the dependence of m on p and q given by Eqs. (2.9a) and (2.9b), one can see that $\exp(ikmz_0)$ is a highly oscillating function for $p^2 + q^2 < 1$ and decreases exponentially for $p^2 + q^2 > 1$; so that in this latter range of p and q, and at any distance z_0 equal or greater than the wavelength λ, it becomes practically zero. For instance, for $z_0 = \lambda$ and $p^2 + q^2 = 2$ this function is $\exp(-2\pi) = 1.87 \times 10^{-3}$. Therefore, $\exp(ikmz_0)$ constitutes *a low pass filter* acting upon $a(p, q)$ by removing the values of $a(p, q)$ at higher spatial frequencies p and q. In other words, *free propagation* from $z = 0$ to $z = z_0 > 0$ amounts to low pass filtering of the angular spectrum $a(p, q)$ with the factor $\exp(ikmz_0)$, m being given by Eqs. (2.9a) and (2.9b). Thus, if one knows $U(x, y, z_0)$, the use of Eq. (2.20) yields $a(p, q)$ for $p^2 + q^2 < 1$, and by means of Eq. (2.15) one can retrieve $f(x, y)$ constrained to the fundamental limit: $p^2 + q^2 = 1$. Details of $f(x, y)$ smaller than λ, namely, with $p^2 + q^2$ larger than the above limit, are lost.

Put in another form, in order to make the factor $\exp(ikmz_0)$ approximately equal to 0.9 for $p^2 + q^2 = 2$, one would require $k\sqrt{p^2 + q^2 - 1}z_0 = 0.1$, which means that the field $U(x, y, z_0)$ must be determined at a distance from the boundary plane $z = 0$ as small as $z_0 = 0.016\lambda$. In Section 9.6 it is seen that this underlines one of the problems associated to inverse diffraction.

2.5. ON THE VALIDITY OF THE ANGULAR SPECTRUM REPRESENTATION

Lalor [2.28], based on an earlier work by Luneburg [2.29], established conditions for the validity of Eq. (2.12): (1) $f(x, y)$ should be square integrable (namely, it should belong to the functional space L^2); (2) $f(x, y)$ should satisfy sectional continuity and $|f| = 0(1/r)$, $|\partial f/\partial r| = 0(1/r)$; (3) $U(\mathbf{r})$ should be regular in $z > 0$ and $|U| = 0(1/r)$, $|\partial U/\partial r| = O(1/r)$; and (4) $U(\mathbf{r})$ should satisfy Sommerfeld's radiation condition, Eq. (2.6), in $z > 0$.

Since the square modulus of the wavefield $U(\mathbf{r})$ is a measure of its energy density in the scalar theory, the norm of $f(x, y)$ in L^2 has an immediate physical meaning in terms of conservation of energy. This is connected with the fact that, in the theory of Fourier integrals (see e.g. Titchmarsh [2.30]) if $f(x, y)$ is in L^2 then $a(p, q)$ is also in L^2, and the norm of this function is equal to that of its transform in L^2. All Fourier transforms in the functional space L^2 should be understood as a limit in the mean, namely:

$$\underset{T\to\infty}{\text{l.i.m.}} \int\int_{-T}^{T} (\cdot)\, dp\, dq.$$

This limit in the mean may be replaced by the ordinary integration:

$$\int\int_{-\infty}^{\infty} (\cdot)\, dp\, dq,$$

if the integral exists almost everywhere.

By strictly analogous arguments, conditions for the validity of the angular spectrum representation, Eq. (2.13), for fields that propagate into $z < 0$, can be established.

Applications of the angular spectrum representation for wavefields whose boundary value $f(x, y)$ is arbitrary in L^2 have been studied by Montgomery [2.31]. Generalizations to wavefields in the space of distributions have been discussed by Sherman and Bremermann [2.32].

2.6. AN ALTERNATIVE REPRESENTATION IN ANGULAR VARIABLES

The angular spectrum representation, Eqs. (2.12) and (2.13), acquires a useful form in terms of the polar angles α and β which characterize the components of the unit vector $\mathbf{s} = (p, q, m)$ in spherical coordinates (Wolf [2.2], Mandel and Wolf [2.3] and Devaney and Wolf [2.10]):

$$p = \sin\alpha\cos\beta, \tag{2.21a}$$

$$q = \sin\alpha\sin\beta, \tag{2.21b}$$

$$m = \cos\alpha. \tag{2.21c}$$

The angle β is real $(-\pi \leqslant \beta \leqslant \pi)$; however, as will be seen shortly, the angle α may be complex.

Let $U(\mathbf{r})$ be a wavefield propagating into $z \geqslant 0$, represented by Eq. (2.12). From Eqs. (2.21) it is straightforward to obtain the following relationship between the surface elements in each set of variables:

$$\frac{dp\,dq}{m} = \sin\alpha\,d\alpha\,d\beta. \tag{2.22}$$

The angle α varies according to its dependence on m through Eq. (2.21c). As $p^2 + q^2$ grows from $p^2 + q^2 = 0$ to $p^2 + q^2 = 1$, due to Eq. (2.21c), α varies from $\alpha = 0$ to $\alpha = \pi/2$. On the other hand, when $p^2 + q^2$ grows, keeping $p^2 + q^2 > 1$, according to Eq. (2.9b) m grows in the form $m = i\sqrt{p^2 + q^2 - 1}$; namely, from $m = 0$ to $m \to i\infty$. However, if $\alpha = \pi/2 - i\delta$, then according to Eq. (2.21c), m should be $m = \cos(\pi/2 - i\delta) = i\sinh\delta$, which also grows from $m = 0$ to $m \to i\infty$; hence, for $p^2 + q^2 > 1$, α is: $\alpha = \pi/2 - i\delta$.

As a result, Eq. (2.12) may be expressed in terms of the angles α and β in the following form:

$$U(\mathbf{r}) = \frac{ik}{2\pi}\int_{-\pi}^{\pi} d\beta \int_{\mathscr{D}^+} d\alpha\,\sin\alpha F(\alpha, \beta)\exp(ik\mathbf{s}\cdot\mathbf{r}), \qquad z > 0; \tag{2.23}$$

where we have written:

$$\frac{ik}{2\pi}F(\alpha, \beta) = \cos\alpha a(p, q), \tag{2.24}$$

with α in $\mathscr{D}^+$. In Eq. (2.23) $\mathscr{D}^+$ is the contour of integration in the complex α-plane of Fig. 2.3a.

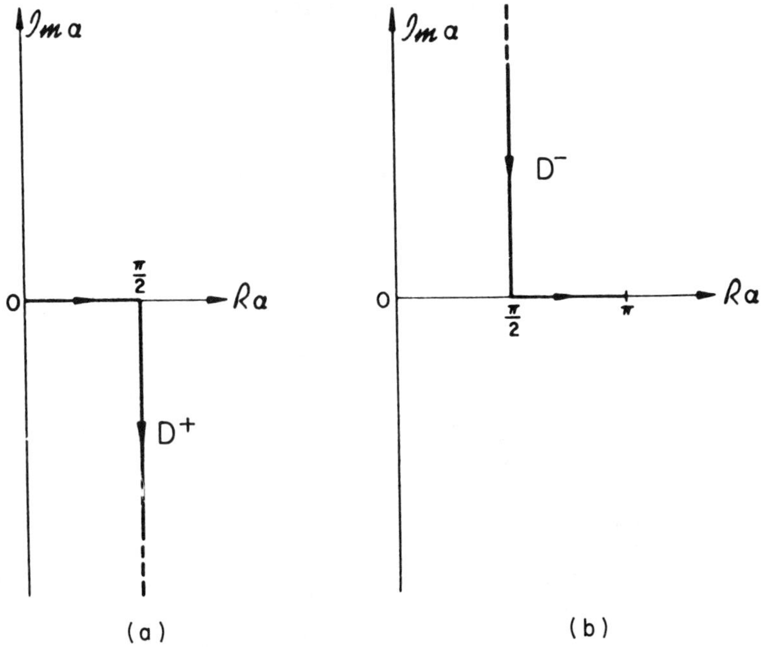

Figure 2.3. (a) The contour $\mathscr{D}^+$ of integration. The contribution of homogeneous waves corresponds to the interval $(0, \pi/2)$, whereas evanescent waves correspond to the line $\alpha = \pi/2 - i\delta$. (b) The contour $\mathscr{D}^-$ of integration. Homogeneous waves contribute in the interval $(\pi/2, \pi)$. Evanescent waves correspond to the line $\alpha = \pi/2 + i\delta$. (Reproduced with permission from paper by A. J. Devaney and E. Wolf, *J. Math. Phys.* **15**, 234 (1974).)

By a strictly similar argument, it can be easily shown that the angular spectrum representation for a wavefield $V(\mathbf{r})$ that propagates into the half-space $z < 0$, according to Eq. (2.13), is given by an integral identical to Eq. (2.23), but now the α integration is performed along the contour $\mathscr{D}^-$ of Fig. 2.3b.

2.7. SOURCE-FREE FIELDS

There are fields that play an important role in diffraction theory (see Refs. 2.5, 2.15, 2.29, 2.33, and 2.34), and that are bandlimited so that they have no evanescent components in their angular spectrum representation. These are known as *source-free fields.*

Two kinds of source-free fields are discussed next:

a. Fields that are solutions of the homogeneous Helmholtz equation:

$$\nabla^2 U + k^2 U = 0, \tag{2.25}$$

throughout the whole space. These fields have contributions of only homogeneous components propagating in all directions in their angular spectrum representation. They were originally studied by Whittaker [2.35], [2.36] and subsequently discussed by Devaney and Wolf [2.10] and by Wolf [2.23].

b. Fields that propagate either into the *source-free* half-space $z \geqslant 0$ or $z < 0$ and have an angular spectrum $a(p, q)$, or $b(p, q)$, that vanishes for $p^2 + q^2 > 1$. These fields were studied by Sherman [2.37], [2.38]. They can also be considered as fields satisfying the Helmholtz equation, either in $z \geqslant 0$ or in $z < 0$, and containing only homogeneous components propagating either towards $z \geqslant 0$ or $z < 0$, respectively. The concept *source-free field* was coined by Sherman in Refs. 2.37 and 2.38.

2.7.1. Source-Free Fields Satisfying the Homogeneous Helmholtz Equation Everywhere and Having Components Propagating in All Directions

Let us assume that there are conditions such that $U(\mathbf{r})$ admits the Fourier representation (Wolf [2.23]):

$$U(\mathbf{r}) = \int \tilde{U}(\mathbf{K}) \exp(i\mathbf{K}\cdot\mathbf{r})\, d^3K; \tag{2.26}$$

where the integral in (2.26) extends over the whole three-dimensional **K**-space. On introducing Eq. (2.26) into the Helmholtz equation, (2.25), we obtain:

$$\int (-K^2 + k^2)\tilde{U}(\mathbf{K}) \exp(i\mathbf{K}\cdot\mathbf{r})\, d^3K = 0. \tag{2.27}$$

Excluding the trivial solution $\tilde{U}(\mathbf{K}) = 0$ one obtains from Eq. (2.27) that $K^2 = k^2$, and therefore $\tilde{U}(\mathbf{K})$ should be of the form:

$$\tilde{U}(\mathbf{K}) = f(\mathbf{K})\, \delta(K - k). \tag{2.28}$$

By expressing **K** in spherical coordinates: $\mathbf{K} = k(p, q, m)$, with p, q and m given by Eqs. (2.21), and introducing:

$$F(\theta, \phi) = 4\pi k f(\mathbf{K}), \tag{2.29}$$

with $0 \leqslant \theta \leqslant \pi$ and $-\pi \leqslant \phi \leqslant \pi$, Eq. (2.26) becomes finally:

$$U(\mathbf{r}) = \frac{k}{4\pi} \int_{-\pi}^{\pi} d\phi \int_{0}^{\pi} d\theta \sin\theta F(\theta, \phi) \exp(ik\mathbf{s}\cdot\mathbf{r}); \tag{2.30}$$

where $\mathbf{s} = (p, q, m)$. Equation (2.30) is the angular spectrum representation for a field that is free of sources throughout the whole space and, hence, satisfies Helmholtz's equation everywhere. The integral expresses $U(\mathbf{r})$ as a superposition of homogeneous plane waves propagating in all directions: those for $0 \leqslant \theta \leqslant \pi/2$ propagate into $z \geqslant 0$, whereas those corresponding to $\pi/2 \leqslant \theta \leqslant \pi$ propagate into $z < 0$.

It is useful to derive the angular spectrum representation of this kind of source-free fields in the (p, q)-variables. To do this let us express Eq. (2.30) in a form similar to Eq. (2.24):

$$\frac{ik}{2\pi} F(\theta, \phi) = \cos\theta a_{sf}(p, q), \qquad 0 \leqslant \theta \leqslant \frac{\pi}{2}, \qquad -\pi \leqslant \phi \leqslant \pi, \tag{2.31}$$

and:

$$\frac{ik}{2\pi} F(\theta, \phi) = \cos\theta b_{sf}(p, q), \qquad \frac{\pi}{2} \leqslant \theta \leqslant \pi, \qquad -\pi \leqslant \phi \leqslant \pi. \tag{2.32}$$

By then introducing Eqs. (2.31) and (2.32) into Eq. (2.30), and making use of Eqs. (2.21), one obtains the following representation in the (p, q)-variables for this kind of source-free field:

$$U(\mathbf{r}) = \frac{1}{2i} \left[\iint_{p^2+q^2 \leqslant 1} a_{sf}(p, q) \exp[ik(px + qy + mz)]\, dp\, dq \right.$$
$$\left. + \iint_{p^2+q^2 \leqslant 1} b_{sf}(p, q) \exp[ik(px + qy - mz)]\, dp\, dq \right]; \tag{2.33}$$

In Eq. (2.33) the complex amplitudes a_{sf} and b_{sf} are the angular spectra, limited to plane waves propagating with $p^2 + q^2 \leqslant 1$, of source free-fields that propagate into $z \geqslant 0$ (cf. Eq. (2.12)), and into $z < 0$ (cf. Eq. (2.13)), respectively. Each of these amplitudes corresponds to the same angular spectrum $F(\theta, \phi)$ evaluated in a different interval of values of the argument θ, as expressed by Eqs. (2.31) and (2.32).

On comparing Eqs. (2.30) and (2.33), one can see that the angular spectrum representation in angular variables is more compact than the one in p, q-variables. The different cases, corresponding to propagation either into $z \geqslant 0$ or into $z \leqslant 0$, can be obtained by considering different paths of integration in the same complex α-plane.

2.7.2. Source-Free Fields Propagating in a Half-Space

For these fields the angular spectrum $a(p, q)$ (cf. Eq. (2.12)), or $b(p, q)$ (cf. Eq. (2.13)), vanishes for $p^2 + q^2 > 1$. They are important because, as will be seen in Chapter 4, these fields have remarkable mathematical properties that make them easier to treat analytically than other wavefields. In addition, fields of this kind are often useful approximations to fields far from sources, including diffraction and scattering obstacles. This is the case, for instance, of optical beams.

As seen from Eq. (2.12), if $U(\mathbf{r})$, propagating into $z \geqslant 0$ is source-free, namely, if:

$$a(p, q) = 0, \qquad p^2 + q^2 > 1, \tag{2.34a}$$

$$a(p, q) = a_{sf}(p, q), \qquad p^2 + q^2 \leqslant 1, \tag{2.34b}$$

in a plane $z = z_0 \geqslant 0$, then it is source-free in every plane $z = \text{constant} \geqslant 0$.

The above result was established by Sherman [2.37], [2.38] who also put forward the following theorem:

Theorem. *If $U(\mathbf{r})$, propagating into $z \geqslant 0$, is source-free, then Eq. (2.12) can be used to extend it into the region $z < 0$ as a bounded function that satisfies the homogeneous Helmholtz equation everywhere.* ■

To show this, let us consider the expression derived from Eq. (2.12) which gives $U(\mathbf{r})$ in terms of its boundary value $f(x, y)$ at $z = 0$, (2.17) (or their equivalent, Eqs. (2.18) and (2.19)), where now we allow z to take on negative values. The kernel $\mathscr{K}$, Eq. (2.19), remains bounded in $z < 0$ providing that m is real, namely, that $p^2 + q^2 < 1$, and therefore that $U(\mathbf{r})$ is source-free. Hence, $U(\mathbf{r})$ exists and is bounded for $z \leqslant 0$. Moreover, by taking derivatives in Eq. (2.17), it is straightforward to check that $U(\mathbf{r})$, given by Eq. (2.17), satisfies the homogeneous Helmholtz equation (2.25) everywhere.

This shows that this kind of source-free field may be also considered throughout all space. There are no sources of these fields anywhere at finite distance from the point in which they are under consideration. These fields will coincide, or approximate, actual physical fields only in regions of space far from real sources (or diffraction or scattering centers). Obviously, similar arguments can be made for source-free fields propagating into $z \leqslant 0$.

Let us emphasize that the two kinds of source-free fields studied above have different sets of angular components, although both satisfy the homogeneous Helmholtz equation everywhere. Those of the first kind, studied in Section 2.7.1, have components propagating in all directions, whereas those of the second kind, studied in this section, have angular components propagating into either $z > 0$ or $z < 0$. In fact, by examining the contours $\mathscr{D}^+$ and $\mathscr{D}^-$ of Figs. 2.3a and 2.3b in the complex α-plane, it is seen that only the real part of these contours contribute to the angular spectrum representation of source-free fields of the second kind. In addition, it is clear that:

Apart from a factor $1/(2i)$, *the sum of two source-free fields, one with angular components propagating into the source-free region* $z \geqslant 0$ *and the other with components propagating into the source-free half-space* $z < 0$, *is equal to a source-free field satisfying the Helmholtz equation everywhere and with angular components propagating in all directions, as given by Eq. (2.30);* [(cf. also Eq. (2.33)].

2.8. ASYMPTOTIC APPROXIMATION TO SOURCE-FREE FIELDS

2.8.1. Fields Propagating in a Half-Space Free of Sources

The asymptotic expression of source-free fields at large distances, in wavelength units, from a reference origin was investigated by Miyamoto and Wolf [2.39] and by Wolf [2.24]. In this context, let us first consider a source-free field that propagates into the half-space $z \geqslant 0$:

$$U(\mathbf{r}) = \iint_{p^2+q^2 \leqslant 1} a_{sf}(p, q) \exp[ik(px + qy + mz)]\, dp\, dq, \tag{2.35}$$

where $m = \sqrt{1 - p^2 - q^2}$. The integral, Eq. (2.35), is evaluated for $kr \to \infty$, $\mathbf{r} = r\mathbf{n} = (x, y, z)$, with the unit vector $\mathbf{n}$ fixed [2.39], by using the principle of the *stationary phase*. (This method of calculating integrals may be found, for instance, in Appendix III of Ref. 2.40.) The result for $z > 0$ is ($kr \to \infty$, $\mathbf{n}$ fixed) (see also Ref. 2.39):

$$U(r\mathbf{n}) \sim -\frac{2\pi i}{k} n_z a_{sf}(n_x, n_y) \frac{\exp(ikr)}{r}, \qquad z > 0,\ (n_z > 0), \tag{2.36a}$$

$$U(r\mathbf{n}) \sim -\frac{2\pi i}{k} n_z a_{sf}(-n_x, -n_y) \frac{\exp(-ikr)}{r}, \qquad z < 0,\ (n_z < 0), \tag{2.36b}$$

where:

$$n_x = \frac{x}{r}, \qquad n_y = \frac{y}{r}, \qquad n_z = \frac{z}{r}. \tag{2.37}$$

Reciprocally, for a source-free field propagating to the left in the half-space $z < 0$:

$$V(\mathbf{r}) = \iint_{p^2+q^2 \leqslant 1} b_{sf}(p, q) \exp[ik(px + qy - mz)]\, dp\, dq, \tag{2.38}$$

(where $m = \sqrt{1 - p^2 - q^2}$), one obtains, in an identical manner as before ($kr \to \infty$, $\mathbf{n}$ fixed):

$$V(r\mathbf{n}) \sim \frac{2\pi i}{k} n_z b_{sf}(-n_x, -n_y) \frac{\exp(-ikr)}{r}, \qquad z > 0, \ (n_z > 0), \tag{2.39a}$$

$$V(r\mathbf{n}) \sim \frac{2\pi i}{k} n_z b_{sf}(n_x, n_y) \frac{\exp(ikr)}{r}, \qquad z < 0, \ (n_z < 0). \tag{2.39b}$$

From Eqs. (2.36) and (2.39) it is observed that the wavefield propagating into $z > 0$ behaves asymptotically in the half-space $z > 0$ as a *diverging spherical wave* with amplitude $a_{sf}(n_x, n_y)$, whereas in the half-space $z < 0$, to which it may be extended as seen in Section 2.7.2, it behaves asymptotically as a *converging spherical wave* with amplitude $a_{sf}(-n_x, -n_y)$.

Reciprocally, a source-free field propagating into $z < 0$ behaves asymptotically in the half-space $z < 0$ as a *diverging spherical wave* of amplitude $b_{sf}(n_x, n_y)$, and in the half-space $z > 0$, to which it may be extended, it behaves asymptotically as a *converging spherical wave*. Therefore, the source-free field $U(\mathbf{r})$ given by Eq. (2.35), propagating into $z > 0$, is *divergent* in $z > 0$ and *convergent* in $z < 0$. Conversely, the source-free field $V(\mathbf{r})$ given by Eq. (2.38), propagating into $z < 0$, is *divergent* in $z < 0$ and *convergent* in $z > 0$.

Equations (2.36) and (2.39) may be easily expressed in angular variables θ and ϕ that define the direction vector $\mathbf{n}$ by using Eqs. (2.31) and (2.32):

$$U(r\mathbf{n}) \sim F(\theta, \phi) \frac{\exp(ikr)}{r}, \qquad z > 0, \tag{2.40a}$$

$$U(r\mathbf{n}) \sim -F(\pi - \theta, \phi + \pi) \frac{\exp(-ikr)}{r}, \qquad z < 0, \tag{2.40b}$$

$$\left(0 \leqslant \theta \leqslant \frac{\pi}{2}, -\pi \leqslant \phi \leqslant \pi\right);$$

$$V(r\mathbf{n}) \sim F(\pi - \theta, \phi + \pi) \frac{\exp(-ikr)}{r}, \qquad z > 0, \tag{2.40c}$$

$$V(r\mathbf{n}) \sim -F(\theta, \phi) \frac{\exp(-ikr)}{r}, \qquad z < 0, \tag{2.40d}$$

$$\left(\frac{\pi}{2} \leqslant \theta \leqslant \pi, -\pi \leqslant \phi \leqslant \pi\right).$$

2.8.2. Fields Satisfying the Helmholtz Equation Everywhere and with Angular Components Propagating in All Directions

The asymptotic expression for these fields as $kr \to \infty$, at $\mathbf{n}$ fixed, may be obtained from Eq. (2.30) by applying the method of the stationary phase quoted in Section 2.8.1, or equivalently, by means of a mathematical lemma due to Jones [2.41]. The result is:

$$U(r\mathbf{n}) \sim \frac{1}{2i}\left[F(\theta, \phi)\frac{\exp(ikr)}{r} - F(\pi - \theta, \phi + \pi)\frac{\exp(-ikr)}{r}\right],$$

$$\left(0 \leqslant \theta \leqslant \frac{\pi}{2}, -\pi \leqslant \phi \leqslant \pi\right). \quad (2.41)$$

The reader can check that Eq. (2.41) is in agreement with Eq. (2.33) and the result established at the end of Section 2.7.2.

2.9. ANGULAR SPECTRUM REPRESENTATION OF ELECTROMAGNETIC FIELDS

The angular spectrum representation for scalar wavefields can be generalized to vector fields. (See, for example, Wolf [2.23].) The electric and magnetic vectors $\mathbf{E}$ and $\mathbf{H}$ propagating in a source-free half space can be represented as:

$$\mathbf{E}(\mathbf{r}) = \iint_{-\infty}^{\infty} \mathbf{e}(\mathbf{s}_\perp) \exp(ik\mathbf{s}^{\pm}\cdot\mathbf{r})\, d^2 s_\perp, \quad (2.42)$$

$$\mathbf{H}(\mathbf{r}) = \iint_{-\infty}^{\infty} \mathbf{h}(\mathbf{s}_\perp) \exp(ik\mathbf{s}^{\pm}\cdot\mathbf{r})\, d^2 s_\perp, \quad (2.43)$$

where $\mathbf{e}(\mathbf{s}_\perp)$ and $\mathbf{h}(\mathbf{s}_\perp)$ are the angular spectra of $\mathbf{E}$ and $\mathbf{H}$, respectively; $\mathbf{s}^{\pm} = (\mathbf{s}_\perp, \pm m)$, $\mathbf{s}_\perp = (p, q, 0)$, with p, q, and m given as in Section 2.2. The sign "plus" or "minus" is chosen according to whether the half-space of propagation is considered in $z \geqslant 0$ or in $z < 0$, respectively.

Like Eqs. (2.12) and (2.13), the representation, Eqs. (2.42) and (2.43), contains both homogeneous and evanescent components.

Maxwell's equations demand that the waves are transverse, namely:

$$\mathbf{s}^{\pm}\cdot\mathbf{e}(\mathbf{s}_\perp) = \mathbf{s}^{\pm}\cdot\mathbf{h}(\mathbf{s}_\perp) = 0, \quad (2.44)$$

and for each angular component the two angular spectra are connected by:

$$\mathbf{h}(\mathbf{s}_\perp) = \mathbf{s}^\pm \times \mathbf{e}(\mathbf{s}_\perp). \tag{2.45}$$

2.10. DIVERGENT AND CONVERGENT SPHERICAL WAVES AND THEIR ANGULAR SPECTRUM REPRESENTATION

It was stated in Chapter 1 that the scalar field $G(|\mathbf{r} - \mathbf{r}'|) = \exp(ik|\mathbf{r} - \mathbf{r}'|)/|\mathbf{r} - \mathbf{r}'|$ represents a *divergent*, or *outgoing*, spherical wave, propagating in free space and originated at the source point $\mathbf{r}'$. Assuming a time dependence $\exp(-i\omega t)$, this function satisfies the wave equation:

$$(\nabla^2 + k^2)G(\mathbf{r}, \mathbf{r}') = -4\pi\, \delta(\mathbf{r} - \mathbf{r}'), \tag{2.46}$$

where $\delta(\mathbf{r} - \mathbf{r}')$ is the three-dimensional δ-function. $G(\mathbf{r}, \mathbf{r}')$ also fulfils Sommerfeld's radiation condition:

$$\lim_{r \to \infty} r\left(\frac{\partial G}{\partial r} - ikG\right) = 0, \tag{2.47}$$

and is the outgoing Green's function for the inhomogeneous Helmholtz equation that governs the wavefield $U(\mathbf{r})$, created by the source density $\rho(\mathbf{r})$:

$$(\nabla^2 + k^2)U(\mathbf{r}) = -4\pi\rho(\mathbf{r}). \tag{2.48}$$

Another solution of Eq. (2.46) is the *convergent*, or *incoming*, spherical wave:

$$G^{(c)}(\mathbf{r}, \mathbf{r}') = \frac{\exp(-ik|\mathbf{r} - \mathbf{r}'|)}{|\mathbf{r} - \mathbf{r}'|}. \tag{2.49}$$

This function satisfies the condition at infinity:

$$\lim_{r \to \infty} r\left(\frac{\partial G^{(c)}}{\partial r} + ikG^{(c)}\right) = 0, \tag{2.50}$$

and represents a spherical wave converging in free space towards the sink point $\mathbf{r}'$.

Convergent spherical waves, and incoming fields in general, are not just mere formal solutions to the propagation equations, but since the advent of *optical phase conjugation* [2.42], [2.43] there exists the actual possibility of generating them, even if with a limited resolution, namely, with angular components propagating with $p^2 + q^2 \leqslant 1$.

The angular spectrum representation of an outgoing spherical wave is of considerable interest in radiation, scattering and diffraction problems. It was originally obtained by Weyl [2.44], and can be also found in several textbooks, e.g. those by Wolf [2.2], Baños [2.45], and Brekhovskikh [2.46]. The discussion outlined next follows the notation of Ref. 2.2.

Writing $\exp(ikr)/r$ in the form of Eq. (2.12), the corresponding angular spectrum $a(p, q)$ is obtained from $f(x, y) = [r^{-1}\exp(ikr)]_{z=0}$ through Eq. (2.16). The resulting integral in the x and y variables is evaluated by changing to polar coordinates ([2.2], [2.3], [2.46]) and it is left as an exercise to prove that it gives the result:

$$\frac{\exp(ikr)}{r} = \frac{ik}{2\pi}\int\int_{-\infty}^{\infty} \frac{1}{m}\exp[ik(px + qy + m|z|)]\,dp\,dq. \tag{2.51}$$

Observe the singularity of the integral, Eq. (2.51), at $z = 0$ which ensures that the factor $\exp(ik|m|z)$ is always convergent for evanescent components.

In a strictly similar manner, the reader can check that the convergent spherical wave is expressed as [2.39]:

$$\frac{\exp(-ikr)}{r} = \frac{ik}{2\pi}\int\int_{-\infty}^{\infty} \frac{1}{m'}\exp[ik(px + qy + m'|z|)]\,dp\,dq, \tag{2.52}$$

where:

$$m' = -\sqrt{1 - p^2 - q^2}, \qquad p^2 + q^2 \leqslant 1, \tag{2.53a}$$

$$m' = i\sqrt{p^2 + q^2 - 1}, \qquad p^2 + q^2 > 1. \tag{2.53b}$$

Note from Eqs. (2.52) and (2.53) that the homogeneous components of the convergent spherical wave and those corresponding to the divergent spherical wave propagate opposite to each other on the z-axis. On the other hand, the evanescent components decay as z increases, both for outgoing and for incoming waves.

By using spherical coordinates α and β defined according to Eqs. (2.21), the reader can verify that $\exp(-ikr)/r$ can be expressed as [2.47]:

$$\frac{\exp(-ikr)}{r} = \frac{ik}{2\pi}\int_{-\pi}^{\pi}\int_{\mathscr{C}^{\pm}} d\alpha \sin\alpha \exp(ik\mathbf{s}\cdot\mathbf{r}), \tag{2.54}$$

where $\mathbf{s} = (p, q, m)$ and is related to α and β through Eqs. (2.21). Like in Eq.

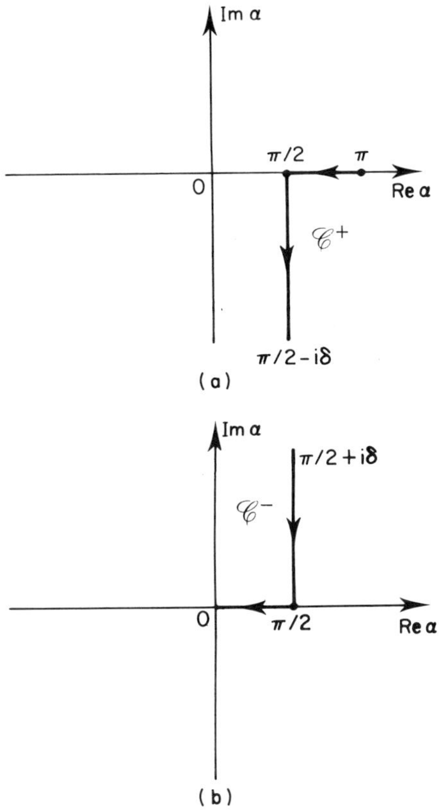

Figure 2.4. (a) The contour $\mathscr{C}^+$ of integration. Homogeneous waves contribute in the interval $(\pi, \pi/2)$. Evanescent waves correspond to the line $\alpha = \pi/2 - i\delta$. (b) The contour $\mathscr{C}^-$ of integration. Homogeneous waves correspond to the interval $(0, \pi/2)$, whereas evanescent waves contribute on the line $\alpha = \pi/2 + i\delta$.

(2.23), the contour $\mathscr{C}^+$ of the α-integration is chosen when $z > 0$ (see Fig. 2.4a), whereas the contour $\mathscr{C}^-$ should be used when $z < 0$ (see Fig. 2.4b).

In fact, the representation, Eq. (2.54), can be generalized to arbitrary *incoming wavefields*, namely, fields that converge towards sink distributions [2.47] and satisfy the condition Eq. (2.50). This generalization reads:

$$U^{(c)}(\mathbf{r}) = \frac{ik}{2\pi} \int_{-\pi}^{\pi} d\beta \int_{\mathscr{C}^+} d\alpha \sin \alpha F(\alpha, \beta) \exp(ik\mathbf{s} \cdot \mathbf{r}), \qquad z > 0, \quad (2.55)$$

and:

$$V^{(c)}(\mathbf{r}) = \frac{ik}{2\pi}\int_{-\pi}^{\pi} d\beta \int_{\mathscr{C}-} d\alpha \sin\alpha F(\alpha, \beta)\exp(ik\mathbf{s}\cdot\mathbf{r}), \qquad z < 0, \quad (2.56)$$

By comparing Eqs. (2.55) and (2.56) with the corresponding Eq. (2.23), we see that the construction of incoming or outgoing fields, either in $z > 0$ or in $z < 0$, is obtained from the same angular spectrum $F(\alpha, \beta)$ just by choosing the appropriate path of integration in the complex α-plane.

2.11. OPTICAL BEAMS: DIFFRACTION-FREE BEAMS

The angular spectrum representation is useful in the characterization of *optical beams* (Kogelnik [2.48]–[2.50], Siegman [2.51]). These are paraxial (namely, very directional) solutions of the homogeneous Helmholtz equation. Details of their study by means of angular spectrum techniques may be found in the works by Wolf [2.2], or Mandel and Wolf [2.3].

Optical beams can be also regarded as source-free fields whose angular spectrum $a(p, q)$ vanishes for $p^2 + q^2 \ll 1$. Therefore the directional cosine m, given by Eq. (2.9a), may be approximated by the expansion: $m \sim 1 - \frac{1}{2}(p^2 + q^2)$. This yields from Eq. (2.12) the following angular spectrum representation for an optical beam propagating into $z > 0$ [2.2], [2.3]:

$$U(\mathbf{r}) = \exp(ikz)\iint_{-\infty}^{\infty} a(p, q)$$
$$\times \exp\left[-\frac{1}{2}ik(p^2 + q^2)z\right]\exp[ik(px + qy)]\, dp\, dq, \qquad z > 0. \quad (2.57)$$

With a similar expression, with negative sign of z, for a beam that propagates into the half-space $z < 0$.

Beam spreading, namely, increasing transversal cross section, is a diffraction phenomenon inherent to its propagation. The reader can check that if $U(x, y, 0) = A\exp[-W^{-2}(x^2 + y^2)]$, A and W being constants representing the amplitude and the *spot size*, respectively, at $z = 0$, then Eq. (2.57) leads to a function $U(x, y, z)$ whose width at $z = z_0$ (z_0 being a constant), increases as z_0 increases. (See also Problem 2.14.) It is possible, however, to obtain solutions of the wave equation that propagate in free space without any significant spreading. Such wavefields are known as *diffraction-free beams* and were investigated by Durnin [2.52] and Durnin *et al.* [2.53]. We briefly discuss now some essential concepts for these beams.

Let us consider the angular spectrum representation of a source-free field propagating into $z \geqslant 0$. This is written as (cf. Section 2.7):

$$U(\mathbf{r}) = \frac{ik}{2\pi} \int_{-\pi}^{\pi} d\phi \int_{0}^{\pi/2} d\theta \sin\theta F(\theta, \phi) \exp(ik\mathbf{s}\cdot\mathbf{r}) \tag{2.58}$$

Consider now the following angular spectrum:

$$F(\theta, \phi) = A(\phi) \frac{\delta(\theta - \theta_0)}{|\sin\theta_0|}, \qquad \theta_0 < \frac{\pi}{2}, \tag{2.59}$$

$A(\theta)$ being an arbitrary complex function.

On introducing Eq. (2.59) into Eq. (2.58) one obtains the following field:

$$U^{(df)}(\mathbf{r}) = \frac{ik}{2\pi} \exp(ikbz) \int_{-\pi}^{\pi} A(\phi) \exp[ika(x\cos\phi + y\sin\phi)]\, d\phi, \tag{2.60}$$

where:

$$a = \sin\theta_0, \tag{2.61a}$$

$$b = \cos\theta_0. \tag{2.61b}$$

It is clear that the wavefield given by Eq. (2.60) has a distribution of intensity across the z-axis which is conserved through propagation into $z \geqslant 0$, namely:

$$|U^{(df)}(x, y, z = constant \geqslant 0)|^2 = |U^{(df)}(x, y, z = 0)|^2, \tag{2.62}$$

where $U^{(df)}(\mathbf{r})$ represents a *diffraction-free beam.* Equations (2.59) and (2.60) show that such a field consists of a linear superposition of plane waves, all propagating at an angle θ_0 to the z-axis.

By writing the scalar product in the exponent of the integrand of Eq. (2.60) as: $x\cos\phi + y\sin\phi = R\cos\alpha$, where:

$$R = \sqrt{x^2 + y^2}, \tag{2.63}$$

and α being the angle between the vectors: (x, y) and $(\cos\phi, \sin\phi)$; Eq. (2.60) may be expressed as:

$$U^{(df)}(\mathbf{r}) = \frac{ik}{2\pi} \exp(ikbz) \int_{0}^{2\pi} A(\alpha) \exp(ikaR\cos\alpha)\, d\alpha. \tag{2.64}$$

We take now into account the well known integral representation of the Bessel functions $J_n(z)$ (see, e.g. Ref. 2.36):

$$2\pi i^n J_n(z) = \int_0^{2\pi} \exp(iz\cos\alpha)\exp(in\alpha)\,d\alpha. \tag{2.65}$$

On comparing Eq. (2.64) with Eq. (2.65) we see that by choosing:

$$A(\alpha) = \frac{i^{-(n+1)}}{k}\exp(in\alpha), \tag{2.66}$$

we obtain a diffraction-free beam which is expressed in terms of Bessel functions as:

$$U^{(df)}(\mathbf{r}) = \exp(ikbz)J_n(kaR). \tag{2.67}$$

Of course, a linear combination of functions of the form Eq. (2.66) will also yield a diffraction-free beam (see Ref. 2.54).

The simplest solution is the one corresponding to $n = 0$, namely:

$$U_0^{(df)}(\mathbf{r}) = \exp(ikbz)J_0(kaR). \tag{2.68}$$

The propagation of this field was studied in detail in Refs. 2.52 and 2.53. Observe that when $\theta_0 = 0$, one has $a = 0$, and then $U_0^{(df)}(\mathbf{r})$ becomes a plane wave.

The Bessel functions are not square integrable and, hence, they are not physically realizable. Therefore, such beams would have infinite energy. It is possible, however, to create real approximations over a finite area. The simulation of such a beam has been done in Ref. 2.52, and its practical realization has been reported in Ref. 2.53. Beams tapered with a Gaussian profile have been also studied [2.55].

The diameter d of the central spot of $U_0^{(df)}$ is given by the position of the first zero of $J_0(z)$ (this is located at $z \sim 3\pi/2$), and thus is: $d = 3\lambda/4a$. The minimum possible value of this diameter is $d = 3\lambda/4$.

Figures 2.5(a)–(e) show the propagation of $U_0^{(df)}$ compared with that of a Gaussian beam [2.2], [2.3], according to the numerical simulation of Durnin in Ref. 2.52. In Fig. 2.5(a) the J_0 beam with $d = 200\ \mu\text{m}$ at $\lambda = 0.5\ \mu\text{m}$, and an aperture radius of 2 mm, is shown together with a Gaussian beam of width $W_0 = 100\ \mu\text{m}$ (see Problem 2.14). Figures 2.5(b)–(e) show the intensity profiles at $z = 25$, 75, 100, and 120 cm, respectively. The intensity of the Gaussian beam appears multiplied by a factor of 10. As seen, the

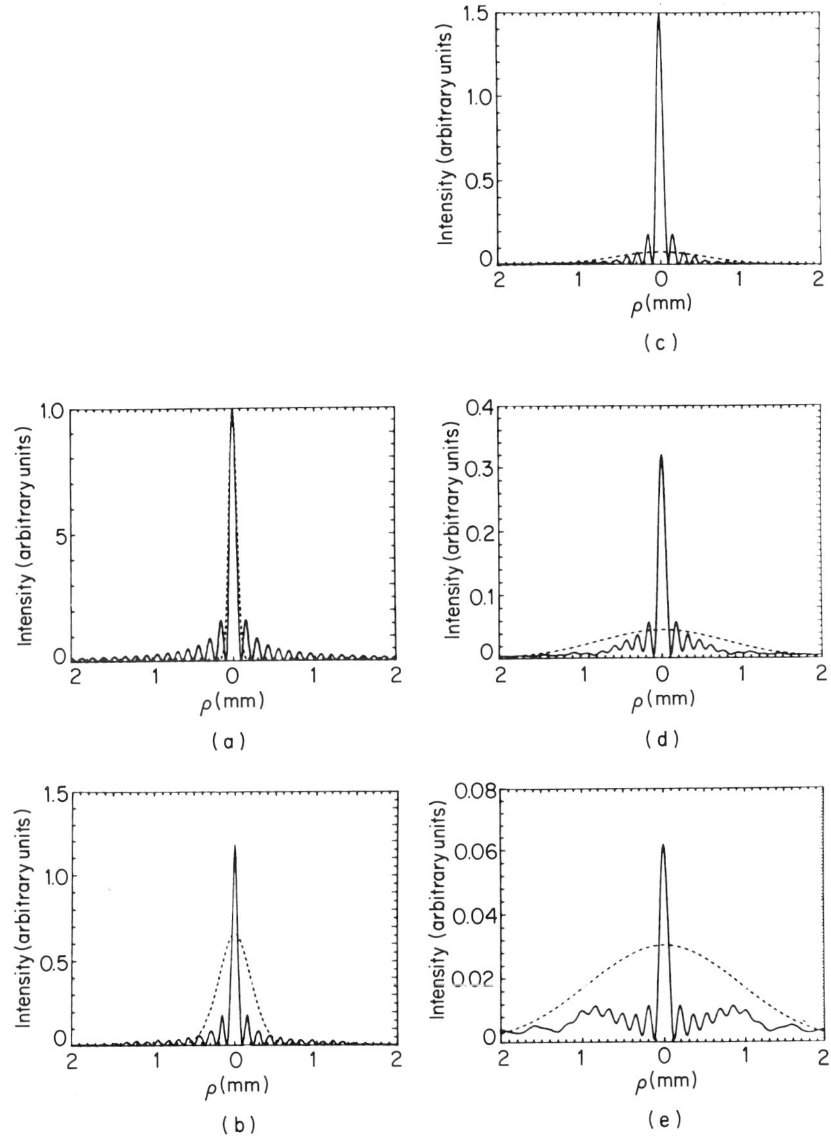

Figure 2.5. Intensity distributions for a J_0 beam (solid line) and a Gaussian beam (broken line), when (a) $z = 0$ (i.e., the initial plane where the beams are assumed to be formed), (b) $z = 25$ cm, (c) $z = 75$ cm, (d) $z = 100$ cm, and (e) $z = 120$ cm. The wavelength is $\lambda = 0.5$ μm. Note that in 2.5(b)–2.5(e) the intensity of the Gaussian beam has been multiplied by 10. (After J. Durnin, *J. Opt. Soc. Am. A* **4**, 651 (1987).)

J_0 beam propagates remarkably larger distances than the Gaussian beam, with its energy highly concentrated about its central maximum. The intensity of $U_0^{(df)}$ at the center $R = 0$ highly oscillates with z and starts to decay near $z = 1$ m; whereas the intensity of the Gaussian beam quickly decays monotonically as z increases. The range over which the J_0 beam remains concentrated increases by making the aperture diameter larger at $z = 0$.

2.12. ASYMPTOTIC APPROXIMATIONS TO ANGULAR SPECTRUM REPRESENTATIONS

In Section 2.8 we discussed the asymptotic approximations for source-free fields. Now we shall present the asymptotic expressions for wavefields containing both homogeneous and evanescent components. Namely, we shall consider a well behaved outgoing field propagating into $z > 0$, satisfying the homogeneous Helmholtz equation (2.1) and Sommerfeld's radiation condition, Eq. (2.6), in $z > 0$. As it is already known, this field admits the angular spectrum representation, Eq. (2.12) (or equivalently, Eq. (2.23) in angular variables), which we shall now write for convenience in the form:

$$U(\mathbf{r}) = U_H(\mathbf{r}) + U_I(\mathbf{r}), \tag{2.69}$$

where:

$$U_H(\mathbf{r}) = \iint_{p^2+q^2 \leqslant 1} a(p, q) \exp[ik(px + qy + mz)]\, dp\, dq, \tag{2.70a}$$

$$U_I(\mathbf{r}) = \iint_{p^2+q^2 > 1} a(p, q) \exp[ik(px + qy + mz)]\, dp\, dq. \tag{2.70b}$$

In Eq. (2.70a) $U_H(\mathbf{r})$ represents the part of $U(\mathbf{r})$ restricted to the contribution of homogeneous waves. Whereas in Eq. (2.70b) $U_I(\mathbf{r})$ constitutes the part of $U(\mathbf{r})$ restricted to inhomogeneous (evanescent) waves, and m is related to p and q through Eqs. (2.9a) and (2.9b).

Let (r, θ, ϕ) be the spherical coordinates of the field point $\mathbf{r}$, with the polar axis $\theta = 0$ taken along the positive x-axis. In order to establish the asymptotic expressions of $U(\mathbf{r})$, the point $\mathbf{r}$ is allowed to vary in a fixed direction (θ, ϕ), where $0 < \theta \leqslant \pi/2$, $0 < \phi \leqslant 2\pi$.

Due to the preferential role of the z-axis, as well as the special case that the plane $z = 0$ plays in the angular spectrum representation, Eqs. (2.70), the cases $0 < \theta < \pi/2$, $\theta = \pi/2$ and $\theta = 0$ are each treated separately.

The following results were obtained by Sherman *et al.* [2.56] by means of the method of the stationary phase. We refer the reader to that reference for their proof.

It is convenient to write:

$$a(p, q) = \frac{W(p, q)}{m}, \tag{2.71}$$

where $W(p, q)$ is bounded for all p and q.

2.12.1. Approximations for $0 < \theta < \pi/2$, $0 \leqslant \phi \leqslant 2\pi$

In this case one has as $kr \to \infty$ the following approximation for the homogeneous and evanescent parts U_H and U_I:

$$U_H(\mathbf{r}) = -\frac{2\pi i}{kr} W(\sin\theta\cos\phi, \sin\theta\sin\phi)\exp(ikr) + U_{H_1}(\mathbf{r}) + 0\left[\frac{1}{(kr)^2}\right], \tag{2.72}$$

$$U_I(\mathbf{r}) = -U_{H_1}(\mathbf{r}) + 0\left[\frac{1}{(kr)^2}\right], \tag{2.73}$$

as $kr \to \infty$ in the (θ, ϕ)-direction with fixed k, θ, and ϕ. Where:

$$\begin{aligned} U_{H_1}(\mathbf{r}) = \left(\frac{2\pi}{\sin\theta}\right)^{1/2} \frac{\exp(i\pi/4)}{(kr)^{3/2}\cos\theta} & [W(\cos\phi, \sin\phi)|_{\theta=0} \\ & \times \exp(ikr\sin\theta) + iW(-\cos\phi, -\sin\phi)|_{\theta=0} \\ & \times \exp(-ikr\sin\theta)] + 0\left[\frac{1}{(kr)^2}\right]. \end{aligned} \tag{2.74}$$

Equations (2.72)–(2.74) show that $U_I(\mathbf{r})$ is of higher order in $1/kr$ than $U_H(\mathbf{r})$.

2.12.2. Approximations for $\theta = \pi/2$

In the plane $z = 0$, for $kr \to \infty$ one has:

$$\begin{aligned} U_H(\mathbf{r}) = -\frac{i\pi}{k}\Bigg[& \frac{\exp(ikr)}{r} W(\cos\phi, \sin\phi)|_{\theta=0} \\ & - \frac{\exp(-ikr)}{r} W(-\cos\phi, -\sin\phi)|_{\theta=0}\Bigg] + 0\left[\frac{1}{(kr)^{3/2}}\right], \end{aligned} \tag{2.75}$$

$$U_I(\mathbf{r}) = -\frac{i\pi}{k}\left[\frac{\exp(ikr)}{r} W(\cos\phi, \sin\phi)|_{\theta=0} + \frac{\exp(-ikr)}{r} W(-\cos\phi, -\sin\phi)|_{\theta=0}\right] + 0\left[\frac{1}{(kr)^{3/2}}\right]. \quad (2.76)$$

In this case $U_H(\mathbf{r})$ and $U_I(\mathbf{r})$ are of the same order in $(1/kr)$.

In the special case in which $W(p, q)$ is of the form:

$$W(p, q) = mM(p, q), \quad (2.77)$$

one has as $kr \to \infty$:

$$U_H(\mathbf{r}) = -U_I(\mathbf{r}) = \frac{\sqrt{2\pi}}{(kr)^{3/2}} \exp\left(i\frac{\pi}{4}\right)[\exp(ikr)M(\cos\phi, \sin\phi)|_{\theta=0} + i\exp(-ikr)M(-\cos\phi, -\sin\phi)|_{\theta=0}] + 0\left[\frac{1}{(kr)^2}\right]. \quad (2.78)$$

Equation (2.78) shows that, again, $U_H(\mathbf{r})$ and $U_I(\mathbf{r})$ are of the same order in $(kr)^{-1}$.

2.12.3. Approximations in the Axial Direction

Along the z-axis, $\theta = 0$, we have the following asymptotic approximation as $kz \to \infty$:

$$U_H(\mathbf{r}) = -2\pi i \frac{\exp(ikz)}{kz} W(0, 0)|_{\theta=0} + \frac{i}{kz} S\left(\frac{\pi}{2}\right) + 0\left[\frac{1}{(kz)^{3/2}}\right], \quad (2.79)$$

$$U_I(\mathbf{r}) = -\frac{i}{kz} S\left(\frac{\pi}{2}\right) + 0\left[\frac{1}{(kr)^{3/2}}\right], \quad (2.80)$$

where:

$$S(\theta) = \int_0^{2\pi} W(\theta, \phi)\, d\phi. \quad (2.81)$$

Hence, U_H and U_I are of the same order in $(kr)^{-1}$.

The leading term in the above asymptotic approximations when $kr \to \infty$ is also known as the *far field*.

Finally, the asymptotic behavior of the sum $U(\mathbf{r}) = U_H(\mathbf{r}) + U_I(\mathbf{r})$ is:

$$U(\mathbf{r}) = -\frac{2\pi i}{k} W(\sin\theta\cos\phi, \sin\theta\sin\phi)\frac{\exp(ikr)}{r} + 0\left[\frac{1}{(kr)^2}\right],$$

$$\text{for } 0 < \theta < \frac{\pi}{2}, 0 < \phi < 2\pi,$$

$$U(\mathbf{r}) = -\frac{2\pi i}{k} W(\cos\phi, \sin\phi)|_{\theta=0}\frac{\exp(ikr)}{r} + 0\left[\frac{1}{(kr)^{3/2}}\right], \qquad \text{for } \theta = \frac{\pi}{2},$$

$$U(\mathbf{r}) = -\frac{2\pi i}{k} W(0, 0)|_{\theta=0}\frac{\exp(ikz)}{z} + 0\left[\frac{1}{(kz)^{3/2}}\right], \qquad \text{for } \theta = 0.$$

Hence, only the second term may differ according to the values of θ. This shows that for $\theta = \pi/2$ or $\theta = 0$ this term decays with $kr \to \infty$ as $(kr)^{-3/2}$ only. The leading term, on the other hand, is of the form of Eqs. (2.40).

2.13. CONTRIBUTION OF THE EVANESCENT COMPONENTS IN THE ASYMPTOTIC EXPRESSIONS

In Section 2.4 we pointed out the loss of spatial frequency information due to the exponential decay of the evanescent wave amplitudes with increasing distance z from the plane $z = 0$. However, whereas the asymptotic expressions for the field $U(\mathbf{r})$ in $0 < \theta < \pi/2$, $0 \leqslant \phi \leqslant 2\pi$ show that the part $U_I(\mathbf{r})$ due to the evanescent wave contribution is one order higher in $1/(kr)$ than the part $U_H(\mathbf{r})$ due to homogeneous waves; in other directions, such as $\theta = \pi/2$ or $\theta = 0$, Eqs. (2.75)–(2.80) show that $U_I(\mathbf{r})$ can be of the same order as $U_H(\mathbf{r})$ in $(kr)^{-1}$; thus, neglecting the contribution of evanescent waves in the angular spectrum representation of $U(\mathbf{r})$ can be inadequate in some cases. Even asymptotically, the contribution of these components may be important. This result seems surprising at first sight due to the exponential decay of these waves with increasing distance z. The following examples aim to clarify this point:

Example 2.1 Let us consider the scalar wavefield $U(\mathbf{r})$ in the half-space $z > 0$, radiated by a time-harmonic source of finite strength with density $\rho(\mathbf{r})$ contained inside a volume V located in $z < 0$. The complex amplitude of this field satisfies the inhomogeneous Helmholtz equation:

$$\nabla^2 U + k^2 U = -4\pi\rho(\mathbf{r}). \tag{2.82}$$

Equation (2.82) is analogous to Eq. (1.21) of Chapter 1; its solution outside the volume V occupied by the source is (cf. Eq. (1.54)):

$$U(\mathbf{r}) = \int_V \rho(\mathbf{r}')G(\mathbf{r}, \mathbf{r}')\, d^3r'. \tag{2.83}$$

By introducing the angular spectrum expansion of $G(\mathbf{r}, \mathbf{r}')$, Eq. (2.51), into Eq. (2.83), and taking into account that $|z - z'| = z - z'$ in $z > 0$, Eq. (2.83) may be written in $z > 0$ as:

$$U(\mathbf{r}) = \iint_{-\infty}^{\infty} a(p, q) \exp[ik(px + qy + mz)]\, dp\, dq, \tag{2.84}$$

with:

$$a(p, q) = \frac{ik}{2\pi m} \int_V \rho(\mathbf{r}') \exp(-ik\mathbf{s}\cdot\mathbf{r}')\, d^3r', \qquad \mathbf{s} = (p, q, m). \tag{2.85}$$

From Eqs. (2.71), (2.81) and (2.85) it is seen that $S(\theta)$ given by Eq. (2.81) does not vanish at $\theta = \pi/2$, therefore both $U_H(\mathbf{r})$ and $U_I(\mathbf{r})$ are necessary to describe $U(\mathbf{r})$ adequately along the positive z-axis. This behavior is due to the singularity on the circle $p^2 + q^2 = 1$ in the factor $1/m$ in (2.85). ■

Example 2.2 Let us consider the divergent spherical wave $G(\mathbf{r}) = \exp(ikr)/r$ originated at a point source $\mathbf{r} = \mathbf{0}$. This is a particular instance of the field of Example 2.1 when $\rho(\mathbf{r}) = \delta(\mathbf{r})$. In this case:

$$a(p, q) = \frac{ik}{2\pi} \frac{1}{m}, \tag{2.86}$$

and one has from Eq. (2.81):

$$S\left(\frac{\pi}{2}\right) = ik. \tag{2.87}$$

The expression for $G(\mathbf{r})$ along the z-axis is easily evaluated from Eqs. (2.69), (2.70a), (2.70b) and (2.86); it coincides with the leading terms of Eq. (2.79) and (2.80) when $kz \to \infty$:

$$G(z) = G_H(z) + G_I(z), \tag{2.88}$$

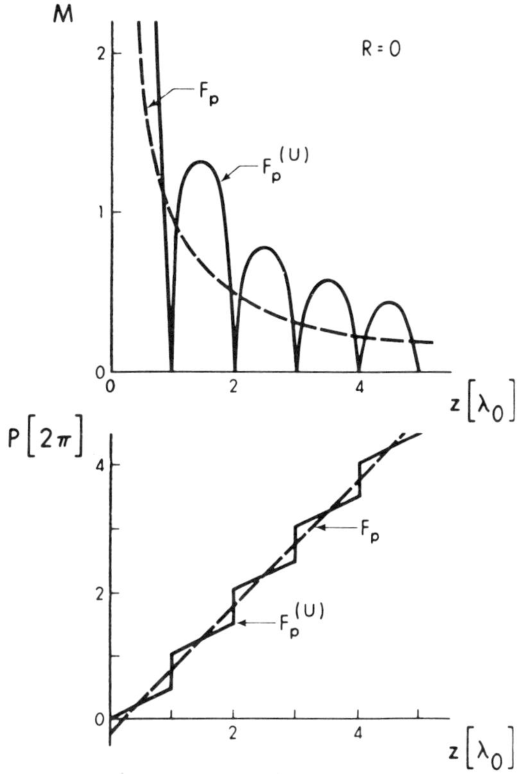

Figure 2.6. Magnitude and phase along the z-axis (in units of λ) of the homogeneous components $G_H(z) \equiv F_P(U)$ and of the total wavefield $G(\mathbf{r}) \equiv F_P$ of a spherical wave created by a point source at the origin. M is the magnitude $|G(\mathbf{r})|$ of the field; P is the phase of $G(\mathbf{r})$. (After H. G. Schmidt-Weinmar, *J. Opt. Soc. Am.* **65**, 1054 (1975).)

with:

$$G_H(z) = \frac{\exp(ikz)}{z} - \frac{1}{z}. \tag{2.89}$$

$$G_I(z) = \frac{1}{z}. \tag{2.90}$$

Equations (2.89)–(2.90) show the error $-1/z$ that would be introduced if the part due to evanescent waves, $G_I(\mathbf{r})$, were neglected in the representation, Eq. (2.88), of $G(\mathbf{r})$.

Figure 2.6, due to Schmidt-Weinmar [2.57], shows the magnitude and the phase along the z-axis, in units of the wavelength λ, of the homogeneous components $G_H \equiv F_p(U)$ and of the sum $G(\mathbf{r}) = G_H(\mathbf{r}) + G_I(\mathbf{r}) \equiv F_p$ for the spherical wave $G(\mathbf{r})$. The difference between $|G(\mathbf{r})|$ which does not oscillate, and $|G_H(\mathbf{r})|$ which has zeros at $z_n = n$, ($n = 1, 2, 3, \ldots$), is evident. Also, Figs. 2.7 and 2.8, and Figs. 2.9 and 2.10, show the contour lines indicating loci of constant magnitude and phase, respectively (Carter [2.58]), for the spherical wave over the xz-plane when it is represented by $G_H(\mathbf{r}) \equiv \mathscr{S}_1$ only, and when it is represented by $G_H^{(1/2)}(\mathbf{r})$ (where $G_H^{(1/2)}(\mathbf{r})$ takes from Eq. (2.70a) only those homogeneous components for which $p^2 + q^2 \leqslant (1/2)^2$), showing that the spherical wave is very modified if, in addition, part of its homogeneous components are removed. ■

The above examples show the error that in some cases may arise by excluding evanescent components in the angular spectrum representation of a wavefield on the z-axis, or in the plane $z = 0$.

Example 2.3 When $U(\mathbf{r})$ is expressible in $z \geqslant 0$ from the boundary value (cf. Section 2.3):

$$U(x, y, 0) = f(x, y), \tag{2.91}$$

satisfying:

$$\iint_{-\infty}^{\infty} |f(x, y)|^2 \, dx \, dy < \infty, \tag{2.92}$$

then, as shown in Eq. (2.16):

$$a(p, q) = \left(\frac{k}{2\pi}\right)^2 \iint_{-\infty}^{\infty} f(x, y) \exp[-ik(px + qy)] \, dx \, dy. \tag{2.93}$$

On comparing Eq. (2.93) and (2.71) one can readily see that in this case $W(p, q)$ is of the form Eq. (2.77) with:

$$a(p, q) = M(p, q). \tag{2.94}$$

It is then straightforward to derive from Eq. (2.81) that $S(\pi/2) = 0$ (this result has been obtained by Foley and Wolf [2.59]). By taking into account Eqs. (2.79) and (2.80) this shows that the asymptotic expression for $U_I(\mathbf{r})$ is of higher order in $(1/kr)$ than the asymptotic expression of $U_H(\mathbf{r})$ and, hence, $U_I(\mathbf{r})$ can be neglected in the *far field* approximation, namely, at $kr \to \infty$. ■

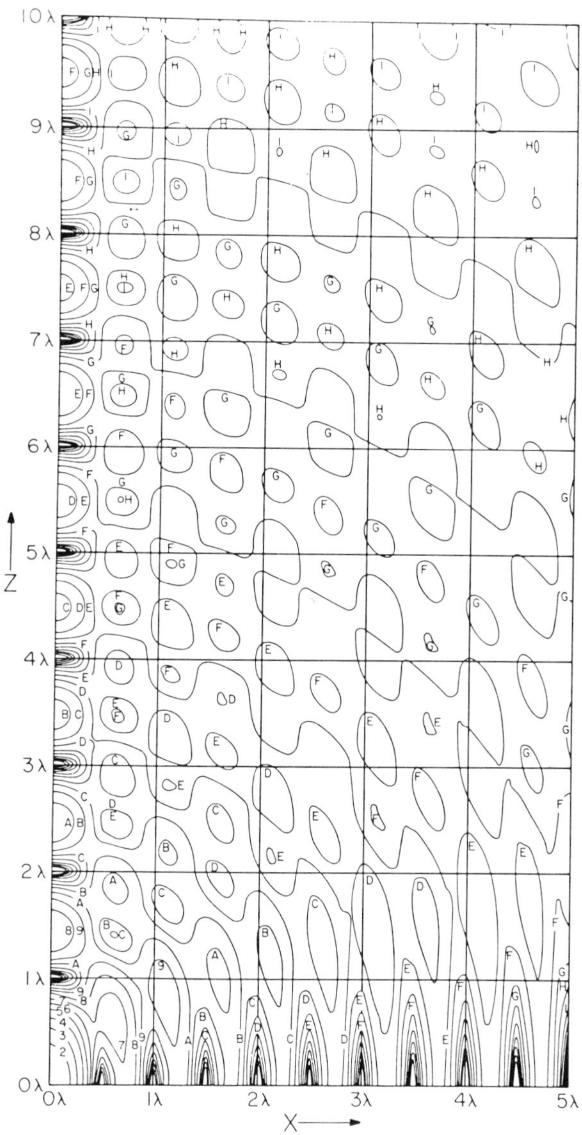

Figure 2.7. Magnitude of $\mathscr{S}_1$ over the xz-plane. The contour lines indicate loci of constant magnitude of $\mathscr{S}_1$ as labeled: $1 = 1.0$, $2 = 0.78$, $3 = 0.606$, $4 = 0.472$, $5 = 0.368$, $6 = 0.286$, $7 = 0.223$, $8 = 0.174$, $9 = 0.135$, $A = 0.105$, $B = 0.082$, $C = 0.064$, $D = 0.050$, $E = 0.039$, $F = 0.030$, $G = 0.023$, $H = 0.018$, $I = 0.014$. These data have been normalized with a maximum value of unity at the origin. The contour levels are also given by the formula $|\mathscr{S}_1| = \exp[-0.25(N-1)]$ for integer values of N (where $A = 10$, $B = 11$, etc.). (After W. H. Carter, *J. Opt. Soc. Am.* **65**, 1054 (1975).)

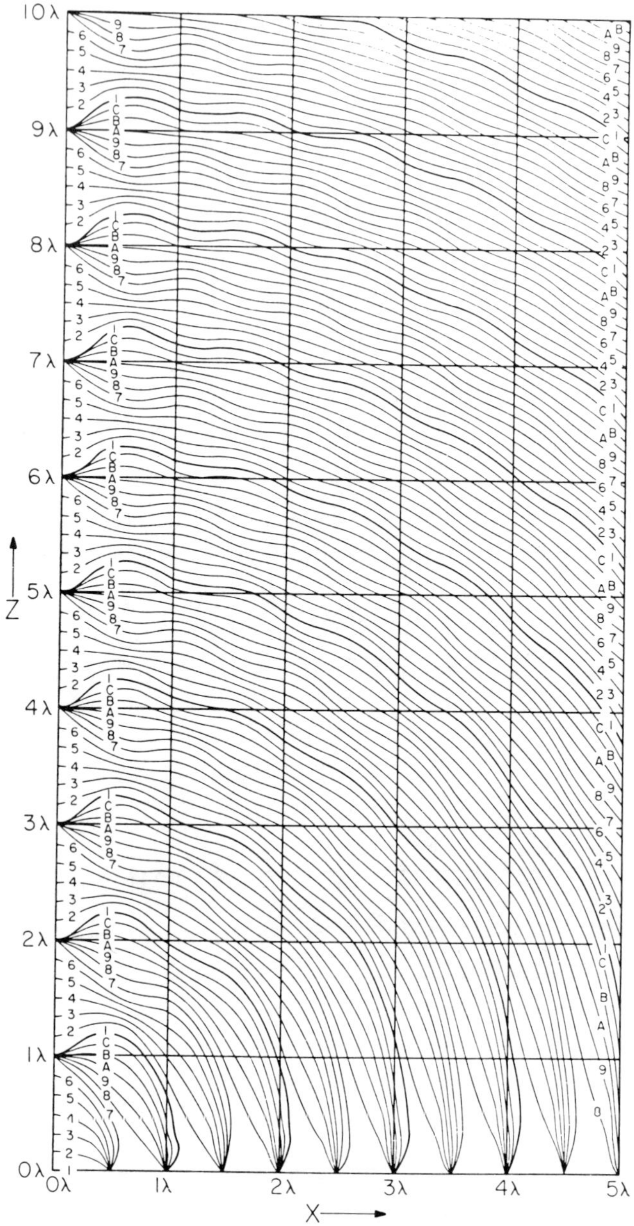

Figure 2.8. Phase of $\mathscr{S}_1$ over the xz-plane, corresponding to the magnitude in Fig. 2.7. The contour lines indicate loci of constant phase 30° apart. The heavy contour lines are 360° apart. (After W. H. Carter, *J. Opt. Soc. Am.* **65**, 1054 (1975).)

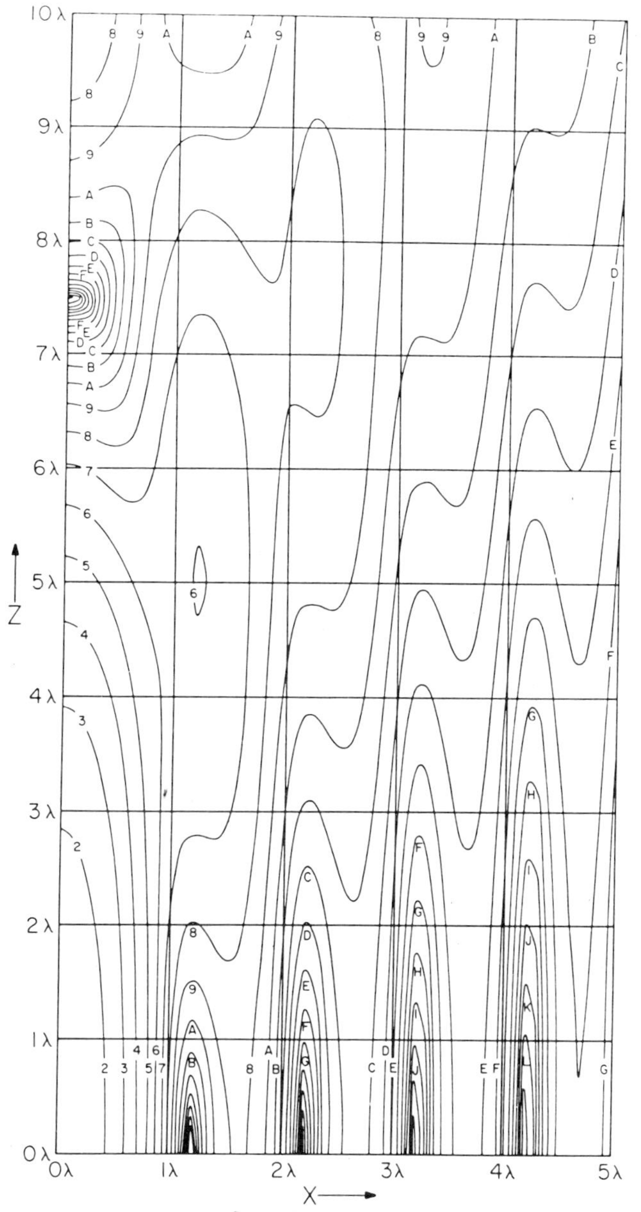

Figure 2.9. Magnitude of $\mathscr{S}_1$ over the xz-plane. This field contains only homogeneous waves that propagate in directions from the origin within a cone that makes an angle of 30° with the z-axis. The format of the contour lines is the same as for Fig. 2.7. (After W. H. Carter, *J. Opt. Soc. Am.* **65**, 1054 (1975).)

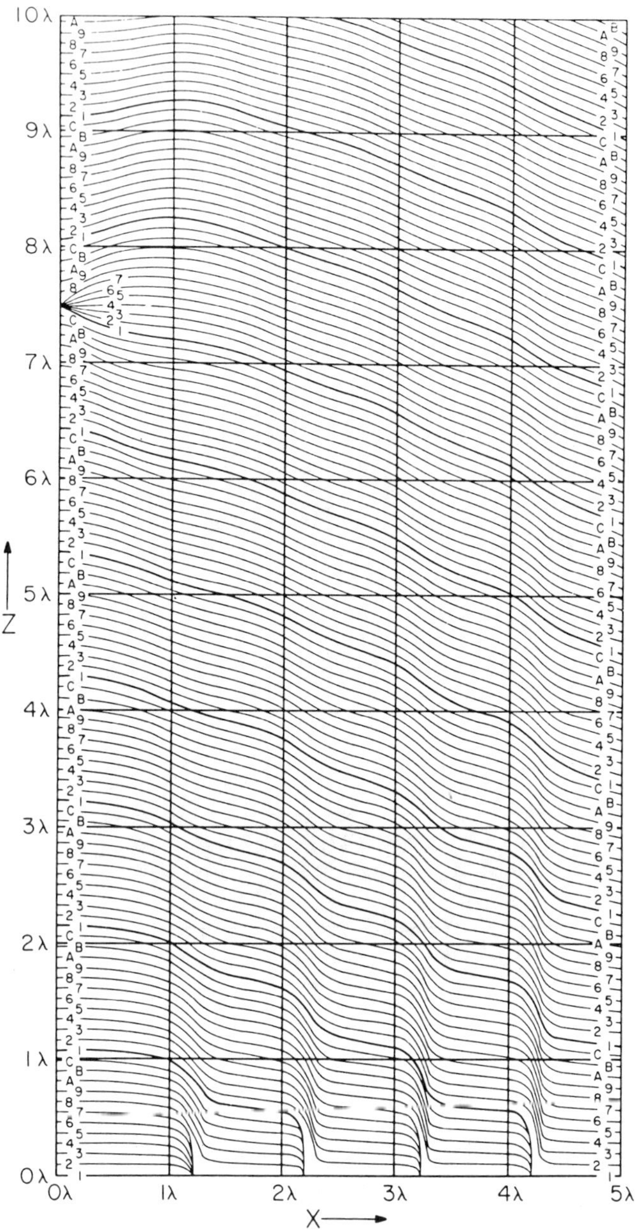

Figure 2.10. Phase of $\mathscr{S}_{1/2}$ over the *xz*-plane, associated to the magnitude of Fig. 2.9. The format for the contour lines is the same as for Fig. 2.9. (After W. H. Carter, *J. Opt. Soc. Am.* **65**, 1054 (1975).)

These examples show that there are instances in which $U_H(\mathbf{r})$ and $U_I(\mathbf{r})$ cannot be treated as independent entities, and thus, in some cases, $U_I(\mathbf{r})$ may be as important as $U_H(\mathbf{r})$, even at large distances from sources. As a matter of fact, it might happen that $U_H(\mathbf{r})$ alone violate Sommerfeld's radiation condition. This occurs for example for the homogeneous part on the z-axis of the field created by a point source, $G_H(\mathbf{r})$, given by Eq. (2.89). This fact has been studied by the authors of Refs. 2.57 and 2.60.

PROBLEMS

2.1. Assuming a field with zero phase and nonzero constant amplitude inside a circular aperture of radius a at the plane $z = 0$. Find:

a. The angular spectrum of this field.

b. The asymptotic expression for the field.

2.2. Show that the angular spectrum of a source-free field $U(\mathbf{r})$ satisfying the Helmholtz equation everywhere and with angular components propagating in all directions, may be expressed as (Wolf [2.23]):

$$F(\theta, \phi) = \frac{k}{2\pi^2} \lim_{\varepsilon \to 0} \int_{k-\varepsilon}^{k+\varepsilon} dK \int U(\mathbf{r}) \exp(-iK\mathbf{s}\cdot\mathbf{r})\, d^3r,$$

where $\mathbf{s}$ is the unit vector $\mathbf{s} = (\sin\theta\cos\phi, \sin\theta\sin\phi, \cos\theta)$.

2.3. Let $U(\mathbf{r}) = \exp(ik\mathbf{n}_0\cdot\mathbf{r})$ be a plane wave with direction of propagation given by the unit vector $\mathbf{n}_0 = (\sin\theta_0\cos\phi_0, \sin\theta_0\sin\phi_0, \cos\theta_0)$.

a. Show that its angular spectrum is:

$$F(\theta, \phi) = \frac{4\pi}{k}\,\Delta(\mathbf{n} - \mathbf{n}_0),$$

where $\Delta(\mathbf{n} - \mathbf{n}_0)$ is the spherical Dirac delta function:

$$\Delta(\mathbf{n} - \mathbf{n}_0) = \frac{1}{|\sin\theta|}\,\delta(\theta - \theta_0)\,\delta(\phi - \phi_0),$$

and $\mathbf{n} = (\sin\theta\cos\phi, \sin\theta\sin\phi, \cos\theta)$.

b. Show that for $kr \to \infty$ one has:

$$U(\mathbf{r}) \sim \frac{2\pi}{ik}\left[\Delta(\mathbf{n} - \mathbf{n}_0)\frac{\exp(ikr)}{r} - \Delta(\mathbf{n} + \mathbf{n}_0)\frac{\exp(-ikr)}{r}\right].$$

2.4. Show that the asymptotic expression for the density of flow of energy $\mathscr{S}(\mathbf{r})$ of a scalar wavefield $U(\mathbf{r})$:

$$\mathscr{S}(\mathbf{r}) = \frac{ck}{8\pi i}(U^* \nabla U - U \nabla U^*),$$

c being a constant, is given by:

$$\mathscr{S}(\mathbf{r}) = \frac{ck^2}{4\pi} U^*(\mathbf{r})U(\mathbf{r})\hat{\mathbf{r}},$$

where $\hat{\mathbf{r}} = \dfrac{\mathbf{r}}{r}$.

2.5. Let $U(\mathbf{r})$ be a wavefield propagating into the half-space $z \geqslant 0$. Show that the evanescent waves do not contribute to the flow of energy entering this half-space; i.e., flowing across the plane $z = 0$.

Show also that the energy is conserved; namely, the flow of energy across the plane $z = 0$ is equal to the flow of energy across the hemisphere at infinity (i.e., with radius $R \to \infty$), in the half-space $z > 0$. This is expressed as:

$$\iint_{z=0} \mathscr{S}\cdot\hat{\mathbf{z}}\, dx\, dy = \int_{R\to\infty} \mathscr{S}(r\mathbf{s})\cdot\hat{\mathbf{r}}\, ds.$$

$$\left(\hat{\mathbf{r}} = \frac{\mathbf{r}}{r}, \mathbf{s} = (p, q, m), ds = r^2\, d\Omega\right),$$

(Ω is the element of solid angle).

2.6. A scalar field $U(\mathbf{r})$ obeys the homogeneous Helmholtz equation (2.1) throughout the half-space $z > 0$ and behaves as an outgoing spherical wave at infinity. It has also the property that in two planes $z = z_1 > 0$ and $z = z_2 > z_1$:

$$U(x, y, z_1) = U(x, y, z_2),$$

for all values of x and y.

Show that:

a. The field distribution in the plane $z = 0$ can have no other angular components of its angular spectrum representation than those that satisfy the condition:

$$u^2 + v^2 = k^2 - \left(\frac{2\pi n}{z_1 - z_2}\right), \qquad (n = 0, 1, 2, \ldots).$$

b. The field distributions will be identical in all planes that are at distances $l(z_2 - z_1)$ from the plane $z = z_1$ throughout the half-space $z > 0$, where l is an integer.

These fields are said to have the property of *self-imaging*.

2.7. Prove that the field $U(\mathbf{r})$ radiated by a spherically symmetric source distribution $\rho(\mathbf{r}) = \rho(r)$, confined to a sphere of radius a, is proportional to a divergent spherical wave, namely:

$$U(\mathbf{r}) = A(k)\,\frac{\exp(ikr)}{r},$$

where:

$$A(k) = 4\pi \int_0^a \rho(r')\sin(kr')r'\,dr'.$$

(Hint: use Eqs. (2.82)–(2.85).)

2.8. Using the orthogonality and closure properties of spherical harmonics Y_l^m (see, e.g. Ref. 2.36), show that the divergent spherical wave can be expanded as [2.10]:

$$\frac{\exp(ikr)}{r} = ik \sum_{l=0}^{\infty} \sum_{m=-l}^{l} a_l^m h_l^{(1)}(kr)\,Y_l^m(\theta, \phi),$$

with:

$$a_l^m = i^l \int_{-\pi}^{\pi} d\beta \int_0^{\pi} d\alpha \sin\alpha\, Y_{lm}^*(\alpha, \beta),$$

and:

$$h_l^{(1)}(kr)\,Y_l^m(\theta, \phi) = (-i)^l \frac{1}{2\pi} \int_{-\pi}^{\pi} d\beta \int_{\mathscr{D}^\pm} d\alpha \sin\alpha\, Y_l^m(\alpha, \beta) \exp(ik\mathbf{s}\cdot\mathbf{r}),$$

$$\mathbf{s} = (\sin\alpha\cos\beta, \sin\alpha\sin\beta, \cos\alpha)$$

$h_l^{(1)}$ are spherical Hankel functions of the first kind.

2.9. Show that a monochromatic outgoing scalar wavefield expressible by the angular spectrum representation, Eq. (2.23), can be expanded as:

$$U(\mathbf{r}) = ik \sum_{l=0}^{\infty} \sum_{m=-l}^{l} a_l^m h_l^{(1)}(kr)\,Y_l^m(\theta, \phi),$$

with:

$$a_l^m = i^l \int_{-\pi}^{\pi} d\beta \int_0^{\pi} d\alpha \sin \alpha F(\alpha, \beta) Y_l^{m*}(\alpha, \beta).$$

This is the multipole expansion of the wavefield (Devaney and Wolf [2.10]).

Find the corresponding multipole expansion for a general incoming wavefield.

2.10. The spherical wave admits the integral representation:

$$G(r) = \frac{1}{2\pi^2} \int_{-\infty}^{\infty} \frac{\exp(i\mathbf{k}\cdot\mathbf{r})}{\mathbf{k}^2 - k^2} d^3k,$$

$$\mathbf{k} = (k_x, k_y, k_z) = k(p, q, m), \quad k = \frac{\omega}{c}.$$

(This expression can be readily derived either from the expansion of the spherical wave in eigenfunctions of the Helmholtz equation: $1/(2\pi)^3 \exp(i\mathbf{k}\cdot\mathbf{r})$, or by direct Fourier transformation of the equation $\nabla^2 G + k^2 G = -4\pi\delta(\mathbf{r})$.)

By performing the k_z-integration in the upper half complex k_z-plane along a contour formed by the real axis and a semicircle of infinite radius, show that the above integral leads to the Weyl expansion, either for a divergent wave, Eq. (2.51), or for a convergent wave, Eq. (2.52), according to whether the k_z-integration contour excludes the pole $k_z = -\sqrt{k^2 - k_x^2 - k_y^2}$ or the pole $k_z = \sqrt{k^2 - k_x^2 - k_y^2}$.

2.11. Show that a source-free field $U(\mathbf{r})$ that satisfies the homogeneous Helmholtz equation everywhere and has an angular spectrum with components propagating in all directions (Eq. (2.30)):

$$U(\mathbf{r}) = \frac{k}{4\pi} \int_{-\pi}^{\pi} d\phi \int_0^{\pi} d\theta \sin \theta F(\theta, \phi) \exp(ik\mathbf{s}\cdot\mathbf{r}),$$

can be expressed as the difference between an outgoing and an incoming field [2.47].

2.12. Show that if the source-free field of Problem 2.11 is identified with an outgoing field of the form Eq. (2.23) in their common domain of validity (namely, in $z > 0$ far enough from sources); then by using angular variables of integration we obtain:

$$\int_{-\pi}^{\pi} d\beta \int_{-\pi/2}^{\pi} d\alpha \sin \alpha F(\alpha, \beta) \exp(ik\mathbf{s}\cdot\mathbf{r})$$

$$= \int_{-\pi}^{\pi} d\beta \int_{\pi/2+i0}^{\pi/2-i\infty} d\alpha \sin \alpha A(\alpha, \beta) \exp(ik\mathbf{s}\cdot\mathbf{r}),$$

i.e., the contribution of the homogeneous waves propagating into $z < \infty$ is equal to the contribution of the inhomogeneous waves of the outgoing wavefield (this result was obtained by Sherman *et al.* [2.61]). Show also that if the result of Problem 2.11 is taken into account, the above equation is equivalent to making the convergent wavefield equal to zero.

2.13. Prove that the solution of the paraxial (or parabolic) wave equation:

$$\frac{\partial^2\psi}{\partial x^2} + \frac{\partial^2\psi}{\partial y^2} + 2ik\frac{\partial\psi}{\partial z} = 0$$

corresponds to a beam, Eq. (2.57), expressed as:

$$U(\mathbf{r}) = \exp(ikz)\psi(\mathbf{r}).$$

2.14. The wavefield of a monochromatic Gaussian scalar beam, propagating into $z > 0$ is the solution to the homogeneous Helmholtz equation in $z > 0$ which satisfies the boundary condition at $z = 0$:

$$f(x, y) = A \exp\left[-\frac{(x^2 + y^2)}{W^2}\right],$$

where W is called the *spot size* of the beam.

a. Show that the angular spectrum of this beam is:

$$a(p, q) = A\frac{k^2W^2}{4\pi}\exp\left[-\frac{1}{4}(kW)^2(p^2 + q^2)\right].$$

b. On introducing into Eq. (2.57) the expression for $a(p, q)$ obtained in (a), find the expression for the beam field at a distance $z = z_0 > 0$.

c. Show that for $kr \to \infty$, and in a fixed direction (θ, ϕ), $0 < \theta \leqslant \pi/2$,

$0 < \phi \leqslant 2\pi$, this field has the asymptotic expression:

$$U(\mathbf{r}) = -iAk\frac{W^2}{2} \times \cos\theta \exp\left[-\frac{1}{4}(kW)^2 \sin^2\theta\right]\frac{\exp(ikr)}{r} + 0\left[\frac{1}{(kr)^2}\right].$$

d. Show that for $kr \to \infty$ the asymptotic expression on the z-axis is:

$$U(\mathbf{r}) = -iAk\frac{W^2}{2}\frac{\exp(ikz)}{z} + O\left[\frac{1}{(kz)^{3/2}}\right].$$

REFERENCES

2.1. J. R. Shewell and E. Wolf, *J. Opt. Soc. Am.* **57**, 1490 (1967).

2.2. E. Wolf, *Lecture Notes Presented at the National Bureau of Standards Under the Wave Optics Competence Program* (1979), (unpublished).

2.3. L. Mandel and E. Wolf, *Optical Coherence and Quantum Optics*, Cambridge University Press, Cambridge, Chapters 3 and 5 (to be published).

2.4. P. M. Woodward and J. D. Lawson, *J.I.E.E.*, Part III **95**, 363 (1948).

2.5. H. G. Booker and P. C. Clemmow, *J.I.E.E.*, Part III **97**, 11 (1950).

2.6. H. G. Booker, J. A. Ratcliffe and D. H. Shinn, *Phil. Trans. Roy. Soc.* A **242**, 579 (1950).

2.7. D. R. Rhodes, *Proc. I.E.E.E.* **52**, 1013 (1964).

2.8. D. M. Kerns, *J. Acoust. Soc. Am.* **57**, 497 (1975).

2.9. E. Wolf. *Proc. Phys. Soc.* **74**, 269 (1959).

2.10. A. J. Devaney and E. Wolf, *J. Math. Phys.* **15**, 234 (1974).

2.11. E. Lalor and E. Wolf, *Phys. Rev. Lett.* **26**, 1274 (1971).

2.12. A. Walther, *J. Opt. Soc. Am.* **57**, 639 (1967); *Am. J. Phys.* **35**, 808 (1967).

2.13. P. C. Clemmow, *Proc. Roy. Soc.* A **205**, 286 (1951).

2.14. C. J. Bouwkamp, *Rep. Prog. Phys.* **17**, 35 (1954).

2.15. J. A. Ratcliffe, *Rep. Prog. Phys.* **19**, 188 (1956).

2.16. D. Gabor, in *Progress in Optics*, Vol. 1, E. Wolf, ed., North-Holland, Amsterdam, 1961, p. 109.

2.17. G. C. Sherman, *J. Opt. Soc. Am.* **57**, 1490 (1967).

2.18. E. Wolf, *Opt. Comm.* **1**, 153 (1969).

2.19. E. Wolf and J. R. Shewell, *J. Math. Phys.* **11**, 2254 (1970).

2.20. M. Nieto-Vesperinas, *J. Opt. Soc. Am.* **72**, 539 (1982).

2.21. R. Petit, ed., *Electromagnetic Theory of Gratings*, Topics in Current Physics, Vol. 22, Springer-Verlag, Berlin, 1980.
2.22. A. Walther, *J. Opt. Soc. Am.* **58**, 1256 (1968).
2.23. E. Wolf, *Phys. Rev. D* **13**, 869 (1976).
2.24. E. Wolf, *J. Opt. Soc. Am.* **70**, 1311 (1980).
2.25. M. Nieto-Vesperinas and E. Wolf, *J. Opt. Soc. Am.* A **2**, 1429 (1985).
2.26. J. W. Goodman, *Introduction to Fourier Optics*, McGraw-Hill, New York, 1968. Chapter 3.
2.27. A. W. Lohmann, *Optical Information Processing*, University Offset, Erlangen, 1978. Chapter 16.
2.28. E. Lalor, *J. Opt. Soc. Am.* **58**, 1235 (1968).
2.29. R. K. Luneburg, *Mathematical Theory of Optics*, University of California Press, Berkeley, 1966.
2.30. E. C. Titchmarsh, *Introduction to the Theory of Fourier Integrals*, Oxford University Press, London, 1948, p. 83.
2.31. W. D. Montgomery, *J. Opt. Soc. Am.* **58**, 1112 (1968).
2.32. G. C. Sherman and H. J. Bremermann, *J. Opt. Soc. Am.* **59**, 146 (1969).
2.33. E. Wolf, *Proc. Roy. Soc.* A **253**, 349 (1959).
2.34. M. Kline and I. W. Kay, *Electromagnetic Theory and Geometrical Optics*, Interscience Publishers, New York, 1965.
2.35. E. T. Whittaker, *Math. Ann.* **57**, 333 (1902).
2.36. E. T. Whittaker and G. N. Watson, *A Course of Modern Analysis*, Cambridge University Press, London, 1963. Section 18.6.
2.37. G. C. Sherman, *Phys. Rev. Lett.* **21**, 1220 (1968).
2.38. G. C. Sherman, *J. Opt. Soc. Am.* **59**, 697 (1969).
2.39. K. Miyamoto and E. Wolf, *J. Opt. Soc. Am.* **52**, 615 (1962).
2.40. M. Born and E. Wolf, *Principles of Optics*. Pergamon Press, Oxford, 1980, 6th edition.
2.41. D. S. Jones, *Proc. Camb. Phil. Soc.* **48**, 733 (1952) (lemma on p. 736).
2.42. C. R. Giuliano, *Applications of Optical Phase Conjugation*, in *Physics Today*, April 1981, p. 27.
2.43. R. A. Fisher, ed., *Optical Phase Conjugation*, Academic Press, New York, 1983.
2.44. H. Weyl, *Ann. Phys.* **60**, 481 (1919).
2.45. A. Baños, *Dipole Radiation in the Presence of a Conducting Half-Space*, Pergamon Press, Oxford, 1966, Chapter 2.
2.46. L. M. Brekhovskikh, *Waves in Layered Media*, Academic Press, New York, 1960. Chapter 4.
2.47. M. Nieto-Vesperinas, *Opt. Comm.* **67**, 391 (1988).
2.48. H. Kogelnik, *Bell Syst. Tech. J.* **44**, 455 (1965).
2.49. H. Kogelnik and T. Li, *Proc. I.E.E.E.* **54**, 1312 (1966).
2.50. H. Kogelnik, *Appl. Opt.* **4**, 1562 (1965).
2.51. A. E. Siegman, *An Introduction to Masers and Lasers*, McGraw-Hill, New York, 1971.
2.52. J. Durnin, *J. Opt. Soc. Am.* A **4**, 651 (1987).

2.53. J. Durnin, J. J. Miceli and J. H. Eberly, *Phys. Rev. Lett.* **58**, 1499 (1987); *Opt. Lett.* **13**, 79 (1988).

2.54. G. Indebetouw, *J. Opt. Soc. Am.* A **6**, 150 (1989).

2.55. F. Gori, G. Guattari and C. Padovani, *Opt. Comm.* **64**, 491 (1987).

2.56. G. C. Sherman, J. J. Stamnes and E. Lalor, *J. Math. Phys.* **17**, 760 (1976).

2.57. H. G. Schmidt-Weinmar, *J. Opt. Soc. Am.* **65**, 1059 (1975).

2.58. W. H. Carter, *J. Opt. Soc. Am.* **65**, 1054 (1975).

2.59. J. T. Foley and E. Wolf, *J. Opt. Soc. Am.* **69**, 761 (1979).

2.60. H. P. Baltes and H. G. Schmidt-Weinmar, *Phys. Lett.* A **60**, 275 (1977).

2.61. G. C. Sherman, A. J. Devaney and L. Mandel, *Opt. Comm.* **6**, 115 (1972).

3

Radiated and Scattered Fields

In this chapter we study time-independent radiated and scattered fields by using the angular spectrum techniques discussed in Chapter 2. This allows us to characterize these fields at any point of space, no matter how close it is to the volume containing the sources. Therefore, it is not necessary to use approximations (e.g. Fresnel or Fraunhofer) usually made in other approaches. Once the angular spectrum is known, the field can be computed at any point exterior to the source volume. The usefulness of this method will be seen in this and subsequent chapters. Also, it provides a new way of dealing with certain radiation problems, such as those on fields from moving particles, as shown by Asby and Wolf [3.1] and Lalor and Wolf [3.2].

We shall deal in this chapter with *direct source* and *scattering problems*; namely, given the sources, or potentials of the medium, we shall study the field that they produce. *Inverse source* and *scattering problems* will be addressed in Chapter 10.

Even the direct scattering problem is usually an involved subject, requiring the solution of (generally coupled) integro-differential equations. Methods on scattering by volumes have been extensively developed in connection with wave propagation in inhomogeneous (often random) media. We shall see that certain approximations useful in the case of weak, or slightly fluctuating, scatterers lead to analytically simplified solutions. This is the case of the first Born, Rytov and eikonal approximations. The approaches suggested for dealing with multiple scattering will not be discussed here, as these have been extensively accounted for in other texts and can be found in the references given at the end of this chapter.

3.1. RADIATED FIELDS FROM A LOCALIZED CHARGE–CURRENT DISTRIBUTION

Let us consider the monochromatic electromagnetic field radiated into free space by a system of charges and currents specified as follows:

$$q(\mathbf{r}, t) = q(\mathbf{r}) \exp(-i\omega t), \tag{3.1a}$$

$$\mathbf{j}(\mathbf{r}, t) = \mathbf{j}(\mathbf{r}) \exp(-i\omega t). \tag{3.1b}$$

We assume that $q(\mathbf{r})$ and $\mathbf{j}(\mathbf{r})$ are continuous and differentiable functions of position and vanish outside a certain volume V. The real physical magnitudes are obtained, as usual, by taking the real part of these quantities.

We shall discuss the characterization of the field radiated from a finite volume distribution by means of angular spectrum methods (Devaney and Wolf [3.3], Friberg and Wolf [3.4], and Carter and Wolf [3.5]).

If the medium inside the volume V is nonmagnetic, the time-independent parts of the vectors **E** and **H** satisfy the inhomogeneous vector equations (cf. Eqs. (1.8) and first term of Eqs. (1.10) of Chapter 1):

$$\nabla \times \nabla \times \mathbf{E}(\mathbf{r}) - k^2\mathbf{E}(\mathbf{r}) = 4\pi i \frac{k}{c} \mathbf{j}(\mathbf{r}), \tag{3.2a}$$

$$\nabla \times \nabla \times \mathbf{H}(\mathbf{r}) - k^2\mathbf{H}(\mathbf{r}) = 4\pi \frac{1}{c} \nabla \times \mathbf{j}(\mathbf{r}), \tag{3.2b}$$

where $k = \omega/c$. Also, the radiated field *outside* the volume V can be written as (cf. Eqs. (1.46) and (1.47) of Chapter 1):

$$\mathbf{E}(\mathbf{r}_>) = \nabla \times \nabla \times \boldsymbol{\Pi}(\mathbf{r}_>), \tag{3.3a}$$

$$\mathbf{H}(\mathbf{r}_>) = -ik\nabla \times \boldsymbol{\Pi}(\mathbf{r}_>), \tag{3.3b}$$

where $\boldsymbol{\Pi}(\mathbf{r})$ is the electric Hertz vector for the radiated field (cf. Eq. (1.42)):

$$\boldsymbol{\Pi}(\mathbf{r}_>) = \frac{i}{kc} \int_V \mathbf{j}(\mathbf{r}')G(\mathbf{r}_>, \mathbf{r}')\, d^3r', \tag{3.4}$$

$G(\mathbf{r}, \mathbf{r}')$ being the scalar outgoing Green function.

Note that in this case, according to Eq. (1.43), the magnetic Hertz vector is zero. For this reason we do not need to use now the subindices e and m.

3.2. ANGULAR SPECTRUM REPRESENTATION OF RADIATED FIELDS

Suppose that the radiating volume V is situated within a strip $0 < z < L$ (Fig. 3.1). Let us introduce into Eq. (3.4) the angular spectrum representation, Eq. (2.51), of the outgoing Green function $G(\mathbf{r}, \mathbf{r}')$; after interchanging the order of integration, one obtains the following plane wave expansion for the electric Hertz vector:

$$\boldsymbol{\Pi}(\mathbf{r}_>) = \iint_{-\infty}^{\infty} \mathbf{a}^{(\pm)}(\mathbf{K}) \exp[i(\mathbf{K}\cdot\mathbf{R}_> \pm k_z z_>)]\, d^2K, \tag{3.5}$$

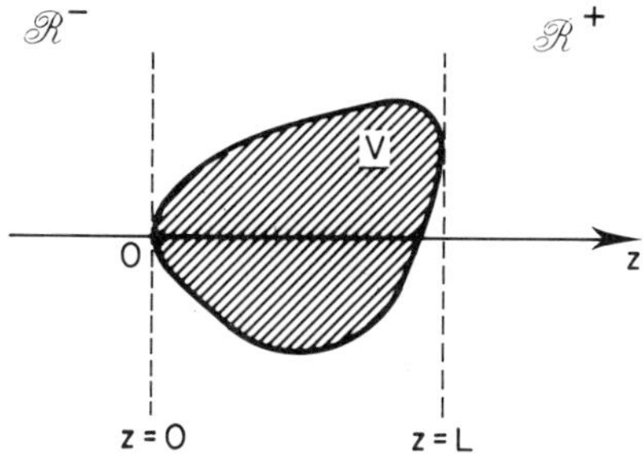

Figure 3.1. The two half-spaces $\mathscr{R}^+$, $z \geqslant L$, and $\mathscr{R}^-$, $z < L$, at each side of the volume V.

where:

$$\mathbf{a}^{(\pm)}(\mathbf{K}) = -\frac{1}{2\pi k c k_z} \int_V \mathbf{j}(\mathbf{r}') \exp[-i(\mathbf{K}\cdot\mathbf{R}'_> \pm k_z z')] \, d^3 r'. \tag{3.6}$$

In Eqs. (3.5) and (3.6) $\mathbf{R}$ and $\mathbf{R}'$ are two-dimensional vectors: $\mathbf{R} = (x, y)$, $\mathbf{R}' = (x', y')$; $\mathbf{r} = (\mathbf{R}, z)$, $\mathbf{r}' = (\mathbf{R}', z')$. Also, $\mathbf{K} = k(p, q)$, $k_z = km$. Where p, q, and m are the directional cosines, satisfying Eqs. (2.9).

The signs "plus" or "minus" in the superscript of a ($\mathbf{K}$), and in the exponent of the integrands in Eqs. (3.5) and (3.6), are chosen according to whether the point $\mathbf{r}_>$ belongs to the source-free half-spaces $\mathscr{R}^+$ or $\mathscr{R}^-$, respectively (see Fig. 3.1). In addition, we have used the fact that $|z_> - z'| = z_> - z'$ when $\mathbf{r}_>$ is in $\mathscr{R}^+$ and $|z_> - z'| = z' - z_>$ when $\mathbf{r}_>$ is in $\mathscr{R}^-$.

On substituting from Eq. (3.5) into Eqs. (3.3a) and (3.3b), we also obtain the angular spectrum representation of the radiated electromagnetic field in the two half-spaces $\mathscr{R}^+$ and $\mathscr{R}^-$:

$$\mathbf{E}(\mathbf{r}_>) = \iint_{-\infty}^{\infty} \mathbf{e}^{(\pm)}(\mathbf{K}) \exp[i(\mathbf{K}\cdot\mathbf{R}_> \pm k_z z_>)] \, d^2 K, \tag{3.7}$$

$$\mathbf{H}(\mathbf{r}_>) = \iint_{-\infty}^{\infty} \mathbf{h}^{(\pm)}(\mathbf{K}) \exp[i(\mathbf{K}\cdot\mathbf{R}_> \pm k_z z_>)] \, d^2 K, \tag{3.8}$$

where:

$$\mathbf{e}^{(\pm)}(\mathbf{K}) = -\mathbf{k}^{(\pm)} \times (\mathbf{k}^{(\pm)} \times \mathbf{a}^{(\pm)}(\mathbf{K})), \tag{3.9}$$

$$\mathbf{h}^{(\pm)}(\mathbf{K}) = k[\mathbf{k}^{(\pm)} \times \mathbf{a}^{(\pm)}(\mathbf{K})], \tag{3.10}$$

$\mathbf{k}^{(\pm)}$ being: $\mathbf{k}^{(\pm)} = (\mathbf{K}, \pm k_z)$. On the other hand, the spectral amplitudes are related by: $\mathbf{h}^{(\pm)}(\mathbf{K}) = 1/k[\mathbf{k}^{(\pm)} \times \mathbf{e}^{(\pm)}(\mathbf{K})]$.

Equations (3.7)–(3.10) permit the determination of the radiated field at *any* point, either in $\mathscr{R}^+$ or $\mathscr{R}^-$. Note the preferential role played by the z-axis. When the radiating volume is finite and has a shape with no particular symmetry, the z-axis can be arbitrarily chosen. If this axis is then rotated, the lines $z = 0, 0'', \ldots$, etc., and $z = L, L', L'', \ldots$, etc., form a convex domain Ω that encloses the volume V, and outside of which Eqs. (3.5), (3.7), and (3.8) are valid with the appropriate rotation of coordinates (see Fig. 3.2).

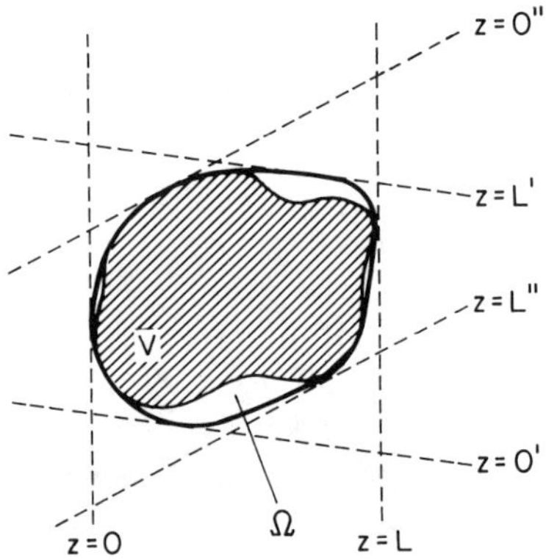

Figure 3.2. Illustration of the convex domain Ω outside which the angular spectrum representation is valid. As the orientation of the z-axis varies, the half-spaces $\mathscr{R}^+$ and $\mathscr{R}^-$ vary accordingly.

3.3. THE FIELD AND THE INTENSITY RADIATED IN THE FAR ZONE

Let the point P of observation be situated in the far zone (Fig. 3.3), the radiation volume V being finite. Denote the position vector of P by $\mathbf{r}_> = r\mathbf{n}$, $\mathbf{n} = (n_x, n_y, n_z)$ being a unit vector. According to Section 2.12, as $kr \to \infty$ the field in a fixed direction specified by $\mathbf{n}$ is given as:

$$\mathbf{E}(r\mathbf{n}) \sim \mp 2\pi i k n_z \mathbf{e}^{(\pm)}(kn_x, kn_y) \frac{\exp(ikr)}{r}, \tag{3.11}$$

$$\mathbf{H}(r\mathbf{n}) \sim \mp 2\pi i k n_z \mathbf{h}^{(\pm)}(kn_x, kn_y) \frac{\exp(ikr)}{r}. \tag{3.12}$$

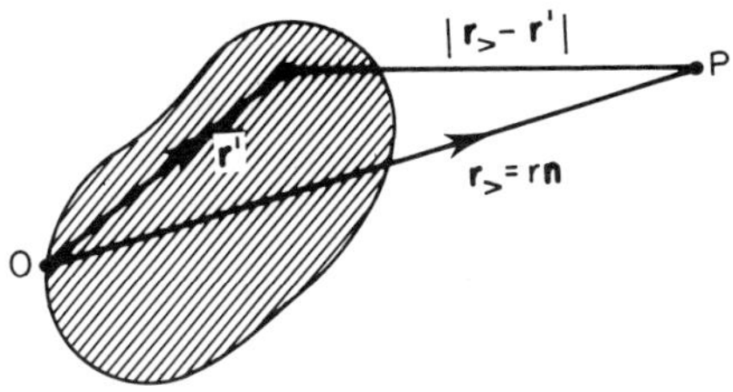

Figure 3.3. Notation for the electromagnetic field in the far zone.

Alternatively, the far zone expressions also can be obtained by making the approximation (Fig. 3.3):

$$|\mathbf{r} - \mathbf{r}'| \sim r - \mathbf{r}'\cdot\mathbf{n}, \tag{3.13}$$

so that the Green function becomes:

$$G(\mathbf{r}, \mathbf{r}') \sim \frac{\exp(ikr)}{r}\exp(-ik\mathbf{r}'\cdot\mathbf{n}). \tag{3.14}$$

Substitution of Eq. (3.14) directly into Eqs. (3.3a), (3.3b) and (3.4) yields again Eqs. (3.11) and (3.12) after making use of Eqs. (3.6), (3.9), and (3.10).

Equations (3.11) and (3.12) show that the far field at a given point $r\mathbf{n}$ is proportional to the amplitude of only one plane wave component. Also, it has the envelope of a divergent spherical wave. If the volume V were constituted by sinks rather than by sources, the envelope of the far field would be a convergent spherical wave.

The radiated intensity in the far zone is given by the magnitude of the time averaged Poynting vector, which as seen in Eq. (1.18), is:

$$\bar{\mathscr{S}}(\mathbf{r}) = \frac{c}{8\pi}\operatorname{Re}[\mathbf{E}(\mathbf{r}) \times \mathbf{H}^*(\mathbf{r})]. \tag{3.15}$$

Substitution of Eqs. (3.11) and (3.12) into Eq. (3.15), together with the use of Eqs. (3.6), (3.9), and (3.10), leads to an expression of the radiated intensity in terms of the three-dimensional Fourier transform of the current density $\mathbf{j}(\mathbf{r})$. We leave its derivation as an exercise for the reader.

3.4. SCALAR THEORY OF RADIATED WAVEFIELDS

A scalar wavefield $U(\mathbf{r})$ radiated by a time-harmonic source density $\rho(\mathbf{r})$ localized in a volume V (Fig. 3.1) satisfies the inhomogeneous Helmholtz equation (Devaney and Wolf [3.3]):

$$\nabla^2 U(\mathbf{r}) + k^2 U(\mathbf{r}) = -4\pi\rho(\mathbf{r}), \tag{3.16}$$

whose solution outside the volume V is known from Section 1.6.3 to be:

$$U(\mathbf{r}_>) = \int_V \rho(\mathbf{r}')G(\mathbf{r}_>, \mathbf{r}')\, d^3r'. \tag{3.17}$$

By introducing the plane wave representation, Eq. (2.51), for $G(\mathbf{r}, \mathbf{r}')$ into Eq. (3.17) and using the same notation as in Section 3.2, one gets the following angular spectrum representation for $U(\mathbf{r})$, either in $\mathscr{R}^+$ or $\mathscr{R}^-$:

$$U(\mathbf{r}_>) = \iint_{-\infty}^{\infty} A^{(\pm)}(\mathbf{K}) \exp[i(\mathbf{K}\cdot\mathbf{R}_> \pm k_z z_>)]\, d^2K, \tag{3.18}$$

with the angular spectrum being:

$$A^{(\pm)}(\mathbf{K}) = \frac{i}{2\pi k_z} \int_V \rho(\mathbf{r}') \exp[-i(\mathbf{K}\cdot\mathbf{R}' \pm k_z z')]\, d^3r'. \tag{3.19}$$

Once again the signs *plus* or *minus* are taken according to whether the point $\mathbf{r}_>$ is in $\mathscr{R}^+$ or $\mathscr{R}^-$, respectively.

The radiated field in the far zone is:

$$U(r\mathbf{s}) \sim (2\pi)^3 \tilde{\rho}(\mathbf{K}) \frac{\exp(ikr)}{r}, \tag{3.20}$$

$\tilde{\rho}(\mathbf{K})$ being the three-dimensional Fourier transform of $\rho(\mathbf{r})$ at values $\mathbf{k} = (\mathbf{K}, k_z)$ of its argument.

3.5. EXAMPLES OF RADIATION FIELDS: CHARGED PARTICLE WITH TWO-DIMENSIONAL MOTION

As an example of the techniques based on the angular spectrum representation of fields, we shall study next the electromagnetic field due to a charged particle. The role of homogeneous and evanescent components appear in this example at the root of the threshold condition marking the existence or absence of radiation. This method was first proposed by Toraldo di Francia [3.6] and later developed by Asby and Wolf [3.1] and by Lalor and Wolf [3.2].

3.5.1. Field Due to a Charged Particle Moving in Vacuum

Let us assume a particle with charge e moving on a trajectory characterized by the position vector $\mathbf{r}_e(t)$, then the electric current density $\mathbf{j}(\mathbf{r}, t)$ and the charge density $\rho(\mathbf{r}, t)$ are:

$$\mathbf{j}(\mathbf{r}, t) = e\mathbf{v}(t)\delta[\mathbf{r} - \mathbf{r}_e(t)], \tag{3.21}$$

$$\rho(\mathbf{r}, t) = e\delta[\mathbf{r} - \mathbf{r}_e(t)], \tag{3.22}$$

where $\mathbf{v}(t) = d\mathbf{r}_e/dt$ is the instantaneous velocity of the particle.

Let us represent the source distributions $\mathbf{j}(\mathbf{r}, t)$ and $\rho(\mathbf{r}, t)$ in terms of their Fourier components $\tilde{\mathbf{j}}(\mathbf{k}, \omega)$ and $\tilde{\rho}(\mathbf{k}, \omega)$:

$$\tilde{\mathbf{j}}(\mathbf{k}, \omega) = \frac{1}{(2\pi)^4}\int_{-\infty}^{\infty} dt' \exp(i\omega t') \int d^3r \exp(-i\mathbf{k}\cdot\mathbf{r})\mathbf{j}(\mathbf{r}, t'), \tag{3.23}$$

$$\tilde{\rho}(\mathbf{k}, \omega) = \frac{1}{(2\pi)^4}\int_{-\infty}^{\infty} dt' \exp(i\omega t') \int d^3r \exp(-i\mathbf{k}\cdot\mathbf{r})\rho(\mathbf{r}, t'), \tag{3.24}$$

where the $\mathbf{r}$-integral is extended to the infinite space. By introducing Eqs. (3.21) and (3.22) into Eqs. (3.23) and (3.24) respectively, we obtain:

$$\tilde{\mathbf{j}}(\mathbf{k}, \omega) = \frac{e}{(2\pi)^4}\int_{-\infty}^{\infty} \exp[i(\omega t' - \mathbf{k}\cdot\mathbf{r}_e(t'))]\mathbf{v}(t')\, dt', \tag{3.25}$$

$$\tilde{\rho}(\mathbf{k}, \omega) = \frac{e}{(2\pi)^4}\int_{-\infty}^{\infty} \exp[i(\omega t' - \mathbf{k}\cdot\mathbf{r}_e(t'))]\, dt'. \tag{3.26}$$

Let us represent the electric and magnetic vectors $\mathbf{E}(\mathbf{r}, t)$ and $\mathbf{H}(\mathbf{r}, t)$ as Fourier integrals:

$$\mathbf{E}(\mathbf{r}, t) = \int_{-\infty}^{\infty} \mathbf{E}(\mathbf{r}, \omega) \exp(-i\omega t)\, d\omega, \tag{3.27}$$

$$\mathbf{H}(\mathbf{r}, t) = \int_{-\infty}^{\infty} \mathbf{H}(\mathbf{r}, \omega) \exp(-i\omega t)\, d\omega. \tag{3.28}$$

Therefore, if $\mathbf{\Pi}(\mathbf{r}, \omega)$ is the Fourier component, for each frequency ω, of the Hertz vector $\mathbf{\Pi}(\mathbf{r}, t)$, we have, analogously to Eqs. (3.3):

$$\mathbf{E}(\mathbf{r}_>, \omega) = \nabla \times \nabla \times \mathbf{\Pi}(\mathbf{r}_>, \omega). \tag{3.29a}$$

$$\mathbf{H}(\mathbf{r}_>, \omega) = -ik\nabla \times \mathbf{\Pi}(\mathbf{r}_>, \omega). \tag{3.29b}$$

All the analysis of previous sections then applies to the Fourier components at a given frequency ω. Hence, $\mathbf{\Pi}(\mathbf{r}, \omega)$ can be represented by means of its angular spectrum $\mathbf{a}^{(\pm)}(\mathbf{K}, \omega)$ as in Eq. (3.5):

$$\mathbf{\Pi}(\mathbf{r}_>, \omega) = \iint_{-\infty}^{\infty} \mathbf{a}^{(\pm)}(\mathbf{K}, \omega) \exp[i(\mathbf{K}\cdot\mathbf{R}_> \pm k_z z_>)]\, d^2K, \tag{3.30}$$

with $\mathbf{a}^{(\pm)}(\mathbf{K}, \omega)$ given according to Eq. (3.6), namely:

$$\mathbf{a}^{(\pm)}(\mathbf{K}, \omega) = -\frac{(2\pi)^2}{kck_z} \tilde{\mathbf{j}}(\mathbf{k}^{(\pm)}, \omega). \tag{3.31}$$

The upper sign *plus* and lower sign *minus* are chosen according to whether the point $\mathbf{r}_>$ is in the half-space to the right, $\mathscr{R}^+$, or to the left, $\mathscr{R}^-$, of the strip $0 < z < L$ assumed to contain the trajectory $\mathbf{r} = \mathbf{r}_e(t)$ (cf. Fig. 3.4).

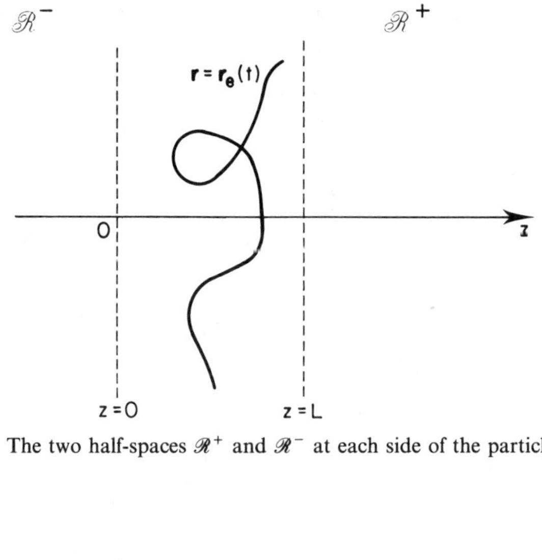

Figure 3.4. The two half-spaces $\mathscr{R}^+$ and $\mathscr{R}^-$ at each side of the particle trajectory.

On introducing Eq. (3.25) into Eq. (3.31) we obtain:

$$\mathbf{a}^{(\pm)}(\mathbf{K}, \omega) = -\frac{e}{(2\pi)^2 kck_z} \iint_{-\infty}^{\infty} \exp[i(\omega t' - \mathbf{k}^{(\pm)} \cdot \mathbf{r}_e(t'))] \mathbf{v}(t')\, dt', \quad (3.32)$$

where, as before, $\mathbf{k}^{(\pm)} = (\mathbf{K}, \pm k_z)$.

Also, from Eqs. (3.6), (3.9), (3.10), and (3.25) we get for the angular spectra $\mathbf{e}^{(\pm)}(\mathbf{K}, \omega)$ and $\mathbf{h}^{(\pm)}(\mathbf{K}, \omega)$ of each component at frequency ω:

$$\mathbf{e}^{(\pm)}(\mathbf{K}, \omega) = \frac{e}{(2\pi)^2 kck_z} \mathbf{k}^{(\pm)} \times \left[\mathbf{k}^{(\pm)} \times \int_{-\infty}^{\infty} \exp[i(\omega t' - \mathbf{k}^{(\pm)} \cdot \mathbf{r}_e(t'))] \mathbf{v}(t')\, dt' \right], \quad (3.33a)$$

$$\mathbf{h}^{(\pm)}(\mathbf{K}, \omega) = -\frac{e}{(2\pi)^2 ck_z} \times \left[\mathbf{k}^{(\pm)} \times \int_{-\infty}^{\infty} \exp[i(\omega t' - \mathbf{k}^{(\pm)} \cdot \mathbf{r}_e(t'))] \mathbf{v}(t')\, dt' \right]. \quad (3.33b)$$

3.5.2. Particle Moving Uniformly in Vacuum

Let the particle move with constant velocity $\mathbf{v}_0$. We shall choose the trajectory along the x-axis so that its parametric equation is:

$$x = x_e(t) = v_0 t, \quad (3.34a)$$

$$y = z = 0. \quad (3.34b)$$

On substituting from Eqs. (3.34) into Eqs. (3.32) and (3.33) we get:

$$\mathbf{a}^{(\pm)}(\mathbf{K}, \omega) = -\frac{e\mathbf{v}_0}{2\pi kck_z} \delta(\omega - K_x v_0), \quad (3.35)$$

where $\mathbf{v}_0 = (v_0, 0, 0)$ and we have taken into account that the δ-function admits the representation:

$$\delta(\omega - K_x v_0) = \frac{1}{2\pi} \int_{-\infty}^{\infty} \exp[i(\omega - K_x v_0)t']\, dt'. \quad (3.36)$$

If the relations $\delta(x) = \delta(-x)$ and $\delta(ax) = \delta(x)/|a|$ are used (a being any real constant), Eq. (3.35) may be written in the form:

$$\mathbf{a}^{(\pm)}(\mathbf{K}, \omega) = -\frac{e}{2\pi kck_z}\,\delta\left(K_x - k\frac{c}{v_0}\right)\hat{\mathbf{x}}, \tag{3.37}$$

$\hat{\mathbf{x}}$ denoting the unit vector in the x-direction in which the particle moves.

Similarly, the angular spectra of the fields $\mathbf{E}(\mathbf{r}, \omega)$ and $\mathbf{H}(\mathbf{r}, \omega)$ at frequency ω are:

$$\mathbf{e}^{(\pm)}(\mathbf{K}, \omega) = \frac{e}{2\pi kck_z}\,\mathbf{k}^{(\pm)} \times [\mathbf{k}^{(\pm)} \times \hat{\mathbf{x}}]\,\delta\left(K_x - k\frac{c}{v_0}\right), \tag{3.38a}$$

$$\mathbf{h}^{(\pm)}(\mathbf{K}, \omega) = -\frac{e}{2\pi ck_z}\,[\mathbf{k}^{(\pm)} \times \hat{\mathbf{x}}]\,\delta\left(K_x - k\frac{c}{v_0}\right). \tag{3.38b}$$

Equations (3.38) imply that the field due to a charged particle moving with constant velocity v_0 in the x-direction consists of plane waves such that the x-component K_x of their wavevectors are:

$$K_x = K_x^{(0)} \equiv k\frac{c}{v_0} \tag{3.39}$$

Since the speed v_0 of the particle is necessarily smaller than the velocity of light in vacuum, $K_x^{(0)}$ is greater than k and hence, according to (2.9b), k_z is purely imaginary. *The field created by a particle in uniform movement consists of evanescent waves only.* With this value of the transversal component $|\mathbf{K}|$ of the wavevector, the contribution of these plane waves to the far field is zero when $kr \to \infty$ (cf. Eqs. (3.11) and (3.12)). The field is static. This agrees with the well known fact that a uniformly moving particle produces nonradiating fields (see, e.g. Refs. 3.7, 3.8 or 3.9); namely, there is no radiated power, as can be seen by evaluating the Poynting vector. (Compare with the result of Problem 2.5.) In fact, by introducing Eq. (3.37) into Eq. (3.30) one obtains for the Hertz vector at frequency ω (see Problem 3.6):

$$\mathbf{\Pi}(\mathbf{r}, \omega) = \frac{ie}{\pi kc}\exp\left[ik\frac{c}{v_0}x\right]K_0(k\gamma d)\hat{\mathbf{x}}, \tag{3.40}$$

where:

$$\gamma = \sqrt{\left(\frac{c}{v_0}\right)^2 - 1}, \tag{3.41}$$

$$d = \sqrt{y^2 + z^2}, \tag{3.42}$$

and (see Problem 3.5):

$$\int_{-\infty}^{\infty} \frac{1}{k_z} \exp[i(K_y y \pm k_z z)]\, dK_y = -2iK_0(k\gamma d), \tag{3.43}$$

where $K_0(k\gamma d)$ is the modified Hankel function of argument $k\gamma d$ and zero order.

As $kd \to \infty$ the function $K_0(k\gamma d)$ has the asymptotic behavior:

$$K_0(k\gamma d) \sim \frac{\pi}{2}\left(\frac{2}{\pi k\gamma d}\right)^{1/2} \exp(-k\gamma d),$$

showing that $\mathbf{\Pi}(\mathbf{r}, \omega)$ (and thus the field) decays exponentially with increasing distance d from the line of motion of the particle.

3.5.3. Čerenkov Radiation

Let us consider now a charged particle moving with constant velocity v_0 in the positive x-direction in a homogeneous, isotropic, nonmagnetic medium, characterized by a refractive index $n(\omega)$. In a way equivalent to that leading to Eqs. (3.37) and (3.38), we obtain the angular spectrum amplitudes:

$$\mathbf{a}^{(\pm)}(\mathbf{K}, \omega) = -\frac{e}{2\pi k(\omega) c k_z}\, \delta\left(K_x - k(\omega)\frac{v(\omega)}{v_0}\right)\hat{\mathbf{x}}, \tag{3.44}$$

and:

$$\mathbf{e}^{(\pm)}(\mathbf{K}, \omega) = \frac{e}{2\pi k(\omega) c k_z}\, \mathbf{k}^{(\pm)} \times [\mathbf{k}^{(\pm)} \times \hat{\mathbf{x}}]\,\delta\left(K_x - k(\omega)\frac{v(\omega)}{v_0}\right), \tag{3.45a}$$

$$\mathbf{h}^{(\pm)}(\mathbf{K}, \omega) = -\frac{e}{2\pi c k_z}\, [\mathbf{k}^{(\pm)} \times \hat{\mathbf{x}}]\,\delta\left(K_x - k(\omega)\frac{v(\omega)}{v_0}\right), \tag{3.45b}$$

where:

$$v(\omega) = \frac{c}{n(\omega)} \tag{3.46}$$

is the phase velocity in the medium for plane waves at frequency ω. Also, in this case:

$$k(\omega) = n(\omega)\frac{\omega}{c} = \frac{\omega}{v(\omega)}. \tag{3.47}$$

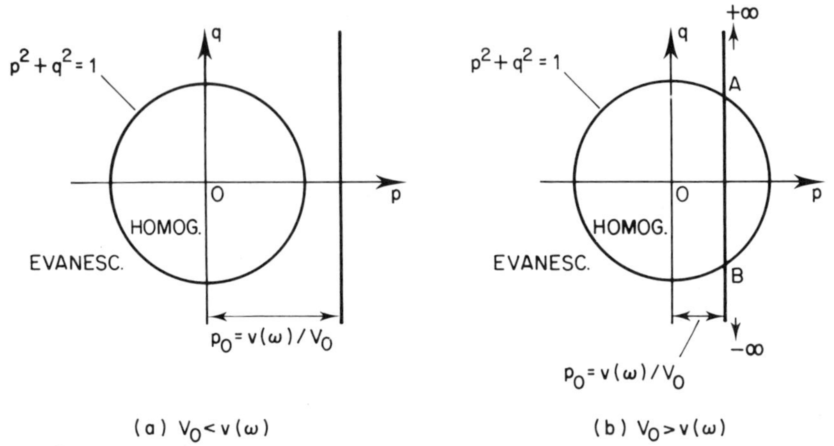

Figure 3.5. Angular spectrum of the field due to a charged particle, moving uniformly with velocity v_0 in the x-direction in a medium of phase velocity $v(\omega)$. The angular spectrum consists of plane waves with $K_x \equiv K_x^{(0)} = v(\omega)/v_0$. Homogeneous waves are represented by points inside the circle, whereas evanescent waves are represented by points outside the circle. (Reproduced with permission from paper by R. Asby and E. Wolf, *J. Opt. Soc. Am.* **61**, 57 (1971).)

Equations (3.44) and (3.45) show that the angular spectrum of the field at frequency ω consists of plane waves such that the transversal component of their wavevector has the value:

$$K_x = K_x^{(0)} \equiv k(\omega)\frac{v(\omega)}{v_0}. \tag{3.48}$$

Two cases can now be distinguished (Asby and Wolf [3.1]):

a. The Speed of the Particle v_0 is Smaller than the Phase Velocity $v(\omega)$
In this case $K_x > k$ (namely, $p > 1$), and hence all waves at frequency ω are evanescent. As in the case of uniform movement in vacuum, the particle *does not radiate* at frequency ω. In Fig. 3.5(a) the point ($K_x^{(0)} = kv(\omega)/c, K_y$) lies outside the circle $K^2 = k^2$.

b. The Speed of the Particle v_0 is Greater than the Phase Velocity $v(\omega)$
In this case $K_x < k$ (which means $p < 1$), and the point ($K_x^{(0)} = kv(\omega)/v_0, K_y$) in Fig. 3.5(b) is on the line that intersects the circle $K^2 = k^2$ at points A and B. These two points limit the range of values of K_y for which $K_x^2 + K_y^2 \leqslant k^2$,

namely, for which the waves are homogeneous (cf. Eqs. (2.9)), i.e., for values:

$$-k(\omega)\left[1-\left(\frac{v(\omega)}{v_0}\right)^2\right]^{1/2} \leqslant K_y \leqslant k(\omega)\left[1-\left(\frac{v(\omega)}{v_0}\right)^2\right]^{1/2} \tag{3.49}$$

the waves are homogeneous; whereas for values:

$$k(\omega)\left[1-\left(\frac{v(\omega)}{v_0}\right)^2\right]^{1/2} < K_y < \infty, \tag{3.50a}$$

$$-\infty < K_y < -k(\omega)\left[1-\left(\frac{v(\omega)}{v_0}\right)^2\right]^{1/2} \tag{3.50b}$$

the waves are evanescent.

The directions of propagation of all homogeneous waves form a circular cone about the x-axis of movement of the particle. The semiangle of this cone is (Fig. 3.6):

$$\theta_0(\omega) = \text{arc cos}\left(\frac{v(\omega)}{v_0}\right). \tag{3.51}$$

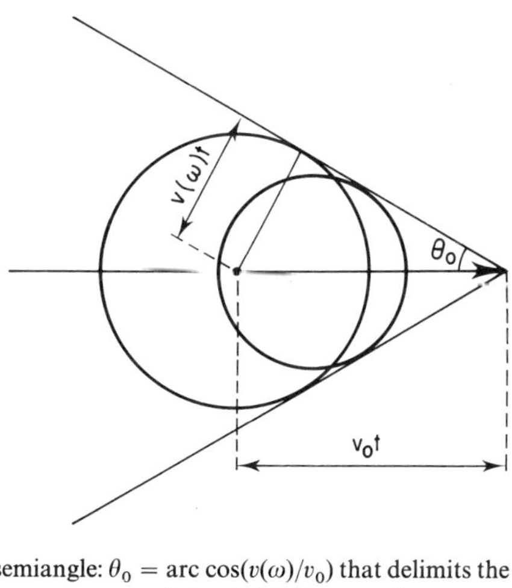

Figure 3.6. Cone of semiangle: $\theta_0 = \text{arc cos}(v(\omega)/v_0)$ that delimits the directions of propagation of the homogeneous components of the field from a charge moving with uniform speed $v_0 > v(\omega)$.

On introducing Eq. (3.44) into Eq. (3.30) the reader can verify that the Hertz vector at frequency ω is given by the following expression:

$$\mathbf{\Pi}(\mathbf{r}, \omega) = -\frac{e}{2kc} \exp\left[ik(\omega) \frac{v(\omega)}{v_0} x \right] H_0^{(1)}(k(\omega)\sigma_\omega d)\hat{\mathbf{x}}, \tag{3.52}$$

where $H_0^{(1)}$ is the Hankel function of the first kind and zero order, and:

$$\sigma_\omega = \sqrt{1 - \left(\frac{v(\omega)}{v_0}\right)^2} = \sin\theta_0, \tag{3.53}$$

$$d = \sqrt{y^2 + z^2}. \tag{3.54}$$

Equation (3.52) represents a conical wave with its axis along the line of motion of the particle (namely, the x-axis).

Asymptotically ($kd \to \infty$), it can be seen (Problem 3.5) that the Hertz vector decays as $1/(\omega d/c)$. This indicates that the particle now gives rise to a radiated field which is known as *Čerenkov radiation* ([3.9] and [3.10]).

Other studies, [3.2], [3.6], and [3.11], show that the evanescent components of the field due to a uniformly moving charge in vacuum can be converted into homogeneous waves after interaction with a dielectric when $v_0 > v(\omega)$, namely, when the speed of the particle exceeds the phase velocity in the medium, *Čerenkov* radiation is produced.

3.6. INTEGRO-DIFFERENTIAL EQUATIONS FOR THE SCATTERED ELECTROMAGNETIC FIELD IN A TIME-INDEPENDENT MEDIUM. ANGULAR SPECTRUM REPRESENTATION OUTSIDE THE STRIP $0 < z < L$

The response of a material to an incident field is described by the *constitutive relations* for the *induced polarization* $\mathbf{P}(\mathbf{r})$ and the *induced magnetization* $\mathbf{M}(\mathbf{r})$. The representation of the scattered field in terms of these quantities is formally equivalent to that used for radiated fields. As a matter of fact, this field can be considered as radiated from sources induced in the material, these sources being characterized by the quantities $\mathbf{P}(\mathbf{r})$ and $\mathbf{M}(\mathbf{r})$. The *direct scattering problem* consists of finding these sources from knowledge of both the incident field and the dielectric and magnetic susceptibilities of the scattering medium.

The angular spectrum representation of scattered fields discussed in this section was established by Friberg and Wolf [3.4], and is analogous to the formalism of Refs. 3.3 and 3.5 discussed in Section 3.2 in connection with radiated fields.

Let $\mathbf{E}^{(i)}(\mathbf{r})$ and $\mathbf{H}^{(i)}(\mathbf{r})$ represent the electric and magnetic vectors, respectively, of a monochromatic field incident on a generally inhomogeneous medium characterized by a refractive index $n(\mathbf{r})$, $\mathbf{r}$ being a position vector. Upon interaction with the medium, a new field $\mathbf{E}(\mathbf{r})$, $\mathbf{H}(\mathbf{r})$ is created. We assume that this medium occupies a volume that is always in practice finite, (see Fig. 3.1). By omitting the time-dependent part $\exp(-i\omega t)$ of the complex amplitudes, it is customary to express the new field as the sum of the incident and the scattered field:

$$\mathbf{E}(\mathbf{r}) = \mathbf{E}^{(i)}(\mathbf{r}) + \mathbf{E}^{(s)}(\mathbf{r}), \tag{3.55a}$$

$$\mathbf{H}(\mathbf{r}) = \mathbf{H}^{(i)}(\mathbf{r}) + \mathbf{H}^{(s)}(\mathbf{r}), \tag{3.55b}$$

$\mathbf{E}^{(s)}$ and $\mathbf{H}^{(s)}$ being the scattered electric and magnetic vectors. It should be remembered that only the real part of the product of these functions by their corresponding harmonic time-dependent factor represent real physical quantities.

As seen in Section 1.2, the vectors $\mathbf{E}(\mathbf{r})$ and $\mathbf{H}(\mathbf{r})$ satisfy the following equations:

$$\nabla \times \nabla \times \mathbf{E}(\mathbf{r}) - k^2\mathbf{E}(\mathbf{r}) = 4\pi k[k\mathbf{P}(\mathbf{r}) + i\nabla \times \mathbf{M}(\mathbf{r})], \tag{3.56a}$$

$$\nabla \times \nabla \times \mathbf{H}(\mathbf{r}) - k^2\mathbf{H}(\mathbf{r}) = 4\pi k[-i\nabla \times \mathbf{P}(\mathbf{r}) + k\mathbf{M}(\mathbf{r})]. \tag{3.56b}$$

Equations (3.56) are analogous to Eqs. (3.2), the only difference being the source term in the right-hand side. In addition, as seen in Section 1.6.2, the scattered electric and magnetic vectors are given at points inside the scattering volume by:

$$\mathbf{E}^{(s)}(\mathbf{r}_<) = \nabla \times \nabla \times \mathbf{\Pi}_e(\mathbf{r}_<) + ik\nabla \times \mathbf{\Pi}_m(\mathbf{r}_<) - 4\pi\mathbf{P}(\mathbf{r}_<), \tag{3.57a}$$

$$\mathbf{H}^{(s)}(\mathbf{r}_<) = \nabla \times \nabla \times \mathbf{\Pi}_m(\mathbf{r}_<) - ik\nabla \times \mathbf{\Pi}_e(\mathbf{r}_<) - 4\pi\mathbf{M}(\mathbf{r}_<). \tag{3.57b}$$

At points outside this volume V the fields are given by expressions identical to Eqs. (3.57) with $\mathbf{P}(\mathbf{r}_>) = \mathbf{M}(\mathbf{r}_>) = 0$.

The electric and magnetic Hertz vectors $\mathbf{\Pi}_e$ and $\mathbf{\Pi}_m$ satisfy the integro-differential equations similar to Eq. (3.4):

$$\mathbf{\Pi}_e(\mathbf{r}) = \int_V \mathbf{P}(\mathbf{r}')G(\mathbf{r}, \mathbf{r}')\, d^3r', \tag{3.58a}$$

$$\mathbf{\Pi}_m(\mathbf{r}) = \int_V \mathbf{M}(\mathbf{r}')G(\mathbf{r}, \mathbf{r}')\, d^3r'. \tag{3.58b}$$

On comparing Eqs. (3.56)–(3.58) with Eqs. (3.2)–(3.4) one can straightforwardly obtain the angular spectrum representation for the scattered vectors in the two half-spaces $\mathscr{R}^+$ and $\mathscr{R}^-$ in the form given by Eqs. (3.7) and (3.8) (see Fig. 3.1), where now the spectral amplitudes $\mathbf{e}^{(\pm)}(\mathbf{K})$ and $\mathbf{h}^{(\pm)}(\mathbf{K})$ are expressed by:

$$\mathbf{e}^{(\pm)}(\mathbf{K}) = -i\frac{(2\pi)^2}{k_z}[\mathbf{k}^{(\pm)} \times [\mathbf{k}^{(\pm)} \times \tilde{\mathbf{P}}(\mathbf{k}^{(\pm)})] + k[\mathbf{k}^{(\pm)} \times \tilde{\mathbf{M}}(\mathbf{k}^{(\pm)})]], \quad (3.59a)$$

$$\mathbf{h}^{(\pm)}(\mathbf{K}) = -i\frac{(2\pi)^2}{k_z}[\mathbf{k}^{(\pm)} \times [\mathbf{k}^{(\pm)} \times \tilde{\mathbf{M}}(\mathbf{k}^{(\pm)})] - k[\mathbf{k}^{(\pm)} \times \tilde{\mathbf{P}}(\mathbf{k}^{(\pm)})]]. \quad (3.59b)$$

Where $\tilde{\mathbf{P}}(\mathbf{k}^{(\pm)})$ and $\tilde{\mathbf{M}}(\mathbf{k}^{(\pm)})$ are the three-dimensional Fourier transforms of $\mathbf{P}(\mathbf{r})$ and $\mathbf{M}(\mathbf{r})$, respectively. Note that since the components of $\mathbf{k}^{(\pm)}$ satisfy: $K^2 + k_z^2 = k^2$, $\tilde{\mathbf{P}}$ and $\tilde{\mathbf{M}}$ are ultimately functions of $\mathbf{K}$.

Equations (3.59a) and (3.59b) characterize the scattered field at any point either in $\mathscr{R}^+$ or $\mathscr{R}^-$, no matter how close to the scattering volume is. Since this volume V is finite, by varying the orientation of the z-axis, a convex domain Ω is built by a procedure similar to that of Section 3.2 (see Fig. 3.2), outside of which this form of representation of the scattered field is valid.

The fields and intensity in the far zone can be easily obtained from Eqs. (3.59) in a way identical to that in Section 3.3.

3.7. ANGULAR SPECTRUM REPRESENTATION OF THE SCATTERED ELECTROMAGNETIC FIELD INSIDE THE STRIP $0 < z < L$

Let us refer again to Fig. 3.1 and address the scattered field at points situated inside the strip $0 < z < L$. Two cases will be considered, depending on whether the point is outside or inside the volume V.

3.7.1. Scattered Field Outside the Scattering Volume

When the point $\mathbf{r}$ is inside the strip $0 < z < L$, but outside the scattering volume V, we introduce the plane wave expansion for $G(\mathbf{r}_>, \mathbf{r}')$, Eq. (2.51), into Eqs. (3.58), then the electric and magnetic vectors are evaluated from Eqs. (3.57). Let us see first the result of this operation for the Hertz vectors. Due to the existence of the factor $|z_> - z'|$ in the exponent of the integrand when Eq. (2.51) is used for $G(\mathbf{r}_>, \mathbf{r}')$, the strip $0 < z < L$ has to be divided into two regions: $z < z'$ and $z > z'$ (see Fig. 3.7), so that:

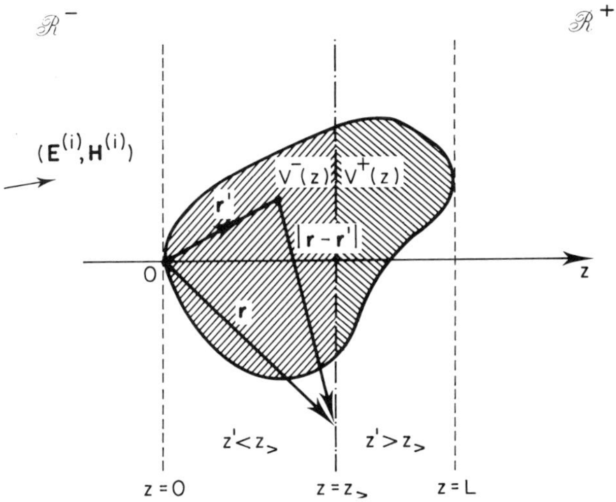

Figure 3.7. Geometry for evaluating the field inside the strip $0 < z < L$.

$$|z_> - z'| = z_> - z', \qquad \text{when } z_> > z', \tag{3.60a}$$

$$|z_> - z'| = z' - z_>, \qquad \text{when } z_> < z'. \tag{3.60b}$$

Let us denote by $V^-(z_>)$ the portion of V that contains points $\mathbf{r}'$ with $z' < z_>$, and by $V^+(z_>)$ the part of V with points $\mathbf{r}'$ for which $z' > z_>$ (Fig. 3.7). Obviously, these two volumes are both functions of the z-component, $z_>$, of the point $\mathbf{r}_> = (\mathbf{R}_>, z_>)$ at which the scattered field is being considered. In addition, it is evident that the sum of these two volumes equals V, namely:

$$V = V^-(z_>) + V^+(z_>). \tag{3.61}$$

On introducing the expansion, Eq. (2.51), for $G(\mathbf{r}_>, \mathbf{r}')$ into Eq. (3.58), and taking Eqs. (3.60) and (3.61) into account, we obtain:

$$\begin{aligned}\mathbf{\Pi}_e(\mathbf{r}_>) = &\iint_{-\infty}^{\infty} \mathbf{a}_e^{(+)}(\mathbf{K}, z_>) \exp[i\mathbf{k}^{(+)}\cdot\mathbf{r}_>]\, d^2K \\ &+ \iint_{-\infty}^{\infty} \mathbf{a}_e^{(-)}(\mathbf{K}, z_>) \exp[i\mathbf{k}^{(-)}\cdot\mathbf{r}_>]\, d^2K,\end{aligned} \tag{3.62a}$$

$$\begin{aligned}\mathbf{\Pi}_m(\mathbf{r}_>) = &\iint_{-\infty}^{\infty} \mathbf{a}_m^{(+)}(\mathbf{K}, z_>) \exp[i\mathbf{k}^{(+)}\cdot\mathbf{r}_>]\, d^2K \\ &+ \iint_{-\infty}^{\infty} \mathbf{a}_m^{(-)}(\mathbf{K}, z_>) \exp[i\mathbf{k}^{(-)}\cdot\mathbf{r}_>]\, d^2K,\end{aligned} \tag{3.62b}$$

where:

$$\mathbf{a}_e^{(\pm)}(\mathbf{K}, z_>) = \frac{i}{2\pi k_z} \int_{V^{\mp}(z_>)} \mathbf{P}(\mathbf{r}') \exp[-i\mathbf{k}^{(\pm)}\cdot\mathbf{r}']\, d^3r', \tag{3.63a}$$

$$\mathbf{a}_m^{(\pm)}(\mathbf{K}, z_>) = \frac{i}{2\pi k_z} \int_{V^{\mp}(z_>)} \mathbf{M}(\mathbf{r}') \exp[-i\mathbf{k}^{(\pm)}\cdot\mathbf{r}']\, d^3r'. \tag{3.63b}$$

Let us introduce now the two-dimensional Fourier transform of the induced polarization $\mathbf{P}(\mathbf{r})$ and magnetization $\mathbf{M}(\mathbf{r})$:

$$\tilde{\mathbf{P}}(\mathbf{K}, z) = \frac{1}{(2\pi)^2} \iint_{-\infty}^{\infty} P(\mathbf{r}) \exp(-i\mathbf{K}\cdot\mathbf{R})\, d^2R, \tag{3.64a}$$

$$\tilde{\mathbf{M}}(\mathbf{K}, z) = \frac{1}{(2\pi)^2} \iint_{-\infty}^{\infty} M(\mathbf{r}) \exp(-i\mathbf{K}\cdot\mathbf{R})\, d^2R. \tag{3.64b}$$

By means of Eqs. (3.64) we easily obtain from Eqs. (3.63):

$$\mathbf{a}_e^{(+)}(\mathbf{K}, z_>) = \frac{2\pi i}{k_z} \int_0^{z_>} \tilde{\mathbf{P}}(\mathbf{K}, z') \exp(-ik_z z')\, dz', \quad z_> > z', \tag{3.65a}$$

$$\mathbf{a}_e^{(-)}(\mathbf{K}, z_>) = \frac{2\pi i}{k_z} \int_{z_>}^{L} \tilde{\mathbf{P}}(\mathbf{K}, z') \exp(ik_z z')\, dz', \quad z_> < z', \tag{3.65b}$$

$$\mathbf{a}_m^{(+)}(\mathbf{K}, z_>) = \frac{2\pi i}{k_z} \int_0^{z_>} \tilde{\mathbf{M}}(\mathbf{K}, z') \exp(-ik_z z')\, dz', \quad z_> > z', \tag{3.66a}$$

$$\mathbf{a}_m^{(-)}(\mathbf{K}, z_>) = \frac{2\pi i}{k_z} \int_{z_>}^{L} \tilde{\mathbf{M}}(\mathbf{K}, z') \exp(ik_z z')\, dz', \quad z_> < z', \tag{3.66b}$$

Equations (3.62), together with Eqs. (3.65) and (3.66), show that now the Hertz vectors $\mathbf{\Pi}_e$ and $\mathbf{\Pi}_m$ contain both forward propagating plane waves with spectral amplitudes $\mathbf{a}_e^{(+)}(\mathbf{K}, z_>)$ and $\mathbf{a}_m^{(+)}(\mathbf{K}, z_>)$, respectively, and backward propagating plane waves with spectral amplitudes $\mathbf{a}_e^{(-)}(\mathbf{K}, z_>)$ and $\mathbf{a}_m^{(-)}(\mathbf{K}, z_>)$, respectively. In this case, these amplitudes depend on the z-component $z_>$ of the point $\mathbf{r}_>$ in which the field is being evaluated.

When Eqs. (3.62), (3.64), (3.65), and (3.66) are introduced into Eq. (3.57), the following expression for the electromagnetic field is obtained:

$$\mathbf{E}^{(s)}(\mathbf{r}_>) = \iint_{-\infty}^{\infty} \mathbf{e}^{(+)}(\mathbf{K}, z_>) \exp(i\mathbf{k}^{(+)}\cdot\mathbf{r}_>)\, d^2K + \iint_{-\infty}^{\infty} \mathbf{e}^{(-)}(\mathbf{K}, z_>) \exp(i\mathbf{k}^{(-)}\cdot\mathbf{r}_>)\, d^2K, \tag{3.67a}$$

$$\mathbf{H}^{(s)}(\mathbf{r}_>) = \iint_{-\infty}^{\infty} \mathbf{h}^{(+)}(\mathbf{K}, z_>) \exp(i\mathbf{k}^{(+)}\cdot\mathbf{r}_>)\, d^2K + \iint_{-\infty}^{\infty} \mathbf{h}^{(-)}(\mathbf{K}, z_>) \exp(i\mathbf{k}^{(-)}\cdot\mathbf{r}_>)\, d^2K, \tag{3.67b}$$

where:

$$\mathbf{e}^{(+)}(\mathbf{K}, z_>) = -i\frac{2\pi}{k_z}\left[\mathbf{k}^{(+)} \times \left[\mathbf{k}^{(+)} \times \int_0^{z_>} \tilde{\mathbf{P}}(\mathbf{K}, z') \exp(-ik_z z')\, dz'\right] + k\left[\mathbf{k}^{(+)} \times \int_0^{z_>} \tilde{\mathbf{M}}(\mathbf{K}, z') \exp(-ik_z z')\, dz'\right]\right], \tag{3.68a}$$

$$\mathbf{e}^{(-)}(\mathbf{K}, z_>) = -i\frac{2\pi}{k_z}\left[\mathbf{k}^{(-)} \times \left[\mathbf{k}^{(-)} \times \int_{z_>}^{L} \tilde{\mathbf{P}}(\mathbf{K}, z') \exp(ik_z z')\, dz'\right] + k\left[\mathbf{k}^{(-)} \times \int_{z_>}^{L} \tilde{\mathbf{M}}(\mathbf{K}, z') \exp(ik_z z')\, dz'\right]\right], \tag{3.68b}$$

$$\mathbf{h}^{(+)}(\mathbf{K}, z_>) = -i\frac{2\pi}{k_z}\left[\mathbf{k}^{(+)} \times \left[\mathbf{k}^{(+)} \times \int_0^{z_>} \tilde{\mathbf{M}}(\mathbf{K}, z') \exp(-ik_z z')\, dz'\right] - k\left[\mathbf{k}^{(+)} \times \int_0^{z_>} \tilde{\mathbf{P}}(\mathbf{K}, z') \exp(-ik_z z')\, dz'\right]\right], \tag{3.69a}$$

$$\mathbf{h}^{(-)}(\mathbf{K}, z_>) = -i\frac{2\pi}{k_z}\left[\mathbf{k}^{(-)} \times \left[\mathbf{k}^{(-)} \times \int_{z_>}^{L} \tilde{\mathbf{M}}(\mathbf{K}, z') \exp(ik_z z')\, dz'\right] - k\left[\mathbf{k}^{(-)} \times \int_{z_>}^{L} \tilde{\mathbf{P}}(\mathbf{K}, z') \exp(ik_z z')\, dz'\right]\right]. \tag{3.69b}$$

Equations (3.67)–(3.69) show that the electric and magnetic vectors contain both forward and backward plane wave components. Note that when the point at which the field is being considered moves to $\mathscr{R}^+$ or $\mathscr{R}^-$ these equations coincide with those of Section 3.6.

3.7.2. Scattered Field Inside the Scattering Volume. The Slowly Varying Amplitude Approximation

If the point at which the scattered field is considered is inside the volume V of the scatterer, the expressions of Section 3.7.1 for the Hertz vectors $\mathbf{\Pi}_e(\mathbf{r})$ and $\mathbf{\Pi}_m(\mathbf{r})$ are still valid, with $\mathbf{r}_>$ being now replaced by $\mathbf{r}_<$, namely:

$$\mathbf{\Pi}_e(\mathbf{r}_<) = \iint_{-\infty}^{\infty} \mathbf{a}_e^{(+)}(\mathbf{K}, z_<) \exp[i\mathbf{k}^{(+)}\cdot\mathbf{r}_<]\, d^2K + \iint_{-\infty}^{\infty} \mathbf{a}_e^{(-)}(\mathbf{K}, z_<) \exp[i\mathbf{k}^{(-)}\cdot\mathbf{r}_<]\, d^2K, \tag{3.70a}$$

$$\mathbf{\Pi}_m(\mathbf{r}_<) = \iint_{-\infty}^{\infty} \mathbf{a}_m^{(+)}(\mathbf{K}, z_<) \exp[i\mathbf{k}^{(+)}\cdot\mathbf{r}_<]\, d^2K + \iint_{-\infty}^{\infty} \mathbf{a}_m^{(-)}(\mathbf{K}, z_<) \exp[i\mathbf{k}^{(-)}\cdot\mathbf{r}_<]\, d^2K, \tag{3.70b}$$

where:

$$\mathbf{a}_e^{(+)}(\mathbf{K}, z_<) = \frac{2\pi i}{k_z} \int_0^{z_<} \tilde{\mathbf{P}}(\mathbf{K}, z') \exp(-ik_z z')\, dz', \quad z_< > z', \tag{3.71a}$$

$$\mathbf{a}_e^{(-)}(\mathbf{K}, z_<) = \frac{2\pi i}{k_z} \int_{z_<}^{L} \tilde{\mathbf{P}}(\mathbf{K}, z') \exp(ik_z z')\, dz', \quad z_< < z', \tag{3.71b}$$

$$\mathbf{a}_m^{(+)}(\mathbf{K}, z_<) = \frac{2\pi i}{k_z} \int_0^{z_<} \tilde{\mathbf{M}}(\mathbf{K}, z') \exp(-ik_z z')\, dz', \quad z_< > z', \tag{3.72a}$$

$$\mathbf{a}_m^{(-)}(\mathbf{K}, z_<) = \frac{2\pi i}{k_z} \int_{z_<}^{L} \tilde{\mathbf{M}}(\mathbf{K}, z') \exp(ik_z z')\, dz', \quad z_< < z', \tag{3.72b}$$

On introducing Eqs. (3.70)–(3.72) into Eqs. (3.57) we get:

$$\mathbf{E}^{(s)}(\mathbf{r}_<) = \iint_{-\infty}^{\infty} \mathbf{e}^{(+)}(\mathbf{K}, z_<) \exp(i\mathbf{k}^{(+)}\cdot\mathbf{r}_<)\, d^2K + \iint_{-\infty}^{\infty} \mathbf{e}^{(-)}(\mathbf{K}, z_<) \exp(i\mathbf{k}^{(-)}\cdot\mathbf{r}_<)\, d^2K, \tag{3.73a}$$

$$\mathbf{H}^{(s)}(\mathbf{r}_<) = \iint_{-\infty}^{\infty} \mathbf{h}^{(+)}(\mathbf{K}, z_<) \exp(i\mathbf{k}^{(+)}\cdot\mathbf{r}_<)\, d^2K + \iint_{-\infty}^{\infty} \mathbf{h}^{(-)}(\mathbf{K}, z_<) \exp(i\mathbf{k}^{(-)}\cdot\mathbf{r}_<)\, d^2K, \tag{3.73b}$$

where:

$$\mathbf{e}^{(+)}(\mathbf{K}, z_<) = -i\frac{2\pi}{k_z}\left[\mathbf{k}^{(+)} \times \left[\mathbf{k}^{(+)} \times \int_0^{z_<} \tilde{\mathbf{P}}(\mathbf{K}, z')\exp(-ik_z z')\,dz'\right]\right.$$
$$\left. + k\left[\mathbf{k}^{(+)} \times \int_0^{z_<} \tilde{\mathbf{M}}(\mathbf{K}, z')\exp(-ik_z z')\,dz'\right]\right]$$
$$- 4\pi\tilde{\mathbf{P}}(\mathbf{K}, z_<)\exp(-ik_z z_<), \tag{3.74a}$$

$$\mathbf{e}^{(-)}(\mathbf{K}, z_<) = -i\frac{2\pi}{k_z}\left[\mathbf{k}^{(-)} \times \left[\mathbf{k}^{(-)} \times \int_{z_<}^{L} \tilde{\mathbf{P}}(\mathbf{K}, z')\exp(ik_z z')\,dz'\right]\right.$$
$$\left. + k\left[\mathbf{k}^{(-)} \times \int_{z_<}^{L} \tilde{\mathbf{M}}(\mathbf{K}, z')\exp(ik_z z')\,dz'\right]\right]$$
$$- 4\pi\tilde{\mathbf{P}}(\mathbf{K}, z_<)\exp(ik_z z_<), \tag{3.74b}$$

$$\mathbf{h}^{(+)}(\mathbf{K}, z_<) = -i\frac{2\pi}{k_z}\left[\mathbf{k}^{(+)} \times \left[\mathbf{k}^{(+)} \times \int_0^{z_<} \tilde{\mathbf{M}}(\mathbf{K}, z')\exp(-ik_z z')\,dz'\right]\right.$$
$$\left. - k\left[\mathbf{k}^{(+)} \times \int_0^{z_<} \tilde{\mathbf{P}}(\mathbf{K}, z')\exp(-ik_z z')\,dz'\right]\right]$$
$$- 4\pi\tilde{\mathbf{M}}(\mathbf{K}, z_<)\exp(-ik_z z_<), \tag{3.75a}$$

$$\mathbf{h}^{(-)}(\mathbf{K}, z_<) = -i\frac{2\pi}{k_z}\left[\mathbf{k}^{(-)} \times \left[\mathbf{k}^{(-)} \times \int_{z_<}^{L} \tilde{\mathbf{M}}(\mathbf{K}, z')\exp(ik_z z')\,dz'\right]\right.$$
$$\left. - k\left[\mathbf{k}^{(-)} \times \int_{z_<}^{L} \tilde{\mathbf{P}}(\mathbf{K}, z')\exp(ik_z z')\,dz'\right]\right]$$
$$- 4\pi\tilde{\mathbf{M}}(\mathbf{K}, z_<)\exp(ik_z z_<). \tag{3.75b}$$

Equations (3.71)–(3.75) can be used to calculate the field inside the medium upon interaction of the incident field with the material. The integral equations (3.71) and (3.72) can easily be written in the alternative form:

$$\frac{\partial \mathbf{a}_e^{(\pm)}(\mathbf{K}, z_<)}{\partial z_<} = \pm\frac{2\pi i}{k_z}\tilde{\mathbf{P}}(\mathbf{K}, z_<)\exp(\mp ik_z z_<), \tag{3.76a}$$

$$\frac{\partial \mathbf{a}_m^{(\pm)}(\mathbf{K}, z_<)}{\partial z_<} = \pm\frac{2\pi i}{k_z}\tilde{\mathbf{M}}(\mathbf{K}, z_<)\exp(\mp ik_z z_<). \tag{3.76b}$$

As an example, let us assume that the response of the medium to the incident field is isotropic, linear, and spatially nondispersive, then according to

Section 1.1, the field inside the volume V is related to the induced polarization $\mathbf{P}(\mathbf{r})$ and magnetization $\mathbf{M}(\mathbf{r})$ by the following constitutive relations:

$$\mathbf{P}(\mathbf{r}) = \chi(\mathbf{r})\mathbf{E}(\mathbf{r}) = \chi(\mathbf{r})[\mathbf{E}^{(i)}(\mathbf{r}) + \mathbf{E}^{(s)}(\mathbf{r})], \quad \text{when } \mathbf{r} \text{ belongs to } V. \tag{3.77a}$$

$$\mathbf{P}(\mathbf{r}) = 0, \quad \text{when } \mathbf{r} \text{ does not belong to } V.$$

$$\mathbf{M}(\mathbf{r}) = \eta(\mathbf{r})\mathbf{H}(\mathbf{r}) = \eta(\mathbf{r})[\mathbf{H}^{(i)}(\mathbf{r}) + \mathbf{H}^{(s)}(\mathbf{r})], \quad \text{when } \mathbf{r} \text{ belongs to } V. \tag{3.77b}$$

$$\mathbf{M}(\mathbf{r}) = 0, \quad \text{when } \mathbf{r} \text{ does not belong to } V.$$

Where $\chi(\mathbf{r})$ and $\eta(\mathbf{r})$ are the *dielectric* and *magnetic susceptibilities*, respectively.

By defining the two-dimensional Fourier transforms of $\chi(\mathbf{r})$ and $\eta(\mathbf{r})$ as:

$$\tilde{\chi}(\mathbf{K}, z) = \frac{1}{(2\pi)^2} \int_{-\infty}^{\infty} \chi(\mathbf{r}) \exp(-i\mathbf{K}\cdot\mathbf{R})\, d^2R, \tag{3.78a}$$

$$\tilde{\eta}(\mathbf{K}, z) = \frac{1}{(2\pi)^2} \int_{-\infty}^{\infty} \eta(\mathbf{r}) \exp(-i\mathbf{K}\cdot\mathbf{R})\, d^2R, \tag{3.78b}$$

and introducing Eqs. (3.73) and (3.78), together with the corresponding angular spectrum representation for the incident field, into Eq. (3.77), and the result into Eq. (3.76), and taking the definitions, Eqs. (3.64), into account, one obtains straightforwardly a differential equation for $\mathbf{a}_e^{(\pm)}(\mathbf{K}, z)$ and $\mathbf{a}_m^{(\pm)}(\mathbf{K}, z)$ that requires to be iteratively solved. The simultaneous existence of both *forward* and *backward* propagating waves complicates the calculation.

Sometimes, however, backward waves can be neglected; this happens when the *slowly varying amplitude approximation* (see, e.g. Ref. 3.12) can be made for $\mathbf{a}_e^{(\pm)}(\mathbf{K}, z)$ and $\mathbf{a}_m^{(\pm)}(\mathbf{K}, z)$. This approximation amounts to assuming that the interaction of the field with the material is significant only after the wave has traveled a large distance in the medium compared with the wavelength. In this case one can consider that $\mathbf{a}_e^{(+)}(\mathbf{K}, z_<)$ and $\mathbf{a}_m^{(+)}(\mathbf{K}, z_<)$ vary so slowly with $z_<$ that the following conditions hold:

$$\left|\frac{\partial^2 \mathbf{a}_e^{(+)}(\mathbf{K}, z_<)}{\partial z_<^2}\right| \ll k \left|\frac{\partial \mathbf{a}_e^{(+)}(\mathbf{K}, z_<)}{\partial z_<}\right|, \tag{3.79a}$$

$$\left|\frac{d^2 \mathbf{a}_m^{(+)}(\mathbf{K}, z_<)}{\partial z_<^2}\right| \ll k \left|\frac{\partial \mathbf{a}_m^{(+)}(\mathbf{K}, z_<)}{\partial z_<}\right|, \tag{3.79b}$$

and:

$$\mathbf{a}_e^{(-)}(\mathbf{K}, z_<) = \mathbf{a}_m^{(-)}(\mathbf{K}, z_<) = 0. \tag{3.80}$$

When the slowly varying amplitude approximation is valid, Eqs. (3.76) can be iteratively solved by using numerical methods like the *finite difference* procedures which use the *Adams–Bashforth algorithm* as a *predictor* and the *Adams–Moulton algorithm* as a *corrector* (see, e.g. Ref. 3.13).

It should be emphasized that Eqs. (3.76) hold for any kind of medium, and irrespective of whether it is linear or nonlinear. When the interaction is nonlinear, the restrictions imposed by the requirements of phase matching between the interacting fields make the slowly varying amplitude approximation accurate, even in the presence of strong interaction, providing that there are no reflections at the boundary surface enclosing the volume V. This is due to the negligible contribution of the backward propagating components because of their phase mismatch. Examples of these calculations for nonlinear wave interactions in both isotropic and anisotropic materials can be found in Refs. 3.14–3.16.

In general, the iterative solution of Eqs. (3.76) for the spectral components of the Hertz vectors leads to the values of $\mathbf{E}(\mathbf{r})$ and $\mathbf{H}(\mathbf{r})$ inside the medium, and then, by means of the constitutive relations, to the computation of the source terms represented by the induced polarization $\mathbf{P}(\mathbf{r})$ and the induced magnetization $\mathbf{M}(\mathbf{r})$. The fields outside V can be subsequently determined according to the analysis of Sections 3.6 or 3.7.1.

3.8. THE FIRST BORN APPROXIMATION

A well known method of solving the integral equations that result from the combination of Eqs. (3.57), (3.58), and (3.77), and whose solutions yield the source terms $\mathbf{P}(\mathbf{r})$ and $\mathbf{M}(\mathbf{r})$, is based on the expansion into a *Neumann series.* This series is also known in scattering theory as the *Born series.*

When the scatterer is sufficiently weak and thin, there is fast convergence and the interaction may be described by the *first Born approximation,* which corresponds to the first term of the Neumann series.

Let us substitute Eqs. (3.77) into (3.58), thus obtaining:

$$\mathbf{\Pi}_e(\mathbf{r}) = \int_V \chi(\mathbf{r}')\mathbf{E}(\mathbf{r}')G(\mathbf{r}, \mathbf{r}')\, d^3r', \tag{3.81a}$$

$$\mathbf{\Pi}_m(\mathbf{r}) = \int_V \eta(\mathbf{r}')\mathbf{H}(\mathbf{r}')G(\mathbf{r}, \mathbf{r}')\, d^3r'. \tag{3.81b}$$

Equations (3.57) and (3.81) constitute a complete system of integral equations for the electric and magnetic vectors. In many cases of practical interest, however, the medium can be assumed to be nonmagnetic so that $\eta(\mathbf{r}) = 0$, namely, $\mathbf{\Pi}_m(\mathbf{r}) = 0$. In this case, the vector Helmholtz equation (3.56a) becomes:

$$\nabla \times \nabla \times \mathbf{E}(\mathbf{r}) - k^2\mathbf{E}(\mathbf{r}) = 4\pi k^2\chi(\mathbf{r})\mathbf{E}(\mathbf{r}), \tag{3.82}$$

which contains the *potential* $4\pi k^2\chi(\mathbf{r})$ in analogy with the scalar Schrödinger equation for the time-independent potential scattering problem in quantum mechanics (see, e.g. Refs. 3.17 or 3.18). Note that in general, when $\mathbf{\Pi}_m(\mathbf{r}) \neq 0$, the two *potentials* of the Helmholtz equations (3.56a) and (3.56b) are the *electric potential* $4\pi k^2\chi(\mathbf{r})$ and the *magnetic potential* $4\pi k^2\eta(\mathbf{r})$.

The Neumann (or Born) series consists of iteratively solving for the field inside V the integral equations obtained from Eqs. (3.55), (3.57), and (3.81). The first term of this iterative series is obtained by substituting $\mathbf{E}$ and $\mathbf{H}$ by the corresponding values $\mathbf{E}^{(i)}$ and $\mathbf{H}^{(i)}$, respectively, in the integrals of Eqs. (3.81). As quoted above, this is the *first Born approximation.* It is seen from Eq. (3.77) that it amounts to considering the interaction between the incident field and the medium so weak that this field remains unperturbed inside the volume V.

There exist no accurate criteria of validity of the first Born approximation. Roughly speaking, the perturbation in the medium upon the incident field will be small if the energy of this field is high and the fluctuations of $\chi(\mathbf{r})$ and $\eta(\mathbf{r})$ are not large. The energy of the scattered field should then be much smaller than that of the incident field. This is in fact the physical meaning of the fast convergence of the Born series. Also, the volume of interaction should be small. A sufficient criterion that covers this is (see, e.g. [3.18] for the quantum-mechanics analogy):

$$\frac{F_0 L}{k} \ll 1, \tag{3.83}$$

F_0 being the *strength* (or standard deviation) of the fluctuation of either the electric or magnetic potential and L representing the maximum linear dimension of the scattering volume. Equation (3.83) suggests that the approximation is valid for high wavenumbers compared with the product of the potential fluctuation and the interaction length. For low angle scattering, the validity of the first Born approximation is less restrictive than Eq. (3.83) imposes.

The fields outside the scatterer adopt a very simple form under this approximation. Let us assume that the incident field is a homogeneous plane wave:

$$\mathbf{E}^{(i)}(\mathbf{r}) = \mathbf{e}_i \exp(i\mathbf{k}_i \cdot \mathbf{r}), \tag{3.84a}$$

$$\mathbf{H}^{(i)}(\mathbf{r}) = \mathbf{h}_i \exp(i\mathbf{k}_i \cdot \mathbf{r}), \tag{3.84b}$$

where the incident wavevector $\mathbf{k}_i$ has magnitude ω/c, and $\mathbf{e}_i$ and $\mathbf{h}_i$ are constant amplitude vectors that satisfy the relations: $\mathbf{h}_i = \mathbf{n}_i \times \mathbf{e}_i$, $\mathbf{n}_i = \mathbf{k}_i/k$. On substituting from Eqs. (3.84) into Eqs. (3.77) one readily obtains the following expressions for the three-dimensional Fourier transforms of $\mathbf{P}(\mathbf{r})$ and $\mathbf{M}(\mathbf{r})$:

$$\tilde{\mathbf{P}}(\mathbf{k}) = \tilde{\chi}(\mathbf{k} - \mathbf{k}_i)\mathbf{e}_i, \tag{3.85a}$$

$$\tilde{\mathbf{M}}(\mathbf{k}) = \tilde{\eta}(\mathbf{k} - \mathbf{k}_i)\mathbf{h}_i \tag{3.85b}$$

$\tilde{\chi}(\mathbf{k})$ and $\tilde{\eta}(\mathbf{k})$ are the three-dimensional Fourier transforms of $\chi(\mathbf{r})$ and $\eta(\mathbf{r})$, respectively. On substituting Eqs. (3.85) into (3.59) one gets the spectral amplitudes of the electromagnetic field in the half-spaces $\mathscr{R}^+$ or $\mathscr{R}^-$ in terms of the Fourier transforms of the electric and magnetic susceptibilities within the first Born approximation, (Friberg and Wolf [3.4]):

$$\begin{aligned}\mathbf{e}_B^{(\pm)}(\mathbf{K}) = -i\,\frac{(2\pi)^2}{k_z}\,&[[\mathbf{k}^{(\pm)} \times [\mathbf{k}^{(\pm)} \times \mathbf{e}_i]]\,\tilde{\chi}(\mathbf{k}^{\pm} - \mathbf{k}_i)\\ &+ k[\mathbf{k}^{(\pm)} \times \mathbf{h}_i]\,\tilde{\eta}(\mathbf{k}^{(\pm)} - \mathbf{k}_i)],\end{aligned} \tag{3.86a}$$

$$\begin{aligned}\mathbf{h}_B^{(\pm)}(\mathbf{K}) = -i\,\frac{(2\pi)^2}{k_z}\,&[[\mathbf{k}^{(\pm)} \times [\mathbf{k}^{(\pm)} \times \mathbf{h}_i]]\,\tilde{\eta}(\mathbf{k}^{\pm} - \mathbf{k}_i)\\ &+ k[\mathbf{k}^{(\pm)} \times \mathbf{e}_i]\,\tilde{\chi}(\mathbf{k}^{(\pm)} - \mathbf{k}_i)].\end{aligned} \tag{3.86b}$$

Equations (3.86) show that these spectral amplitudes are expressed in terms of the three-dimensional Fourier transforms of χ and η evaluated at the *wavevector transfer* $\mathbf{K} = \mathbf{k}^{(\pm)} - \mathbf{k}_i$.

The far fields are easily obtained from Eqs. (3.86) in a manner similar to that used in Section 3.3.

3.9. SCATTERING FROM A WEAKLY FLUCTUATING RANDOM MEDIUM

In this section we shall consider scattering by nonmagnetic, isotropic, linear, spatially nondispersive, time-independent *random continuum media*. These are systems characterized by a dielectric permittivity $\varepsilon(\mathbf{r})$ which is a continuous random function of position $\mathbf{r}$. Fundamental accounts on the theory of wave scattering and propagation in random media, as well as on the mathematical grounds of the theory of random variables, can be found, e.g. in the works by Tatarskii [3.19], Chernov [3.20], Ishimaru [3.21], Doob [3.22], and Papoulis [3.23]. An excellent summary of statistical concepts in connection with this problem can be also found in the work by Ross [3.24].

We shall consider media that are *statistically homogeneous*, by this it is meant that the ensemble average $\langle \varepsilon \rangle$ of the permittivity over different realizations of the material (namely, over different samples) is a constant independent of the position. Without loss of generality, this quantity can be normalized to unity. We shall then write $\varepsilon(\mathbf{r})$ as the sum of the average and the fluctuating part:

$$\varepsilon(\mathbf{r}) = 1 + \delta\varepsilon(\mathbf{r}), \tag{3.87}$$

where $\delta\varepsilon(\mathbf{r})$ represents the random fluctuating part of $\varepsilon(\mathbf{r})$.

However, since:

$$\varepsilon(\mathbf{r}) = 1 + 4\pi\chi(\mathbf{r}), \tag{3.88}$$

we obtain from Eqs. (3.87) and (3.88):

$$4\pi\chi(\mathbf{r}) = \delta\varepsilon(\mathbf{r}). \tag{3.89}$$

On the other hand, since $\varepsilon(\mathbf{r}) = n^2(\mathbf{r})$, $n(\mathbf{r})$ being the refractive index, if the fluctuations are weak, by using Eq. (3.89) one can make the following approximation:

$$\delta\varepsilon(\mathbf{r}) \sim 2\delta n(\mathbf{r}). \tag{3.90}$$

$\delta n(\mathbf{r})$ being the fluctuating part of $n(\mathbf{r})$.

The *extent* of the fluctuation is characterized by the *covariance* $C_n(\mathbf{r}_1, \mathbf{r}_2)$ [3.19]–[3.24]:

$$C_n(\mathbf{r}_1, \mathbf{r}_2) = \langle \delta n(\mathbf{r}_1) \delta n(\mathbf{r}_2) \rangle, \tag{3.91}$$

where, once again, the bracket $\langle \cdot \rangle$ means ensemble average over many realizations of the medium. A consequence of the statistical homogeneity of ε is that:

$$C_n(\mathbf{r}_1, \mathbf{r}_2) = C_n(\boldsymbol{\rho}), \qquad \boldsymbol{\rho} = \mathbf{r}_1 - \mathbf{r}_2 \tag{3.92}$$

In addition, if the medium is *statistically isotropic* [3.19]–[3.24]:

$$C_n(\boldsymbol{\rho}) = C_n(\rho), \qquad \rho = |\mathbf{r}_1 - \mathbf{r}_2|. \tag{3.93}$$

The covariance function $C_n(\rho)$ is an even, monotonically decreasing function of ρ; more details on its properties can be found for instance in Refs. 3.22 and 3.23. The effective width T of $C_n(\rho)$ constitutes the *scale of the fluctuation* and is called the *correlation length*. On the other hand, its height: $\sigma_n^2 = C_n(0) = \langle (\delta n)^2 \rangle$ is the variance, or *strength* of the fluctuation. The function $B_n(\rho) = [C_n(0)]^{-1} C_n(\rho)$ is the *autocorrelation function* of the fluctuation. If $n(\mathbf{r})$ follows Gaussian statistics, then the second order moment $C_n(\boldsymbol{\rho})$ is sufficient to characterize it completely [3.21], [3.22].

Since the strength of the fluctuation is assumed to be small, we can use the first Born approximation, as indicated by Eq. (3.83). According, then, to Section 3.3, Eqs. (3.11), (3.86), (3.89), and (3.90), the electric vector of the scattered field in the far zone is given by:

$$\mathbf{E}_B^{(s)}(r\mathbf{n}) \sim -\frac{k^2}{2\pi} \mathbf{n} \times (\mathbf{n} \times \mathbf{e}_i) \int_V \exp(-i\mathbf{K}\cdot\mathbf{r})\delta n(\mathbf{r})\, d^3r, \tag{3.94}$$

with $\mathbf{K}$ being the wavevector transfer: $k\mathbf{n} - \mathbf{k}_i$; and the scattered magnetic vector being determined from Eq. (3.94) through:

$$\mathbf{H}_B^{(s)}(r\mathbf{n}) = \mathbf{n} \times \mathbf{E}_B^{(s)}(r\mathbf{n}). \tag{3.95}$$

By introducing Eqs. (3.94) and (3.95) into Eq. (3.15) one gets the ensemble average of the time averaged Poynting vector:

$$\langle \bar{\mathscr{S}}^{(s)}(r\mathbf{n}) \rangle = \frac{ck^4}{64\pi^3 r^2} |\mathbf{n} \times (\mathbf{n} \times \mathbf{e}_i)|^2$$
$$\times \int_V \int_V \exp(-i\mathbf{K}\cdot(\mathbf{r} - \mathbf{r}'))\langle \delta n(\mathbf{r})\delta n(\mathbf{r}')\rangle\, d^3r\, d^3r'. \tag{3.96}$$

Let us write:

$$\mathbf{n} \times (\mathbf{n} \times \mathbf{e}_i) = \mathbf{n} \times \mathbf{t} \sin\Theta = \mathbf{e}_\mathbf{n} \sin\Theta, \tag{3.97}$$

Θ being the angle between the direction of polarization $\mathbf{e}_i$ of the incident wave and the direction of observation $\mathbf{n}$; $\mathbf{t}$ and $\mathbf{e}_\mathbf{n}$ are unit vectors such that:

$$\mathbf{t} \sin\Theta = \mathbf{n} \times \mathbf{e}_i, \tag{3.98}$$

and:

$$\mathbf{e}_\mathbf{n} = \mathbf{n} \times \mathbf{t}. \tag{3.99}$$

Also, let us transform the variables $\mathbf{r}$ and $\mathbf{r}'$ of the integrand in Eq. (3.96) into the difference $\boldsymbol{\rho}$ and average $\mathbf{r}_a$:

$$\boldsymbol{\rho} = \mathbf{r} - \mathbf{r}', \tag{3.100a}$$

$$\mathbf{r}_a = \tfrac{1}{2}(\mathbf{r} + \mathbf{r}'). \tag{3.100b}$$

On introducing Eq. (3.97) into (3.96), taking the definition, Eq. (3.91), and Eqs. (3.92) and (3.93) into account, and using the transformation, Eq. (3.100), we easily obtain from Eq. (3.96):

$$\langle \bar{\mathcal{S}}^{(s)}(r\mathbf{n}) \rangle = \mathbf{n}\,\frac{ck^4}{64\pi^3 r^2}\sin^2\Theta \int_V d^3 r_a \int d^3\rho\, \exp(-i\mathbf{K}\cdot\boldsymbol{\rho}) C_n(\rho), \quad (3.101)$$

where the integral in $\boldsymbol{\rho}$ is extended to all space, whereas the integral in $\mathbf{r}_a$ is done over the scattering volume V. In establishing these limits of integration we have made use of the fact that the correlation distance T of the fluctuation is much smaller than the linear dimension L of the scattering volume, and therefore $C_n(\rho)$ is negligibly small for $r_a \gg T$.

The integral over $\mathbf{r}_a$ is equal to V. On the other hand, the integral in ρ represents the three-dimensional Fourier transform of the covariance $C_n(\rho)$. This is the *spectral density* Φ_n:

$$\Phi_n(\mathbf{K}) = \frac{1}{(2\pi)^3}\int d^3\rho\, \exp(-i\mathbf{K}\cdot\boldsymbol{\rho}) C_n(\boldsymbol{\rho}). \quad (3.102)$$

Sometimes the spectral density is defined as the Fourier transform of the correlation function $B_n(\rho)$ [3.23]–[3.24]. In this case the only difference with Eq. (3.102) is the normalization factor $C_n(0)$.

With the above considerations, and using the definition, Eq. (3.102), we can express Eq. (3.101) as:

$$\langle \bar{\mathcal{S}}^{(s)}(r\mathbf{n}) \rangle = \mathbf{n}\,\frac{cVk^4}{8r^2}\sin^2\Theta\,\Phi_n(\mathbf{K}). \quad (3.103)$$

Therefore, the mean scattered intensity in the far zone per unit volume and per unit of solid angle is:

$$\frac{1}{V}\frac{d\langle I^{(s)}(r\mathbf{n})\rangle}{d\Omega} = \frac{ck^4}{8}\sin^2\Theta\,\Phi_n(\mathbf{K}). \quad (3.104)$$

Note that for a statistically homogeneous and isotropic fluctuation the spectral density can be expressed after performing the angle integration in Eq. (3.102):

$$\Phi_n(\mathbf{K}) = \frac{4\pi}{|\mathbf{K}|}\int_0^\infty C_n(\rho)\sin(|\mathbf{K}|\rho)\rho\, d\rho. \quad (3.105)$$

It is easy to see that:

$$|\mathbf{K}| = 2k \sin\frac{\theta}{2}, \tag{3.106}$$

where θ is the angle between the direction of observation $\mathbf{n}$ and the direction of propagation $\mathbf{k}_i$ of the incident field.

Equation (3.104) shows that the mean scattered intensity is proportional to the spectral density of the refractive index fluctuation, has the $\sin^2 \Theta$ dependence and, due to a well known property of Fourier transforms, is more concentrated about the direction of incidence the larger the correlation distance T of the refractive index fluctuation is.

3.10. SCALAR APPROACH TO SCATTERED SCALAR WAVEFIELDS

Let us assume that the scatterer is nonmagnetic, namely, $\eta = 0$, so that $\mathbf{M}(\mathbf{r}) = 0$. Equations (3.56) then become:

$$\nabla \times \nabla \times \mathbf{E}(\mathbf{r}) - k^2\mathbf{E}(\mathbf{r}) = 4\pi k^2 \mathbf{P}(\mathbf{r}), \tag{3.107a}$$

$$\nabla \times \nabla \times \mathbf{H}(\mathbf{r}) - k^2\mathbf{H}(\mathbf{r}) = -4\pi i k \nabla \times \mathbf{P}(\mathbf{r}), \tag{3.107b}$$

with:

$$\mathbf{P}(\mathbf{r}) = \chi(\mathbf{r})\mathbf{E}(\mathbf{r}). \tag{3.108}$$

Taking into account the vector identity:

$$\nabla \times \nabla \times \mathbf{E} = \nabla(\nabla \cdot \mathbf{E}) - \nabla^2 \mathbf{E}, \tag{3.109}$$

and introducing Eq. (3.109) into (3.107a) we obtain the following differential equation for the electric vector:

$$\nabla^2\mathbf{E}(\mathbf{r}) + k^2\mathbf{E}(\mathbf{r}) = -k^2[n^2(\mathbf{r}) - 1]\mathbf{E}(\mathbf{r}) + \nabla[\nabla \cdot \mathbf{E}(\mathbf{r})], \tag{3.110}$$

where we have made use of the expression:

$$4\pi\chi(\mathbf{r}) = \begin{cases} n^2(\mathbf{r}) - 1 & \text{if } \mathbf{r} \text{ belongs to } V, \\ 0 & \text{if } \mathbf{r} \text{ is outside } V, \end{cases} \tag{3.111}$$

$n(\mathbf{r})$ denoting the refractive index of the scatterer.

Equation (3.110) shows that the change in polarization of the electric vector, as a result of the scattering described by the right-hand side, is due to the *source term* $\nabla[\nabla\cdot\mathbf{E}(\mathbf{r})]$. When the scale over which $n(\mathbf{r})$ varies is much larger than the wavelength λ, this term, and hence the depolarization, may be neglected (see also, e.g. Strohben [3.25]). One is then left with a differential equation that for each Cartesian component of the electric vector has the form:

$$\nabla^2 U(\mathbf{r}) + k^2 U(\mathbf{r}) = F(\mathbf{r})U(\mathbf{r}), \tag{3.112}$$

where the *potential* $F(\mathbf{r})$ is:

$$F(\mathbf{r}) = \begin{cases} -k^2[n^2(\mathbf{r}) - 1] & \text{if } \mathbf{r} \text{ belongs to } V, \\ 0 & \text{if } \mathbf{r} \text{ is outside } V, \end{cases} \tag{3.113}$$

Note that Eq. (3.112) is formally identical to the time-independent Schrödinger equation of *potential scattering* of quantum mechanics [3.17], [3.18]. Its integral form, discussed in Section 1.6.3, Eq. (1.54), is also similar to Eq. (3.17), where the *source distribution* $\rho(\mathbf{r})$ is given by Eq. (1.49) and, hence, it contains the wavefunction $U(\mathbf{r})$.

The solution of Eq. (3.112) can be written at points outside the scattering volume (cf. Section 1.6.3):

$$U(\mathbf{r}_>) = U^{(i)}(\mathbf{r}_>) + U^{(s)}(\mathbf{r}_>), \tag{3.114}$$

and for points outside V:

$$U(\mathbf{r}_<) = U^{(i)}(\mathbf{r}_<) + U^{(s)}(\mathbf{r}_<), \tag{3.115}$$

where the scattered field is:

$$U^{(s)}(\mathbf{r}) = -\frac{1}{4\pi}\int_V F(\mathbf{r}')U(\mathbf{r}')G(\mathbf{r}, \mathbf{r}')\, d^3r'. \tag{3.116}$$

Equations (3.115) and (3.116) constitute a Fredholm integral equation of the second kind for the field $U(\mathbf{r}_<)$ inside the scattering volume. Once this field has been found, the field outside the scatterer can be obtained by means of Eqs. (3.114) and (3.116) substituting $\mathbf{r}$ by $\mathbf{r}_>$.

With reference to Fig. 3.1, the fields inside the strip $0 < z < L$ (either inside or outside V), and the fields in $\mathscr{R}^+$ or $\mathscr{R}^-$, can be found from Eqs. (3.114)–(3.116) in a way identical to that developed in Sections 3.6 and 3.7. This is left as an exercise for the reader.

It should be remarked, however, that although Eq. (3.112) represents an adequate way of describing the strong multiple interaction of a scalar wavefield with a scattering medium (like, e.g. an acoustic wave), when Eq. (3.112) is applied to the electromagnetic field, it is meaningful only within the domain of validity of the first Born approximation, or the Rytov or the eikonal approximations to be discussed next. Therefore, it is only within the domain of validity of these three models when one can assume that the scatterer does not depolarize the wave. The three approximations belong to the range of *small fluctuations* [3.19]–[3.21], [3.25]. Due to its importance in many problems, we shall derive explicitly the expressions for scattered scalar wavefields within the first Born approximation. The Rytov and the eikonal procedures will then be discussed.

3.10.1. The First Born Approximation for Scalar Wavefields

Let the incident field be a plane monochromatic wave of unit amplitude and wavevector $\mathbf{k}_i$:

$$U^{(i)}(\mathbf{r}) = \exp(i\mathbf{k}_i \cdot \mathbf{r}), \tag{3.117}$$

then, the first Born approximation assumes that $U(\mathbf{r})$ in the integrand of Eq. (3.116) can be replaced by $U^{(i)}(\mathbf{r})$. As mentioned before, this constitutes the first term of the Neumann series for the integral equation represented by Eqs. (3.115) and (3.116). When one makes this operation, and Eq. (2.51) is introduced into Eq. (3.116), the following expression for the scattered field is obtained (Wolf, [3.26]):

$$U_B^{(s)}(\mathbf{r}_>) = \iint_{-\infty}^{\infty} A_B^{(\pm)}(\mathbf{K}) \exp[i(\mathbf{K} \cdot \mathbf{R}_> \pm k_z z_>)]\, d^2K, \tag{3.118}$$

with the angular spectrum being given by:

$$A_B^{(\pm)}(\mathbf{K}) = -i\frac{\pi}{k_z}\tilde{F}(\mathbf{k}^{(\pm)} - \mathbf{k}_i), \tag{3.119}$$

where, as before, $\mathbf{k}^{(\pm)} = (\mathbf{K}, \pm k_z)$, $\mathbf{r}_> = (R_>, z_>)$, and the sign "plus" or "minus" is considered according to whether the point $\mathbf{r}_>$ is assumed to be in $\mathscr{R}^+$ or $\mathscr{R}^-$, respectively.

Equations (3.118) and (3.119) permit the scattered field to be found either in $\mathscr{R}^+$ or $\mathscr{R}^-$, no matter how close the point $\mathbf{r}_>$ is to the scattering volume V. Equation (3.119) shows that the angular spectrum of this scattered field is given by the three-dimensional Fourier transform of the *potential* $F(\mathbf{r})$ evaluated at the wavevector transfer $\mathbf{K}^{(\pm)} = \mathbf{k}^{(\pm)} - \mathbf{k}_i$.

On the other hand, the scattered field in the far zone is:

$$U(r\mathbf{n}) \sim -2\pi^2 \tilde{F}(\mathbf{K}^{(\pm)}) \frac{\exp(ikr)}{r}. \tag{3.120}$$

3.10.2. The Rytov Approximation

Another procedure useful in the range of *small fluctuations* of the refractive index $n(\mathbf{r})$ is the *Rytov approximation* [3.19]–[3.21]. It is often used in problems of optical and microwave propagation in turbulence [3.25], optical tomography, and acoustic wave propagation in biological media. The scale of the fluctuation must be large compared with the wavelength. This means that the field is scattered in a cone of small angle about the direction of incidence, and, therefore, there is no backscattering into the region $\mathscr{R}^+$ of Fig. 3.1. A discussion on the relative advantages of the Rytov approximation over the first Born approximation may be found, e.g. in Refs. 3.19 and 3.20.

The Rytov method consists of writing the wavefield $U(\mathbf{r})$ as:

$$U(\mathbf{r}) = \exp[\phi(\mathbf{r})], \tag{3.121}$$

then developing a series solution for $\phi(\mathbf{r})$.

Using Eq. (3.121) we obtain:

$$\nabla^2 U = U[|\nabla\phi|^2 + \nabla^2\phi]. \tag{3.122}$$

Introducing Eq. (3.122) into (3.112) one is led to:

$$\nabla^2\phi(\mathbf{r}) + |\nabla\phi(\mathbf{r})|^2 + k^2 = F(\mathbf{r}). \tag{3.123}$$

On the other hand, the incident field may be written as:

$$U^{(i)}(\mathbf{r}) = \exp[\phi^{(i)}(\mathbf{r})], \tag{3.124}$$

where $\phi^{(i)}(\mathbf{r})$ satisfies the homogeneous equation:

$$\nabla^2\phi^{(i)}(\mathbf{r}) + |\nabla\phi^{(i)}(\mathbf{r})|^2 + k^2 = 0. \tag{3.125}$$

Let us subtract Eq. (3.125) from (3.123) and write:

$$\phi(\mathbf{r}) = \phi^{(i)}(\mathbf{r}) + \phi^{(1)}(\mathbf{r}), \tag{3.126}$$

then we obtain:

$$\nabla^2\phi^{(1)}(\mathbf{r}) + 2\nabla\phi^{(i)}(\mathbf{r})\cdot\nabla\phi^{(1)}(\mathbf{r}) = -|\nabla\phi^{(1)}(\mathbf{r})|^2 + F(\mathbf{r}). \tag{3.127}$$

Using the identity:

$$\nabla^2(U^{(i)}\phi^{(1)}) = (\nabla^2 U^{(i)})\phi^{(1)} + 2U^{(i)}\nabla\phi^{(i)}\cdot\nabla\phi^{(1)} + U^{(i)}\nabla^2\phi^{(1)}, \tag{3.128}$$

and taking into account that:

$$\nabla^2 U^{(i)} + k^2 U^{(i)} = 0, \tag{3.129}$$

we can write Eq. (3.127) as:

$$(\nabla^2 + k^2)(U^{(i)}\phi^{(1)}) = -[|\nabla\phi^{(1)}|^2 - F(\mathbf{r})]U^{(i)}. \tag{3.130}$$

Equation (3.130) is of the form Eq. (3.112) and, hence, the solution $\phi^{(1)}$ can be expressed by means of Eq. (3.116) as:

$$\phi^{(1)}(\mathbf{r}) = \frac{1}{4\pi}\frac{1}{U^{(i)}(\mathbf{r})}\int_V G(\mathbf{r},\mathbf{r}')[|\nabla\phi^{(1)}(\mathbf{r}')|^2 - F(\mathbf{r}')]U^{(i)}(\mathbf{r}')\, d^3r'. \tag{3.131}$$

Equation (3.131) can be iteratively solved like Eq. (3.116). The first iteration $\phi^{(1)}_{(0)}$ consists of setting $\phi^{(1)} = 0$ in the integrand of Eq. (3.131), then using Eq. (3.126), we obtain:

$$U(\mathbf{r}) = \exp[\phi^{(i)}(\mathbf{r}) + \phi^{(1)}_{(0)}(\mathbf{r})], \tag{3.132}$$

where, according to Eq. (3.131):

$$\phi^{(1)}_{(0)}(\mathbf{r}) = -\frac{1}{4\pi}\frac{1}{U^{(i)}(\mathbf{r})}\int_V G(\mathbf{r},\mathbf{r}')F(\mathbf{r}')U^{(i)}(\mathbf{r}')\, d^3r'. \tag{3.133}$$

Therefore, from Eq. (3.124) the first Rytov approximation may be written as:

$$U(\mathbf{r}) = U^{(i)}(\mathbf{r})\exp[\phi^{(1)}_{(0)}(\mathbf{r})], \tag{3.134}$$

with $\phi^{(1)}_{(0)}(\mathbf{r})$ given by Eq. (3.133).

When the exponential of Eq. (3.134) is expanded into a series, the first two terms are:

$$U(\mathbf{r}) = U^{(i)}(\mathbf{r})[1 + \phi^{(1)}_{(0)}(\mathbf{r})]. \tag{3.135}$$

By taking Eq. (3.133) into account, we see that Eq. (3.135) constitutes the first Born approximation for $U(\mathbf{r})$. Therefore, the first Born approximation is included in the first Rytov approximation, thus being adequate when, in addition to λ being very small compared with the scale of the refractive index fluctuation, the series, Eq. (3.134), converges so fast that only the first terms contribute. This happens when the scattered intensity is small.

In problems dealing with propagation in random media, it is customary to write:

$$\phi(\mathbf{r}) = \chi(\mathbf{r}) + iS(\mathbf{r}). \tag{3.136}$$

The real part of $\phi(\mathbf{r})$, represented by $\chi(\mathbf{r})$, constitutes the logarithm of the amplitude of $U(\mathbf{r})$ and is usually called the *log amplitude fluctuation.* The imaginary part of $\phi(\mathbf{r})$, denoted by $S(\mathbf{r})$, represents the *phase* of $U(\mathbf{r})$. The reason for using the representation (3.136) is that in many practical situations the log amplitude fluctuation follows normal statistics. This is the so-called *log normal model*; then, as the field propagates in the random medium, the fluctuations of $U(\mathbf{r})$ can be easily accounted for analytically (see discussions on this subject matter in, e.g. Refs. 3.21, 3.25, 3.27, and 3.28). (See Problem 3.14.)

3.10.3. The Eikonal Approximation

For weak scatterers, the geometrical optics limit [3.20], [3.21] can be employed when the total length L of the scattering volume traversed by the passing radiation is such that $\lambda L \ll T^2$, T being the scale of fluctuation of $n(\mathbf{r})$. (A full discussion on the foundations and validity of this limit may be found for instance in Chapter 3 of Ref. 3.29.) In this approximation the wavefield inside the medium is expressed as:

$$U(\mathbf{r}) = A(\mathbf{r}) \exp[iS(\mathbf{r})], \tag{3.137}$$

where $A(\mathbf{r})$ is the amplitude and $S(\mathbf{r})$ is the phase, *optical path*, or *eikonal* [3.29]. The function $S(\mathbf{r})$ of the ray that can be associated in this case, and which describes a path between two points $\mathbf{r}_1$ and $\mathbf{r}_2$, is given by:

$$S(\mathbf{r}_1) - S(\mathbf{r}_2) = k \int_{\mathbf{r}_1}^{\mathbf{r}_2} n(\mathbf{r})\, ds, \tag{3.138}$$

ds being the path element.

However, in order to evaluate Eq. (3.138) the ray path has to be known. In a weak scatterer this path can be approximated by a straight line along the direction of incidence. Let this direction be along the z-axis, then Eq. (3.138) may be written as:

$$S(\mathbf{R}, L) - S(\mathbf{R}, 0) = k \int_0^L n(\mathbf{R}, z)\, dz, \qquad \mathbf{R} = (x, y). \tag{3.139}$$

The geometrical optics limit can be straightforwardly derived from the first Rytov approximation, Eq. (3.134), for a weak scatterer when $\lambda L \ll T^2$. In this case one can assume that the angle θ of scattering is very small. We shall use the angular spectrum representation of the wavefield $U(\mathbf{r})$. Let the incident field $U^{(i)}(\mathbf{r})$ be a plane wave of unit amplitude, propagating along the z-axis:

$$U^{(i)}(\mathbf{r}) = \exp(ikz). \tag{3.140}$$

Since the propagation is now rectilinear inside V, $\phi^{(1)}$ (Eqs. (3.131) and (3.140)) will be nonzero only at points inside the geometrical projection $\mathbf{S}$ of the scatterer over a plane $z = constant$ (Fig. 3.8).

From Eq. (3.133) and using Eqs. (2.51) and (3.140), one has at points $\mathbf{r}_>$ in $\mathscr{R}^+$ lying in the projection $\mathbf{S}$:

$$\phi_{(0)}^{(1)}(\mathbf{r}_>) = -\frac{i}{8\pi^2} \frac{1}{\exp(ikz_>)} \iint_{-\infty}^{\infty} d^2K \exp[i(\mathbf{K}\cdot\mathbf{R}_> + k_z z_>)]$$
$$\times \frac{1}{k_z} \int_V \exp[-i(\mathbf{K}\cdot\mathbf{R}' + k_z z')] \exp(ikz') F(\mathbf{r}')\, d^3r'. \tag{3.141}$$

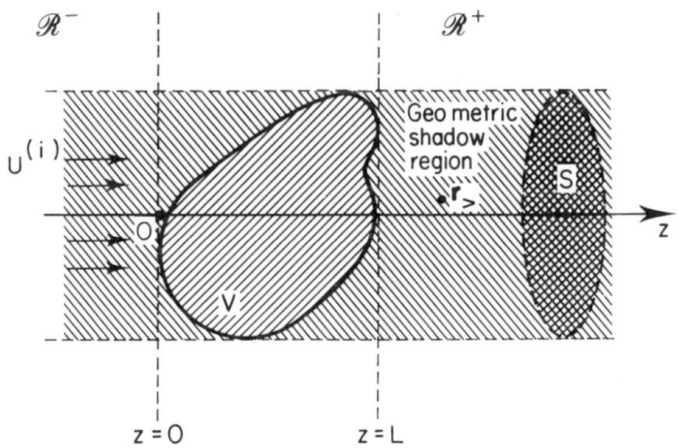

Figure 3.8. The geometric shadow of the scatterer.

The geometrical optics limit amounts to assuming that the angular spectrum:

$$A(\mathbf{K}) = \frac{1}{k_z}\int_V \exp[-i(\mathbf{K}\cdot\mathbf{R}' + k_z z')] \exp(ikz')F(\mathbf{r}')\, d^3r' \qquad (3.142)$$

is different from zero only at spatial frequencies **K** very nearly zero. That is, such that:

$$k_z = \sqrt{k^2 - K^2} \sim k. \qquad (3.143)$$

Equation (3.141) may then be written as:

$$\phi^{(1)}_{(0)}(\mathbf{r}_>) = -\frac{i}{8\pi^2 k}\iint_{-\infty}^{\infty} d^2K \exp(i\mathbf{K}\cdot\mathbf{R}_>) \int_V \exp(-i\mathbf{K}\cdot\mathbf{R}')F(\mathbf{R}', z')\, d^2R'\, dz'. \qquad (3.144)$$

Rearranging the order of integration in Eq. (3.144) and recalling that:

$$\iint_{-\infty}^{\infty} \exp[-i\mathbf{K}\cdot(\mathbf{R} - \mathbf{R}')]\, d^2K = (2\pi)^2\delta(\mathbf{K} - \mathbf{K}'). \qquad (3.145)$$

From Eq. (3.144) one gets:

$$\phi^{(1)}_{(0)}(\mathbf{r}_>) = -\frac{i}{2k}\int_0^L F(\mathbf{R}, z)\, dz, \qquad \mathbf{r}_> = (\mathbf{R}_>, z_>), \qquad (3.146)$$

R being in **S**.

Let us write:

$$n(\mathbf{r}) = 1 + \delta n(\mathbf{r}), \qquad (3.147)$$

where $\delta n(\mathbf{r})$ represents the fluctuation of the refractive index. One has:

$$n^2(\mathbf{r}) - 1 = 2\delta n(\mathbf{r}) + [\delta n(\mathbf{r})]^2. \qquad (3.148)$$

For a weakly fluctuating random medium the second term of (3.148) can be neglected, (cf. Eqs. (3.87) and (3.90)), so that:

$$n^2(\mathbf{r}) - 1 \sim 2\delta n(\mathbf{r}). \qquad (3.149)$$

Thus, by introducing Eq. (3.149) into (3.113), and the result into Eq. (3.146), we obtain finally:

$$\phi_{(0)}^{(1)}(\mathbf{r}_>) = ik \int_0^L \delta n(\mathbf{r})\, dz. \tag{3.150}$$

Equation (3.150) constitutes the *eikonal approximation* for $\phi_{(0)}^{(1)}(\mathbf{r})$. From Eqs. (3.150), (3.140), and (3.134) we obtain the result:

$$U(\mathbf{r}_>) = \exp\left[ik\left(z_> + \int_0^L \delta n(\mathbf{r})\, dz\right)\right]. \tag{3.151}$$

Note that if $U(\mathbf{r})$ were evaluated at a point $\mathbf{r}_<$ inside the volume V of the scattering medium, the upper limit L of the integral in Eq. (3.151) should be replaced by $z_<$.

Equation (3.151) constitutes the basis of a wide class of computer tomography methods (see a review of these procedures, e.g. in Ref. 3.30). It is also of wide application in techniques on wave propagation in random media (see, e.g. Refs. 3.21 and 3.25).

3.11. MULTIPLE SCATTERING THEORIES

The approximations discussed in this chapter are useful for describing field interactions with weak scatterers and, in particular, with weakly fluctuating media. However, they become inadequate as the strength of the interaction, or the refractive index fluctuations, become larger. As a matter of fact, the range of validity of the Rytov approximation, for example, has been found to be less broad than expected from the theoretical predictions [3.31]. Hence, for multiple scattering resulting from stronger interactions, other methods are required. These have been largely developed for wave propagation in random media. They are not studied here, but the reader is referred to specific references quoted below.

One group of such methods contains a phenomenological approach and is based on the *radiative transfer equation* (see, e.g. Refs. 3.32 and 3.33). This model also gives rise to the well known *diffusion approximation* [3.21] for very dense random media. The connection of this rather heuristic model with the theory based on Maxwell's equations encounters several difficulties that have generated a vast literature (see, e.g. Refs. 3.34–3.37 and references cited therein).

Another group of methods contains rigorous diagrammatic procedures that solve iteratively the wave equation for the total field by using Feynman graphs, [3.38]–[3.40]. These lead to the *Dyson equation* [3.41] for the mean field, and to the *Bethe–Salpeter equation*, [3.42] for its correlation function. These techniques are extensively used. Excellent reviews can be found in Refs. 3.21, 3.25, and 3.43–3.45. Reviews and references on light scattering by dense media and the associated effect of *photon localization* manifested by the phenomenon of *enhanced backscattering* are given in Refs. 3.46 and 3.47.

PROBLEMS

3.1. A Hertzian dipole consists of two opposite charges, $+e$ and $-e$, separated by a distance l and oscillating with time-harmonic frequency as $\exp(-i\omega t)$. The current I and the dipole moment of this system are $I = -i\omega e$ and $P = Il$, respectively. Assuming the dipole to be at the origin and the charges to be on the z-axis, the electric current density is:

$$\mathbf{j}(\mathbf{r}) = Il\delta(\mathbf{r})\hat{\mathbf{z}},$$

$\hat{\mathbf{z}}$ being the unitary vector along the z-axis.

Find the angular spectra for both the electromagnetic field radiated by this dipole and the time averaged Poynting vector.

3.2. Making use of Section 3.3 show that:

$$\lim_{r\to\infty} r(\nabla \times \mathbf{E} - ik\hat{\mathbf{r}} \times \mathbf{E}) = 0,$$

$$\lim_{r\to\infty} r(\nabla \times \mathbf{H} - ik\hat{\mathbf{r}} \times \mathbf{H}) = 0,$$

$\hat{\mathbf{r}}$ denoting the unit vector in the direction of $\mathbf{r}$. These are the radiation conditions for electromagnetic fields [3.48], [3.49].

3.3. Assuming fields with time dependence $\exp(-i\omega t)$, Maxwell's equations imply the possibility of introducing a *vector potential* $\mathbf{A}(\mathbf{r})$, such that:

$$\mathbf{B}(\mathbf{r}) = \nabla \times \mathbf{A}(\mathbf{r}),$$

as well as a *scalar potential* $\phi(\mathbf{r})$, defining:

$$\mathbf{E}(\mathbf{r}) = ik\mathbf{A}(\mathbf{r}) - \nabla\phi(\mathbf{r}).$$

Show that these potentials are related by the *Lorentz condition*:

$$\nabla \cdot \mathbf{A} - ik\phi = 0,$$

and both satisfy the inhomogeneous Helmholtz equations:

$$\nabla^2 \mathbf{A} + k^2 \mathbf{A} = -\frac{4\pi}{c}\mathbf{j},$$

$$\nabla^2 \phi + k^2 \phi = -4\pi\rho.$$

$\mathbf{A}$ and ϕ are not uniquely defined since the same fields $\mathbf{E}$ and $\mathbf{H}$ are obtained from any other potentials $\mathbf{A}'$ and ϕ' that are given by the *gauge transformation*:

$$\mathbf{A}' = \mathbf{A} + \nabla\chi,$$

$$\phi' = \phi + ik\chi,$$

χ being an arbitrary function.

Show that the Lorentz condition imposes on χ to satisfy the homogeneous Helmholtz equation:

$$\nabla^2 \chi + k^2 \chi = 0.$$

3.4. Find the angular spectra of the scalar and vector potentials for a charge moving uniformly in vacuum.

3.5. Show that:

$$\int_{-\infty}^{\infty} \frac{1}{k_z} \exp[i(K_y y \pm k_z z)]\, dK_y = -2iK_0(k\gamma d),$$

$$\gamma = \sqrt{\left(\frac{c}{v_0}\right)^2 - 1}, \qquad d = \sqrt{y^2 + z^2}, \qquad k_z = i\sqrt{K_y^2 + k^2\gamma^2},$$

$K_0(k\gamma d)$ being the modified Hankel function of zero order and argument $k\gamma d$.

3.6. Prove Eq. (3.40) for the Hertz vector $\mathbf{\Pi}(\mathbf{r}, \omega)$ of radiation at frequency ω from a charge with uniform movement in vacuum.

3.7. Consider a spherical shell of radius R with a source density distribution:

$$\rho(\mathbf{r}) = \frac{1}{4\pi r^2} e\delta(r - R),$$

where e is a constant.

Show that the angular spectrum of the field radiated by this distribution is zero if $kR = l\pi$, l being a positive integer (Schott [3.50], see also Ref. 3.51).

3.8. Find the angular spectrum and far field within the first Born approximation for a plane monochromatic wave scattered by a homogeneous sphere of radius a.

3.9. Find the angular spectrum, far field and Poynting vector, within the first Born approximation, for the scattering of a plane wave by a nonmagnetic periodic structure of spacing a with permittivity:

$$\varepsilon(\mathbf{r}) = \sum_{j=1}^{\infty} f_j \delta(\mathbf{r} - \mathbf{r}_j),$$

where the vector $\mathbf{r}_j$ is: $\mathbf{r}_j = (l, m, n)\mathbf{a}$ (l, m, and n being integers, and f_j representing positive constants).

3.10. A nonabsorbing *phase screen* is a layer of dielectric material that introduces changes in the phase of the incident wavefield $U^{(i)}$ so that the wave emerging from the screen in a plane immediately behind it, is:

$$U(\mathbf{r}) = U^{(i)}(\mathbf{r}) \exp[i\varphi(\mathbf{r})],$$

where the real function $\varphi(\mathbf{r})$ represents the phase added to that of the incident field after traversing the screen.

Assume that $\varphi(\mathbf{r})$ is a homogeneous and isotropic random function with normal statistics, zero mean, and correlation function $B_\varphi(\rho)$:

$$B_\varphi(\rho) = \frac{\langle \varphi(\mathbf{r})\varphi(\mathbf{r} + \boldsymbol{\rho})\rangle}{\sigma_\varphi^2},$$

where σ_φ^2 is the variance of $\varphi(\mathbf{r})$. The width of $B_\varphi(\rho)$ is negligible compared with the linear dimension of the illuminated area. Show that the mean scattered intensity in the far zone, per unit of solid angle and per unit of illuminated area, is:

$$\frac{dI^{(s)}(r\mathbf{n})}{d\Omega} = c\pi \cos^2 \theta \langle |A(\mathbf{s}_\perp)|^2 \rangle.$$

Where $\mathbf{n}$ is a unit vector characterizing the scattering direction: $\mathbf{n} = (n_x, n_y, n_z)$, $U^{(i)}$ is a plane wave with wavevector: $k\mathbf{n}^{(i)} = k(n_x^{(i)}, n_y^{(i)}, n_z^{(i)})$, and the average of the square modulus of the spectral amplitude $A(\mathbf{s}_\perp)$ is:

$$\langle |A(\mathbf{s}_\perp)|^2 \rangle = \frac{ck^2}{4\pi} \cos^2\theta \int_0^\infty d^2\rho \exp(-ik\mathbf{s}_\perp \cdot \boldsymbol{\rho}) \exp[-\sigma_\varphi^2(1 - B_\varphi(\rho))],$$

$\mathbf{s}_\perp$ being the vector transfer: $\mathbf{s}_\perp = (n_x - n_x^{(i)}, n_y - n_y^{(i)})$; $\mathbf{n}^{(i)}$ is assumed along the z-axis and $n_z = \cos\theta$.

Show that if $B_\varphi(\rho) = \exp[-\rho^2/T^2]$ the scattered intensity in the far zone defined above is:

$$\frac{dI^{(s)}(r\mathbf{n})}{d\Omega} = \frac{ck^2}{4\pi} \cos^2\theta \exp(-\sigma_\varphi^2)\left[\frac{(2\pi)^2}{k^2}\delta(s_\perp) + \pi T^2 \sum_{m=1}^{\infty} \frac{\sigma_\varphi^{2m}}{m!} \exp[-k^2T^2s_\perp^2/4m]\right].$$

The δ-function in the first term represents the *forward* or *coherent* component of the scattered intensity, whereas the second term involving the series expansion describes the *diffuse* part scattered at all angles. (This result was originally derived by Bramley [3.52].)

Find the limits of the above expression when: (a) $\sigma_\varphi^2 \ll 1$, and (b) $\sigma_\varphi^2 \gg 1$.

3.11. Consider a slab of ground glass of uniform refractive index n. One of its surfaces is rough with a random profile: $z = D(x, y)$ (see Fig. 3.9).

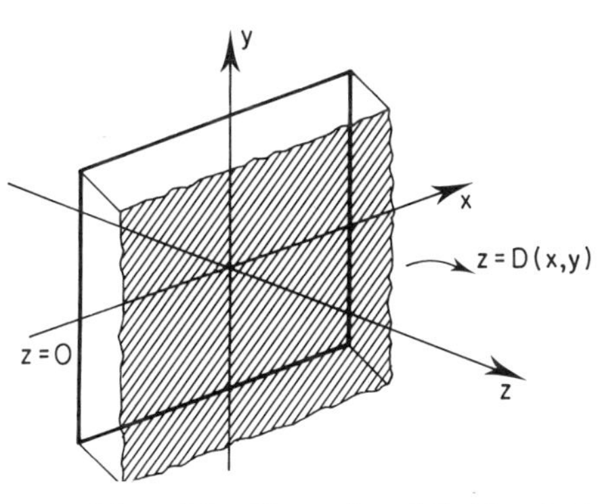

Figure 3.9. Illustration of Problem 3.11.

Show by means of the eikonal approximation that this slab behaves as a phase screen, so that if $U^{(i)}(\mathbf{r})$ denotes an incident wave, the wavefield emerging from the screen in a plane immediately behind it, is:

$$U(\mathbf{r}) = U^{(i)}(\mathbf{r})A \exp[ik(n - 1)D(x, y)],$$

A being a complex constant. Hence, the variance σ_φ^2 of the phase φ due to the screen is: $\sigma_\varphi^2 = k^2(n - 1)^2\sigma_D^2$, σ_D^2 being the variance of $D(x, y)$.

Apply to this case the results of Problem 3.10.

3.12. A *Goldfischer model* [3.53] of the phase screen is that for which the random phases φ are uncorrelated: $B_\varphi(\rho) = \delta(\rho)$. Suppose that this phase screen is superimposed on a nonrandom planar object of finite extent that emits with a field amplitude distribution $o(x, y)$ ($o(x, y)$ being a nonrandom real and positive function). The field then emerging from the screen in a plane immediately behind it, is, apart from a complex constant that can be omitted:

$$U(\mathbf{r}) = o(x, y) \exp[i\varphi(x, y)].$$

Show that the mean scattered intensity distribution behind the screen is:

$$\frac{dI^{(s)}(r\mathbf{n})}{d\Omega} = \frac{ck^2}{4\pi} \cos^2\theta \exp(-\sigma_\varphi^2)[(2\pi)^4|\tilde{o}(\mathbf{s}_\perp)|^2 + K],$$

where $\tilde{o}(\mathbf{s}_\perp)$ is the two-dimensional Fourier transform of $o(x, y)$, and K is a constant.

The above result shows that the mean scattered intensity in the far zone is equal to the power spectrum of the object amplitude distribution plus a constant background. This has several applications in image synthesis [3.54].

3.13. Show that the mean scattered intensity from an inhomogeneous medium with covariance:

$$B_n(\rho) = \langle n^2 \rangle \exp\left(-\frac{\rho}{T}\right),$$

($\langle n^2 \rangle$ being the variance and T denoting the correlation distance of the refractive index fluctuations) is given by:

$$\frac{d\langle I^{(s)}(\theta)\rangle}{d\Omega} = \frac{Vc}{8\pi^2} \frac{k^4 T^3 \sin^2\Theta \langle n^2 \rangle}{(1 + 4k^2T^2 \sin^2(\theta/2))^2}.$$

Show also that:

a. When the correlation distance l of the refractive index fluctuations is much smaller than the wavelength, one has:

$$\frac{d\langle I^{(s)}(\theta)\rangle}{d\Omega} = \frac{Vc}{8\pi^2} k^4 T^3 \sin^2\Theta\langle n^2\rangle,$$

so that the scattering is isotropic and proportional to λ^{-4}. This is known as *Rayleigh scattering* and describes the blue color of the sky.

b. When T is much larger than the wavelength the scattering is concentrated about the direction of incidence, and one has:

$$\frac{d\langle I^{(s)}(\theta)\rangle}{d\Omega} = \frac{Vc}{128\pi^2} \frac{\sin^2\Theta\langle n^2\rangle}{T\sin^4(\theta/2)}.$$

3.14. The perturbation of a turbulent medium on a wavefield $U(\mathbf{r}) = \exp[\phi(\mathbf{r})]$ is often characterized in terms of the *structure function* $\mathscr{D}_\phi$ (Tatarskii, [3.19]):

$$\mathscr{D}_\phi(\mathbf{r}, \mathbf{r}') = \langle[\phi(\mathbf{r}) - \phi(\mathbf{r}')]^2\rangle.$$

a. On expressing $\phi(\mathbf{r})$ by Eq. (3.136), prove that:

$$\mathscr{D}_\phi = \mathscr{D}_\chi + \mathscr{D}_S,$$

where $\mathscr{D}_\chi$ and $\mathscr{D}_S$ are the *log* and *phase* structure functions, respectively.

Assuming homogeneous and isotropic Gaussian statistics for ϕ, find the relationship between $\mathscr{D}_\phi$, σ_ϕ, and $B_\phi(\rho)$.

b. In a *Kolmogorov model* ([3.19],-[3.21]) the structure function is: $\mathscr{D}_\phi(\rho) = Ar^{5/3}$. Find the mean intensity in the far zone for a field that, emerging from a turbulent layer, and expressed in the form of Eq. (3.121), is characterized by such a structure function.

REFERENCES

3.1. R. Asby and E. Wolf, *J. Opt. Soc. Am.* **61**, 52 (1971).

3.2. E. Lalor and E. Wolf, *Phys. Rev. Lett.* **26**, 1274 (1971).

3.3. A. J. Devaney and E. Wolf, *J. Math. Phys.* **15**, 234 (1974).

3.4. A. T. Friberg and E. Wolf, *J. Opt. Soc. Am.* **73**, 26 (1983).

3.5. W. H. Carter and E. Wolf, *Phys. Rev. A* **36**, 1258 (1987).

3.6. G. Toraldo di Francia, *Nuovo Cimento* **16**, 61 (1960).

3.7. L. D. Landau and E. M. Lifshitz, *The Classical Theory of Fields*, Pergamon Press, Oxford, 1962, 2nd edition.

3.8. W. K. H. Panofsky and M. Phillips, *Classical Electricity and Magnetism*, Addison Wesley, Reading, Massachusetts, 1969, 3rd printing.

3.9. J. D. Jackson, *Classical Electrodynamics*, J. Wiley, New York, 1965.

3.10. J. V. Jelley, *Cherenkov Radiation and its Applications*, Pergamon Press, London, 1958.

3.11. J. W. Van Laak, *Am. J. Phys.* **41**, 636 (1973).

3.12. Y. R. Shen, *The Principles of Non Linear Optics*, J. Wiley, New York, 1984.

3.13. B. Carnahan, H. A. Luther and J. D. Wilkes, *Applied Numerical Methods*, J. Wiley, New York, 1969.

3.14. J. R. Morris and Y. R. Shen, *Phys. Rev. A* **15**, 1143 (1977).

3.15. S. C. Sheng and A. E. Siegman, *Phys. Rev. A* **21**, 599 (1980).

3.16. M. Nieto-Vesperinas and G. Lera, *Opt. Comm.* **69**, 329 (1989); *J. Opt. (Paris)* **20**, 169 (1989).

3.17. K. Gottfried, *Quantum Mechanics*, W. A. Benjamin, New York, 1966.

3.18. P. Roman, *Advanced Quantum Theory*, Addison Wesley, Reading, Massachusetts, 1965.

3.19. V. I. Tatarskii, *Wave Propagation in a Turbulent Medium*, Dover, New York, 1967.

3.20. L. A. Chernov, *Wave Propagation in a Random Medium*, Dover, New York, 1967.

3.21. A. Ishimaru, *Wave Propagation and Scattering in Random Media*, Academic Press, New York, 1978.

3.22. J. L. Doob, *Stochastic Processes*, J. Wiley, New York, 1953.

3.23. A. Papoulis, *Probability Random Variables and Stochastic Processes*, McGraw-Hill, New York, 1965.

3.24. G. Ross, *Opt. Acta* **15**, 451 (1968).

3.25. J. W. Strohben, ed., *Laser Beam Propagation in the Atmosphere*, Topics in Applied Physics, Vol. 25, Springer-Verlag, Berlin, 1978.

3.26. E. Wolf, *Opt. Comm.* **1**, 153 (1969).

3.27. F. Roddier, in *Progress in Optics*, Vol. 19, E. Wolf, ed., North-Holland, Amsterdam, 1981.

3.28. J. W. Goodman, *Statistical Optics*, J. Wiley, New York, 1985.

3.29. M. Born and E. Wolf, *Principles of Optics*, Pergamon Press, Oxford, 1982, 6th edition.

3.30. A. C. Kak, *Proc. I.E.E.E.* **67**, 1245 (1979).

3.31. S. F. Clifford, in *Laser Beam Propagation in the Atmosphere*, J. W. Strohben, ed., Topics in Applied Physics, Vol. 25, Springer-Verlag, Berlin, 1978, Chapter 2.

3.32. S. Chandrasekhar, *Radiative Transfer*, Dover, New York, 1960.

3.33. D. M. Menzel, ed., *Selected Papers on the Transfer of Radiation*, Dover, New York, 1966.

3.34. G. I. Ovchinnikov and V. I. Tatarskii, *Radiophys. Quantum Electron.* **15**, 1087 (1972).

3.35. E. Wolf, *Phys. Rev. D* **13**, 869 (1979).

3.36. E. Wolf, *J. Opt. Soc. Am.* **68**, 6 (1978).

3.37. M. Nieto-Vesperinas, *J. Opt. Soc. Am. A* **3**, 1354 (1986).

3.38. R. C. Bourret, *Nuovo Cimento* **26**, 1 (1962); *Can. J. Phys.* **40**, 782 (1962).

3.39. V. Twersky, *Proc. Am. Math. Soc. Symp. Stochast. Proc. Math. Phys. Eng.* **16**, 84 (1964).

3.40. V. I. Tatarskii, *The Effects of the Turbulent Atmosphere on Wave Propagation*, U.S. Department of Commerce TT-68-50464, Springfield, Virginia, 1971.

3.41. F. Dyson, *Phys. Rev.* **75**, 1736 (1949).

3.42. E. E. Salpeter and H. A. Bethe, *Phys. Rev.* **84**, 1232 (1951).

3.43. U. Frisch, *Ann. d'Astrophys.* **29**, 645 (1966); **30**, 565 (1967).

3.44. Yu. N. Barabanenkov, Y. A. Kravtsov, S. M. Rytov and V. I. Tatarskii, *Soviet Phys. Usp.* **13**, 551 (1971).

3.45. S. M. Rytov, Y. A. Kravtsov and V. I. Tatarskii, *Principles of Statistical Radiophysics*, Vol. 4: *Wave Propagation through Random Media*, Springer-Verlag, Berlin, 1989.

3.46. P. Sheng, ed., *Scattering and Localization of Classical Waves in Random Media*, World Scientific, Singapore, 1990.

3.47. M. Nieto-Vesperinas and J. C. Dainty, eds., *Scattering in Volumes and Surfaces*, North-Holland, Amsterdam, 1990.

3.48. C. Müller, *Foundations of the Mathematical Theory of Electromagnetic Waves*, Springer-Verlag, Berlin, 1969.

3.49. J. A. Kong, *Theory of Electromagnetic Waves*, J. Wiley, New York, 1975.

3.50. G. A. Schott, *Phil. Mag. Suppl.* 7 **15**, 752 (1933).

3.51. G. H. Goedecke, *Phys. Rev.* **135**, B281 (1964).

3.52. E. N. Bramley, *Proc. I.E.E. B* **102**, 533 (1955).

3.53. L. I. Goldfischer, *J. Opt. Soc. Am.* **55**, 247 (1965).

3.54. P. S. Idell, J. R. Fienup and R. S. Goodman, *Opt. Lett.* **12**, 858 (1987).

4

Mathematical Properties of Radiated and Scattered Fields

Radiated and scattered fields from media of finite extent have important properties of analyticity, both for the angular spectrum and far field amplitude as for their wavefunction, at any point outside the medium. These characteristics are studied in this chapter. A brief review of the basic mathematical background concerning entire functions is given in Appendices 4.1–4.3.

4.1. INTRODUCTION

Wavefields scattered or emitted from finite sources have remarkable mathematical properties. They are analytic functions, and more specifically, they are usually extended in the complex space associated with their arguments as entire functions of exponential type. This has fundamental consequences concerning the information content of these fields. This, in turn, is relevant for the determination of the characteristics of sources and scattering potentials.

Entire functions are encoded by their zeros in the complex space of their arguments. Also, their real and imaginary parts are related by dispersion relations that are of great importance in cases involving only one variable. In more than one complex variable, however, it is difficult to take advantage of these properties, given the difficulties of characterizing the *zero lines*, and to solve the singular integral equations that arise from the dispersion relations. The analytic properties of these functions constitute, nonetheless, attributes of basic conceptual relevance for the fields.

4.2. THE ANGULAR SPECTRUM OF WAVEFIELDS IN FREE SPACE AS THE BOUNDARY VALUE OF AN ENTIRE FUNCTION

We have seen in previous chapters that the angular spectrum $A(\mathbf{K})$ of wavefields, $\mathbf{K} = k(p, q)$, propagating in free-space, may be given either by the two-dimensional Fourier transform of the boundary value $f(x, y)$ of this field at a plane $z = z_0$ (z_0 can be arbitrarily chosen to be zero) (Eq. (2.16)), or as the three-dimensional Fourier transform of the source distribution. (This distribution can correspond to either a *radiating source*, like an oscillating current of density $\mathbf{j}(\mathbf{r})$; or to a *scatterer*, as the induced polarization $\mathbf{P}(\mathbf{r})$, or the induced magnetization $\mathbf{M}(\mathbf{r})$.)

In the first case, in which $A(\mathbf{K})$ is described from a function $f(x, y)$ at a plane $z = 0$, one has a Fourier transform relationship between two functions that are dependent on two variables. In the second case, corresponding to radiation and scattering, the three-dimensional Fourier transform of either $\mathbf{j}(\mathbf{r})$ or $\mathbf{M}(\mathbf{r})$ is evaluated at wavevectors $\mathbf{k}^{(\pm)} = (\mathbf{K}, \pm k_z)$, $\mathbf{K} = k(p, q)$, $k_z = km$, where p, q, and m are given by Eqs. (2.9), and $\mathbf{k}^{(\pm)}$ satisfies (cf. Section 3.2):

$$|\mathbf{k}^{(\pm)}|^2 = k^2 = \frac{\omega^2}{c^2}, \tag{4.1}$$

and:

$$k_z = \sqrt{k^2 - |\mathbf{K}|^2}, \qquad |\mathbf{K}|^2 \leqslant k^2, \tag{4.2a}$$

$$k_z = i\sqrt{|\mathbf{K}|^2 - k^2}, \qquad |\mathbf{K}|^2 > k^2, \tag{4.2b}$$

Due to the *Plancherel–Polya theorem* (Appendix 4.1), if $f(x, y)$ is zero outside a finite area of the plane $z = 0$, the Fourier transform $F(\mathbf{K})$ of $f(x, y)$ is the boundary value of an *entire function of exponential type* of the two complex variables: $z_1 = K_1 + iK'_1$, $z_2 = K_2 + iK'_2$ for $\mathbf{K} = k(p, q)$. Similarly, in radiation and scattering from a medium of finite extent, the three-dimensional Fourier transforms $\tilde{\mathbf{j}}(\mathbf{z})$, $\tilde{\mathbf{P}}(\mathbf{z})$, and $\tilde{\mathbf{M}}(\mathbf{z})$ are entire functions of exponential type in the complex space of the $\mathbf{z}$-argument: $\mathbf{z} = (z_1, z_2, z_3)$, $z_1 = K_1 + iK'_1$, $z_2 = K_2 + iK'_2$, $z_3 = K_3 + iK'_3$. Thus according to Sections 3.2 and 3.6, the functions $\tilde{\mathbf{j}}(\mathbf{k}^{(\pm)})$, $\tilde{\mathbf{P}}(\mathbf{k}^{(\pm)})$, and $\tilde{\mathbf{M}}(\mathbf{k}^{(\pm)})$ are boundary values of these entire functions on the sphere given by Eq. (4.1). For homogeneous waves, the component k_z is real, and therefore, the functions $\tilde{\mathbf{j}}(\mathbf{k}^{(\pm)})$, $\tilde{\mathbf{P}}(^{k(\pm)})$, and $\tilde{\mathbf{M}}(\mathbf{k}^{(\pm)})$ are evaluated for real values of $\mathbf{k}^{(\pm)}$. However, for evanescent waves, k_z is imaginary, and hence, these functions are expressed in terms of their analytic continuations for complex values of $\mathbf{k}^{(\pm)}$.

4.3. CONSEQUENCES FOR RADIATED AND SCATTERED FIELDS

In this section we address some theorem derived by Wolf and Nieto-Vesperinas [4.1]. For the main purpose of the discussion that follows,

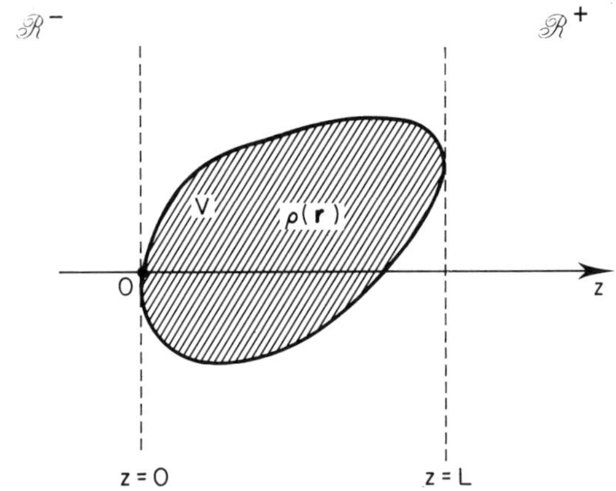

Figure 4.1. The regions $\mathscr{R}^+$ and $\mathscr{R}^-$.

we can disregard the vector nature of the electromagnetic field, and establish the results for each Cartesian component of either the electric or the magnetic vector. Hence, the rest of this analysis will be carried on within the formalism for scalar wavefields.

It will be convenient to work with the angular spectrum in angular variables α and β (cf. Section 2.6). Thus, we shall write:

$$p = \sin\alpha\cos\beta, \tag{4.3a}$$

$$q = \sin\alpha\sin\beta, \tag{4.3b}$$

$$m = \cos\alpha, \tag{4.3c}$$

$$\mathbf{n} = (p, q, m).$$

So that the field at the half spaces $\mathscr{R}^+$ and $\mathscr{R}^-$ (Fig. 4.1), emitted by a source with distribution $\rho(\mathbf{r})$ contained in a volume V, will be given by (cf. [4.2], and Sections 2.6 and 3.4):

$$U(\mathbf{r}) = \frac{ik}{2\pi}\int_{-\pi}^{\pi} d\beta \int_{\mathscr{D}^\pm} d\alpha \sin\alpha A(\alpha, \beta)\exp(ik\mathbf{n}\cdot\mathbf{r}). \tag{4.4}$$

The angular spectrum $A(\alpha, \beta)$ being:

$$A(\alpha, \beta) = \int_V \rho(\mathbf{r}')\exp[ik(x'\sin\alpha\cos\beta + y'\sin\alpha\sin\beta + z'\cos\alpha)]\, d^3r'. \tag{4.5}$$

As before, the integral in Eq. (4.5) is extended to infinity, but $\rho(\mathbf{r}')$ is zero outside V. Also, the integration contours $\mathscr{D}^+$ and $\mathscr{D}^-$ are those of Figs. 2.3(a) and 2.3(b), respectively.

We shall assume that V constitutes a convex domain of the three-dimensional space, so that Eq. (4.4) is valid at any point outside V with the appropriate rotation of the z-axis as remarked at the end of Section 3.2 (see also Fig. 3.2).

The source distribution $\rho(\mathbf{r}')$ can either be radiating or correspond to a scatterer; in the latter case, it is related to the scattering potential $F(\mathbf{r}')$ through Eq. (1.49).

So far the angular spectrum $A(\alpha, \beta)$ has been introduced for values of α on the contours $\mathscr{D}^+$ and $\mathscr{D}^-$ of the complex α-plane, and for real values of β in the range $0 \leqslant \beta \leqslant 2\pi$. We shall now broaden the definition of this function.

Assuming that $\rho(\mathbf{r}')$ is continuous in $\mathbf{r}$ throughout V, the integrand of Eq. (4.5) is also continuous in $\mathbf{r}'$ through V, being also a continuous function of α and β for all values (real or complex) of these variables. In addition, for every value of $\mathbf{r}'$ in V, and for every value of α, this integrand is an entire function of β. Also, it is an entire function of α for every value of $\mathbf{r}'$ in V, and for every value of β.

As a consequence, and from the theory of analytic functions (see, e.g. Ref. 4.20), $A(\alpha, \beta)$ is, for each value of α, an entire function of β, and for each value of β, it is an entire function of α. This implies, according to *Hartogs' theorem* (see Appendix 4.1), that *the angular spectrum $A(\alpha, \beta)$ is an entire function of the two complex variables α, β.*

This result has some interesting consequences. In particular:

The amplitude $A(\theta, \phi)$ of the far field from a source of finite volume V is the boundary value on the unit sphere ($\sin\theta\cos\phi$, $\sin\theta\sin\phi$, $\cos\theta$), $0 \leqslant \theta \leqslant \pi$, $0 \leqslant \phi \leqslant 2\pi$, *of an entire function $A(\alpha, \beta)$ of the two complex variables α and β. (θ and ϕ are real quantities corresponding to the polar and azimuthal angles, respectively, of the direction of observation* $\mathbf{n}$.)

From this second result, and taking into account that, according to the principle of analytic continuation, an analytic function cannot vanish in a region of its domain of analyticity unless it vanishes everywhere ([4.19], [4.20]), we obtain the following theorem:

Theorem 1. *If the amplitude $A(\theta, \phi)$ of the far field, either radiated or scattered from a finite volume with a continuous source distribution, vanishes for all directions (θ, ϕ) within a finite solid angle $d\Omega$, however small, it vanishes for all directions; and more generally, $A(\alpha, \beta)$ is zero for all values α, β real or complex.* ■

The first part of this theorem was established by Müller [4.3]. Its present generalization has been given in Ref. 4.1.

Observe also that the same consequence could have been obtained for the wavefield given in terms of its boundary value $f(x, y)$ at a plane $z = 0$ if $f(x, y)$ is of finite extent.

4.4. CONSEQUENCES FOR HOMOGENEOUS AND EVANESCENT COMPONENTS

The analyticity of the angular spectrum $A(\alpha, \beta)$, or similarly the analyticity of $\tilde{\rho}(\mathbf{z})$, has also some important consequences as regards to the structure of the wavefield. To show this, let us recall that Eq. (4.4) represents the field, radiated or scattered, in the half-spaces $\mathscr{R}^+$ and $\mathscr{R}^-$ (Fig. 4.1), as a superposition of homogeneous and evanescent plane waves. We are going to pose next the following question: is it possible to generate a field $U(\mathbf{r})$, scattered or radiated from a finite volume, that does not contain either homogeneous or evanescent waves?

Suppose first that the field $U(\mathbf{r})$ does not contain homogeneous waves, either in the half-space $\mathscr{R}^+$ or in the half-space $\mathscr{R}^-$; then $A(\alpha, \beta) = 0$ for all α and β such that:

$$0 \leqslant \alpha \leqslant \frac{\pi}{2}, \qquad 0 \leqslant \beta \leqslant 2\pi, \tag{4.6}$$

if there are no homogeneous components in $\mathscr{R}^+$, and $A(\alpha, \beta) = 0$ for all α and β such that:

$$\frac{\pi}{2} \leqslant \alpha \leqslant 2\pi, \qquad 0 \leqslant \beta \leqslant 2\pi, \tag{4.7}$$

if there are no homogeneous waves in $\mathscr{R}^-$. In either case, it follows at once from *Theorem 1* that $A(\alpha, \beta) = 0$ for all values α and β; hence, according to Eq. (4.4), the field $U(\mathbf{r})$ vanishes identically throughout both half-spaces $\mathscr{R}^+$ and $\mathscr{R}^-$.

An additional result follows from this conclusion. To see it, let us first address the following theorem:

Theorem 2. *The solution $U(\mathbf{r})$ of the inhomogeneous Helmholtz equation*

$$\nabla^2 U(\mathbf{r}) + k^2 U(\mathbf{r}) = -4\pi\rho(\mathbf{r}), \tag{4.8}$$

(where $\rho(\mathbf{r})$ is a localized source distribution in a finite volume V; i.e., $\rho(\mathbf{r})$ is zero outside V) is an analytic function in the space of complex variables $x + ix'$,

$y + iy'$, $z + iz'$; where x, y, and z are the Cartesian coordinates of a point field outside V. ■

The proof of this theorem is simple: according to Section 3.4, $U(\mathbf{r})$ can be expressed *outside* V as:

$$U(\mathbf{r}) = \int_V \rho(\mathbf{r}')G(\mathbf{r}, \mathbf{r}')\, d^3r'. \tag{4.9}$$

The Green function $G(\mathbf{r}, \mathbf{r}')$ is a continuous differentiable function for points $\mathbf{r}$ outside V and, therefore, the integrand of Eq. (4.9) is a differentiable function with respect to each Cartesian component of $\mathbf{r}$. As a consequence, it is possible to extend analytically $U(\mathbf{r})$ to each complex variable $x + ix'$, $y + iy'$, $z + iz'$, U being separately analytic with respect to each of these complex arguments. Hence, according to Hartogs' theorem, U is analytic in the complex space of these three complex variables.

It should be remarked that if the volume V is not a convex domain, then one cannot guarantee the analyticity of the exterior field in the concavities of the boundary surface. This point, which has been studied in connection with the validity of the angular spectrum representation in these concavities, which is known as the Rayleigh hypothesis (See Section 7.7), has led to the result that the exterior field is analytic outside the convex domain Ω that contains V, discussed at the end of Section 3.2 (see also Fig. 3.2). In addition, if the boundary surface S of V is analytic, then the exterior field can be analytically continued up to S, i.e., inside the concavities. (This is a necessary but not sufficient condition for the validity of the Rayleigh hypothesis. A necessary and sufficient condition for its validity is that the singularities of the field lie inside a sphere inscribed in V.)

Theorem 2 has other ways of being formulated (see, e.g. Refs. 4.4 and 4.5), and constitutes a particular case of a more general result which is referred to as the *Cauchy–Kowalewski theorem*:

Cauchy–Kowalewski theorem. *There exists one analytic solution of a partial differential equation:*

$$\nabla^2 U(\mathbf{r}) = g\left(\mathbf{r}, U, \frac{\partial U}{\partial x}, \frac{\partial U}{\partial y}, \frac{\partial U}{\partial z}\right), \tag{4.10}$$

that satisfies the Cauchy boundary conditions on a smooth boundary $\mathscr{C}$ of a bounded simply connected domain:

$$\Phi = U(\mathbf{r})|_{\mathscr{C}}$$

$$\Psi = \left.\frac{\partial U}{\partial \mathbf{n}}\right|_{\mathscr{C}}$$

n *being the outward normal to* $\mathscr{C}$, *g being an analytic function of its five arguments, and* $U(\mathbf{r})$ *being twice continuously differentiable.* ■

We shall not give here the proof of this theorem which can be found in the text by Colton and Kress [4.6].

A first corollary of Theorem 2 is that *a function* $U(\mathbf{r})$ *satisfying the homogeneous Helmholtz equation:*

$$\nabla^2 U(\mathbf{r}) + k^2 U(\mathbf{r}) = 0 \tag{4.11}$$

everywhere, is analytic throughout the whole space, and hence, it can be extended analytically to complex values of its arguments for any **r**. ■

We shall discuss this consequence in greater detail in Section 4.5.

A second corollary of Theorem 2 is that *due to the principle of analytic continuation, the field at any finite domain of the space outside the source volume V determines uniquely the field everywhere outside V.*

Another consequence follows immediately from *Theorem 2*:

If $U(\mathbf{r})$ *has no homogeneous components, and therefore, vanishes both in* $\mathscr{R}^+$ *and* $\mathscr{R}^-$, *then* $U(\mathbf{r})$ *vanishes everywhere outside the volume V.* ■

Now we pass on to analyze whether $U(\mathbf{r})$ can have no evanescent waves outside V. Let us assume that the field outside V does not contain evanescent components, either in $\mathscr{R}^+$ or in $\mathscr{R}^-$, then $A(\alpha, \beta) = 0$ for all α and β such that:

$$\alpha = \frac{\pi}{2} - i\delta, \qquad 0 \leqslant \delta < \infty, \qquad 0 \leqslant \beta \leqslant 2\pi, \tag{4.12}$$

if there are no evanescent waves in $\mathscr{R}^+$.

On the other hand, $A(\alpha, \beta) = 0$ for all α and β such that:

$$\alpha = \frac{\pi}{2} + i\delta, \qquad \infty > \delta \geqslant 0, \qquad 0 \leqslant \beta \leqslant 2\pi, \tag{4.13}$$

if there are no evanescent components in $\mathscr{R}^-$. Since, then, $A(\alpha, \beta)$ is an entire function of the complex arguments $\alpha + i\alpha'$ and $\beta + i\beta'$, it follows from the principle of analytic continuation that $A(\alpha, \beta) = 0$ for all α and β either real or complex. Hence, by an argument identical to the one given above in connection with the absence of homogeneous waves, the field $U(\mathbf{r})$ vanishes everywhere outside the volume V.

Both the case of absence of homogeneous waves and the case of absence of evanescent waves can be summarized in the following theorem (Wolf and Nieto-Vesperinas [4.1]):

Theorem 3. *The field $U(\mathbf{r})$ generated by a source or scatterer that occupies a finite volume V, and whose distribution $\rho(\mathbf{r})$ is continuous, contains necessarily both homogeneous and evanescent waves outside V, unless $U(\mathbf{r})$ vanishes everywhere outside V. In addition, $U(\mathbf{r})$ cannot vanish in any finite domain of points outside V unless it vanishes everywhere outside V.* ■

It should also be remarked that some aspects concerning the analyticity of radiation patterns were studied by Rhodes [4.7].

4.5. CONSEQUENCES FOR SOURCE-FREE FIELDS

We have seen that a function $U(\mathbf{r})$ that satisfies the homogeneous Helmholtz equation everywhere can be extended analytically in the complex space $x + ix'$, $y + iy'$, $z + iz'$, as an entire function. Such a function corresponds to a *source-free field* (cf. Section 2.7). For example, the plane wave $\exp[ik(px + qy + mz)]$ has the analytic continuation:

$$\exp[-k(px' + qy' + mz')] \exp[ik(px + qy + mz)].$$

Sherman [4.8], [4.9] defined *source-free fields* as those for which their angular spectrum $a_{sf}(p, q)$ vanishes for $p^2 + q^2 > 1$ (see also Section 2.7). At first glance, such fields seem to contradict *Theorem 3* if the field arises from a continuous finitely extended source distribution, as will always be the case in practice. However, as seen in Section 2.4, at optical frequencies, and at distances not very close to the source, any determination of the field will make the assessment of $A(p, q)$ for $p^2 + q^2 > 1$ impossible, because of the unavoidable presence of noise. Therefore, the field could be approximated by a source-free field. Under the conditions studied in Section 2.13 in connection with the importance of the evanescent wave contribution, this can be a useful approximation to the actual field.

As mentioned before, if $U(\mathbf{r})$ is source-free, then $a_{sf}(p, q) = 0$ for $p^2 + q^2 > 1$. This, according to the Plancherel–Polya theorem (Appendix 4.1), implies that $U(\mathbf{r})$ can be extended analytically as an *entire function of exponential type* to complex values of its three arguments $x + ix'$, $y + iy'$, $z + iz'$. Hence, far enough from V, and under the conditions established in Section 2.13, the emitted field may be approximately considered an entire function of exponential type. This property is more specific than the one established in Theorem 2.

For situations in which the field emitted from V can be represented by a function $U(x)$ of one argument only, the whole theory of entire functions of one complex variable, summarized in Appendices 4.1 and 4.2, is applicable to $U(x)$. This is the case in one-dimensional problems, when there exists circular symmetry, or when the function $U(\mathbf{r})$ is separable.

In one dimension, since the source-free wavefunction $U(x + iy)$ has a density of zeros that, according to Appendix 4.1, is $\mathscr{D} = k/\pi = 2/\lambda$, this function cannot represent the field exterior and close to the source domain. In fact, as a consequence of *Carlson's theorem* (Appendix 4.1), such density cannot be increased if the field is an entire function of exponential type, namely, if it is source-free, unless this field is zero everywhere. This impossibility of recovering higher zero density for the exterior field, once it has propagated a certain distance away from the sources, underlines the irreversible loss of structural information beyond the evanescent wave limit, as the field propagates, and it is analogous to the low pass filtering of the spatial frequencies corresponding to these inhomogeneous components of the angular spectrum, as discussed in Section 2.4. On the other hand, analytic continuation of the angular spectrum beyond the accessible domain of spatial frequencies (corresponding to homogeneous waves) is unstable in practice due to the presence of noise. This will be further addressed in Chapter 9.

4.6. CONCLUSIONS AND EXTENSIONS

We have seen that the finitude of the volume enclosing the sources produces remarkable properties of analyticity in the angular spectrum of the emitted fields (either radiated or scattered). Specifically, these characteristics relate the wavefunction to entire functions of exponential type. Due to the loss of evanescent components in the field determination at a certain distance from the sources, one can, in certain circumstances, approximate this field by a source-free field that can also be analytically continued as an entire function of exponential type. This attribute is even more specific than the mere analyticity of the field, which stems from being a solution to the Helmholtz equation.

The characterization of the wavefield as an entire function permits encoding of the information, conveyed by the field, about the source. This can be done by means of a Hadamard product (Appendix 4.1) in terms of the zeros of its analytic continuation, and constitutes a procedure alternative to the one based on the *Shannon sampling theorem* [4.10], [4.17], [4.18] (see Appendix 4.3), containing certain advantages and some disadvantages over Shannon's method. The conceptual advantages stem from the detailed knowledge that this scheme provides about the analytic properties of the wavefunction, and the consequences established. That is to say, we conclude that it is theoretically possible to characterize the field, everywhere outside the sources, from its values in a finite domain of this region. We also learn that it cannot be zero in any domain of this region unless it is zero everywhere. However, the description in terms of the zeros of its analytic continuation, and the exploitation of the dispersion relations between the real and imaginary parts, or equivalently, between the modulus and the phase, as described in Appendix 4.2, is uniquely possible in geometries dealing with *one* complex variable only.

The interpretation of the field propagation in the free space outside the source volume in terms of *Huygens' principle* is equivalent to a conservation of the structural information of the field (i.e., of the number of its zeros) [4.11]–[4.13], and underlines its analytic properties [4.14].

PROBLEMS

4.1. Prove that the zeros of the far field diffraction pattern from a circular aperture are real. Find dispersion relations for the real and imaginary parts, and for the phase and the modulus.

4.2. Prove Eq. (A4.17).

4.3. Find the dispersion relations for the real and imaginary parts of a function $F(x_1, x_2)$ that is separable, namely, it verifies Eq. (A4.18).

4.4. Find the dispersion relations for the modulus and the phase of the function of Problem 4.3.

4.5. Find the zeros of the function $\cos(2\pi z) + K$ for different values of the constant K.

4.6. Show that the zeros of the analytic continuation of the intensity $I(x)$ of the wavefield $F(x)$ (assuming $F(x)$ to be an entire function of exponential type) are the same as those of $F(z)$ plus their complex conjugate. Find the zeros of the intensity corresponding to the far field arising from a constant source distribution confined to a rectangular aperture.

4.7. What happens when a coherent background is added to the wavefield of Problem 4.6? What are the consequences for the dispersion relations between the real and imaginary parts of this field?

REFERENCES

4.1. E. Wolf and M. Nieto-Vesperinas, *J. Opt. Soc. Am. A* **2**, 886 (1985).

4.2. A. J. Devaney and E. Wolf, *J. Math. Phys.* **15**, 234 (1974).

4.3. C. Müller, *J. Rat. Mech. Anal.* **4**, 235 (1955).

4.4. P. R. Garabedian, *Partial Differential Equations*, J. Wiley, New York, 1964.

4.5. D. L. Colton, *Theory of Partial Differential Equations*, Pitman, London, 1980.

4.6. D. L. Colton and R. Kress, *Integral Equations Methods in Scattering Theory*, J. Wiley, New York, 1983.

4.7. D. R. Rhodes, *Proc. I.E.E.E.* **52**, 1013 (1964).

4.8. G. C. Sherman, *Phys. Rev. Lett.* **21**, 761 (1968).

4.9. G. C. Sherman, *J. Opt. Soc. Am.* **59**, 697 (1969).

4.10. J. W. Goodman, *Introduction to Fourier Optics*, McGraw-Hill, New York, 1968.

4.11. G. Ross, *Phil. Trans. Roy, Soc.* **268**, 177 (1970).

4.12. D. Gabor, in *Optical and Acoustical Holography*, E. Camatini, ed., Proc. NATO Advanced Study Institute, Milan, 1971; Plenum Press, New York, 1972.

4.13. D. Gabor, *J. Phys.* **E8**, 73 (1975).

4.14. M. Nieto-Vesperinas, *J. Math. Phys.* **25**, 2109 (1984).

4.15. M. Plancherel and G. Polya, *Comm. Math. Helv.* **9**, 224 (1937); *Comm. Math. Helv.* **10**, 110 (1938).

4.16. B. A. Fuks, *Introduction to the Theory of Analytic Functions of Several Complex Variables*, Am. Math. Soc. Translations of Mathematical Monographs, Vol. 8, Providence, Rhode Island, 1963.

4.17. A. Papoulis, *The Fourier Integral and its Applications*, McGraw-Hill, New York, 1962.

4.18. R. N. Bracewell, *The Fourier Transform and its Applications*, McGraw-Hill, New York, 1978, 2nd edition.

4.19. E. C. Titchmarsh, *The Theory of Functions*, Oxford University Press, London, 1975, 2nd edition.

4.20. E. T. Copson, *An Introduction to the Theory of Functions of a Complex Variable*, Oxford University Press, London, 1978, 14th printing.

4.21. L. Hormander, *An Introduction to Complex Analysis in Several Variables*. North-Holland, Amsterdam, 1973.

4.22. R. P. Boas, *Entire Functions*, Academic Press, New York, 1954.

4.23. G. Ross, M. A. Fiddy, M. Nieto-Vesperinas and I. Manolitsakis, *Opt. Acta* **26**, 229 (1979).

4.24. G. Ross, M. A. Fiddy and M. Nieto-Vesperinas, in *Inverse Scattering Problems in Optics*, H. P. Baltes, ed., Topics in Current Physics, Vol. 20, Springer-Verlag, Berlin, 1980.

4.25. M. A. Fiddy, G. Ross, M. Nieto-Vesperinas and A. M. J. Huiser, *Opt. Acta* **29**, 23 (1982).

4.26. B. J. Hoenders, *J. Math. Phys.* **16**, 1719 (1975).

4.27. M. A. Fiddy, in *Image Recovery: Theory and Application*, H. Stark, ed., Academic Press, New York, 1987.

4.28. W. F. Osgood, *Topics in the Theory of Several Complex Variables*, Dover, New York, 1966.

4.29. W. F. Osgood, *Lehrbuch der Funktionentheorie*, Vol. 2, Teubner, Leipzig, 1929.

4.30. M. H. Hayes and J. H. McClellan, *Proc. I.E.E.E.* **70**, 197 (1982).

4.31. M. H. Hayes, *I.E.E.E. Trans. Acoust. Speech Sig. Proc.* **ASSP 30**, 140 (1982).

4.32. E. C. Titchmarsh, *Introduction to the Theory of Fourier Integrals*, Oxford University Press, London, 1948.

4.33. H. M. Nussenzveig, *Causality and Dispersion Relations*, Academic Press, New York, 1972.

4.34. P. Roman, *Advanced Quantum Theory*, Addison-Wesley, New York, 1965.

4.35. L. Schiff, *Quantum Mechanics*, McGraw-Hill, New York, 1968.

4.36. E. Wolf, *Proc. Phys. Soc.* **80**, 1269 (1962).

4.37. H. M. Nussenzveig, *J. Math. Phys.* **8**, 561 (1967).

4.38. J. S. Toll, *Phys. Rev.* **104**, 1918 (1956).

4.39. R. H. T. Bates, *Mon. Not. Roy. Astron. Soc.* **142**, 413 (1969).

4.40. R. E. Burge, M. A. Fiddy, A. H. Greenaway and G. Ross, *Proc. Roy. Soc. A* **350**, 191 (1976).

4.41. R. L. Weaver and Y. H. Pao, *J. Math. Phys.* **22**, 1909 (1981).

4.42. M. Nieto-Vesperinas, *Optik* **56**, 377 (1980).

4.43. F. D. Gakhov, *Boundary Value Problems*, Pergamon Press, London, 1966.

4.44. N. I. Muskhelishvili, *Singular Integral Equations*, Noordhoff, Leyden, 1977.

Appendix 4.1

Entire Functions of Exponential Type

In this appendix we outline some basic concepts pertaining to entire functions of one and several complex variables.

Plancherel–Polya Theorem [4.15], [4.16] *Let* $F(\mathbf{x})$ *be* square integrable *(namely, it belongs to the functional space* L^2*),* $\mathbf{x} = (x_1, \ldots, x_N)$*, and be the* N*th dimensional Fourier transform of a function* $f(\mathbf{t})$, $\mathbf{t} = (t_1, \ldots, t_N)$*:*

$$F(\mathbf{x}) = \frac{1}{(2\pi)^N} \int_{-\infty}^{\infty} f(\mathbf{t}) \exp(-i\mathbf{x}\cdot\mathbf{t})\, d\mathbf{t}. \tag{A4.1}$$

In order that $F(\mathbf{x})$ *can be extended to the whole space* $\mathbf{C}^N$ *of the complex variables* $z_j = x_j + iy_j$, $(j = 1, \ldots, N)$ *as an* entire function of exponential type, *it is necessary and sufficient that* $f(\mathbf{t})$ *be of* finite support D*, namely, that* $f(\mathbf{t})$ *be zero outside some finite domain* D *of the* $\mathbf{t}$*-space. If* $f(\mathbf{t})$ *is square integrable, the integral* (A4.1) *should be replaced by:*

$$\operatorname*{l.i.m.}_{T\to\infty} \int_{-T}^{T} (\cdot)\, d\mathbf{t}. \quad \blacksquare$$

A function $F(\mathbf{x})$ with a Fourier transform of finite support is also called a *bandlimited function* [4.17], [4.18]. Thus, bandlimited functions can be extended as entire functions of exponential type to complex values of their arguments.

A function $F(\mathbf{z})$ is named *entire* when it is *analytic* for all $\mathbf{z}$ finite, i.e., it has a derivative $dF/d\mathbf{z}$ for this $\mathbf{z}$. Sometimes these are also called *integral functions* [4.19], [4.20]. As $F(\mathbf{z})$ has derivative for $\mathbf{z}$ finite, it satisfies the Cauchy–Riemann equations, [4.16], for all $\mathbf{z}$ finite:

$$\frac{\partial \operatorname{Re} F}{\partial x_j} = \frac{\partial \operatorname{Im} F}{\partial y_j}, \qquad \frac{\partial \operatorname{Re} F}{\partial y_j} = -\frac{\partial \operatorname{Im} F}{\partial x_j}, \qquad (j = 1, \ldots, N). \tag{A4.2}$$

The concept *exponential type* characterizes the rate of growth of the function. Namely, an entire function of exponential type is bounded by an exponential:

$$|F(\mathbf{z})| \leqslant A \exp[rH(\mathbf{y})], \qquad \mathbf{y} = (y_1, \ldots, y_N). \tag{A4.3}$$

where A is a constant, r is the modulus of $\mathbf{z}$, and $H(\mathbf{y})$, known as the *P-indicator* of $F(\mathbf{z})$ is given by [4.16]:

$$H(\mathbf{y}) = \sup H(\mathbf{x}, \mathbf{y}), \tag{A4.4}$$

sup meaning *least upper bound*, and:

$$H(\mathbf{x}, \mathbf{y}) = \limsup_{r \to \infty} \frac{\ln|F(\mathbf{z})|}{r}. \tag{A4.5}$$

It can be shown from Eq. (A4.3) that the growth of $F(\mathbf{z})$ in the space $\mathbf{C}^N$ is determined by the shape of the support of its inverse Fourier transform $f(\mathbf{t})$.

Hartogs' theorem [4.16], [4.21]. *Let $F(\mathbf{z})$ be a function that is separately analytic with respect to each complex variable $\mathbf{z}_i$, $i = 1, \ldots, N$; namely, $F(\mathbf{z})$ is analytic with respect to each variable $\mathbf{z}_i$ in the corresponding complex plane $\mathbf{C}_i$ of that variable, while the others: $\mathbf{z}_j$, $j \neq i$, are kept constant. $F(\mathbf{z})$, then, is analytic with respect to the N independent variables $\mathbf{z}_1, \ldots, \mathbf{z}_N$ in the complex space $\mathbf{C}^N$.* ■

Observe that a corresponding statement for functions of several *real* variables would be false.

For $N = 1$ the well known *Paley–Wiener theorem* holds:

Paley–Wiener theorem [4.22]. *The function:*

$$F(x) = \frac{1}{2\pi} \int_{-\infty}^{\infty} f(t) \exp(-ixt)\, dt \tag{A4.6}$$

can be extended in the complex z-plane $\mathbf{C}$: $z = x + iy$, as an entire function of exponential type if and only if $f(t)$ is of finite support. ■

Once again, if $f(t)$ is square integrable, the integral in Eq. (A4.6) should be replaced by the limit in the mean.

Let us choose the interval of integration to be $(-T, T)$, the support thus being $2T$. The rate of growth of $F(z)$ in the complex plane is now given by the bound:

$$|F(z)| \leqslant A \exp(Ty), \tag{A4.7}$$

The maximum value of $H(y)$ lies on the y-axis, $z = y$, either with $y > 0$ or with $y < 0$. On the other hand, $H(z = x) = H(z = -x) = 0$. Hence, $F(z)$ has the lower rate of growth on the real axis. This is expected because $F(x)$ has to decrease as $|x| \to \infty$ since $F(x)$ is a square integrable function.

Entire functions are generalizations of polynomials ([4.19]; Chapter 8), and like these, they may be described everywhere by their zeros [4.19], [4.22]. According to theorems by Weierstrass and Hadamard [4.19], [4.22], when $N = 1$, an entire function, like a polynomial, always factorizes. A function given by Eq. (A4.6) can be represented in terms of its zeros z_n in $\mathbf{C}$ by means of the *Hadamard product*:

$$F(z) = \prod_{n=1}^{\infty} \left(1 - \frac{z}{z_n}\right). \tag{A4.8a}$$

Therefore, the zeros of $F(z)$ encode the necessary information concerning this function [4.23]–[4.25]. There exists a numerable infinity of zeros. Asymptotically they are regularly spaced by an interval π/T corresponding to the *Nyquist frequency*, and lie either on the real axis, or on a line parallel to the real axis. Their position thus being given by [4.26]:

$$z_n = \frac{1}{2T}\left[2\pi n - i \log\left(\frac{f(T)}{f(-T)}\right)\right]. \tag{A4.8b}$$

For $N > 1$ a representation of $F(z)$ analogous to (A4.8a) is still possible (see, e.g. Fiddy [4.27]). An extension of Eq. (A4.8a) to $N > 1$ has been given by Osgood [4.28], [4.29], according to whom $F(\mathbf{z})$ can be decomposed as a product of nonfactorizable functions $G_n(\mathbf{z})$:

$$F(\mathbf{z}) = \prod_{n=1}^{L} G_n(\mathbf{z}). \tag{A4.9}$$

The difference between Eqs. (A4.8a) and (A4.9) is that whereas every factor in Eq. (A4.8a) has a well-defined form, in (A4.9) the functions $G_n(\mathbf{z})$ can be quite complicated and there is no generally obvious number L of factors $G_n(\mathbf{z})$ that can be determined from (A4.1). It may happen even that $L = 1$, since, like polynomials in several variables, not every function $F(\mathbf{z})$ is necessarily factorizable (see, e.g. Refs. 4.30 and 4.31). The problem of factorizability into a number of factors with either $L = 1$ or $L > 1$, as stems from the Osgood product, is intimately connected with the *uniqueness* of the *phase retrieval problem* associated with inverse diffraction, source, and scattering problems in optics, and will be discussed in Section 9.10.

Example. Let

$$f(t) = \begin{cases} \pi & \text{if } |t| \leqslant T \\ 0 & \text{if } |t| > T, \end{cases}$$

then, according to (A4.6):

$$F(z) = \frac{\sin Tz}{z}, \qquad z = x + iy.$$

this function behaves for $y = 0$ as:

$$F(x) = \frac{\sin Tx}{x},$$

which decreases to zero as $|x| \to \infty$ and is square integrable. Also, as $x \to \infty$, it becomes:

$$F(iy) = \frac{\sinh Ty}{y} = \frac{\exp(Ty) + \exp(-Ty)}{2y},$$

which is a function that grows as $\exp(Ty)$ as $|y| \to \infty$, consistently with Eq. (A4.7).

The zeros of $\sin(Tz/z)$ are located at $n\pi/T$ ($n = \pm 1, \pm 2, \ldots$) and therefore:

$$\frac{\sin Tz}{z} = \prod_{n=1}^{\infty} \left(1 - \frac{Tz}{n\pi}\right).$$

The p-indicator also provides information on the density of zeros in any region around the origin of the complex space. For $N = 1$ the *density of zeros* is defined as:

$$\mathscr{D} = \lim_{|z| \to \infty} \frac{\mathscr{N}(z)}{|z|},$$

where $\mathscr{N}(z)$ is the number of zeros for which $|z_n| \leqslant |z|$. Since asymptotically these zeros are regularly spaced at intervals π/T on a line parallel to the real axis, in this limit the density is proportional to $2T$.

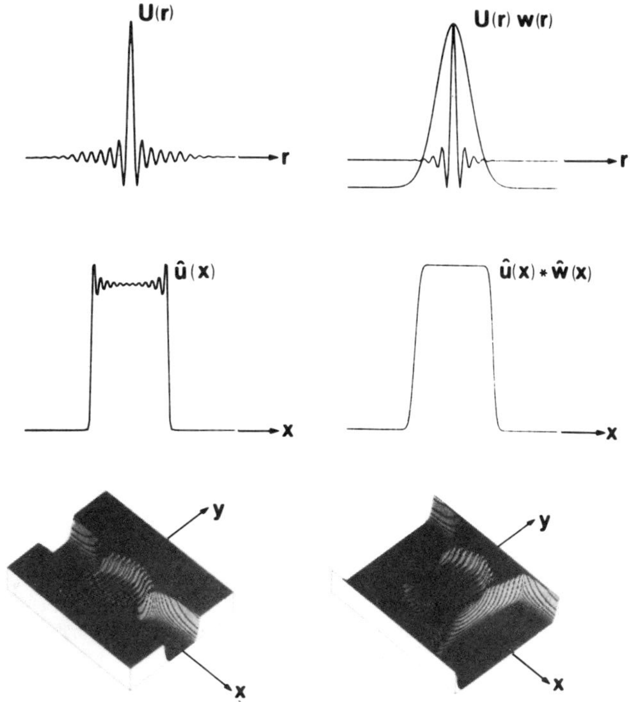

Figure 4.2. The distribution of zeros in the complex plane of the Fourier transform of a truncated Gaussian. The amplitude of the Fourier transform represented has been truncated to facilitate the observation of the zeros. (After G. Ross, M. A. Fiddy, and M. Nieto-Vesperinas, in *Inverse Scattering Problems in Optics*, H. P. Baltes, ed., Topics in Current Physics, Vol. 20, Springer-Verlag, Berlin, 1980.)

The basic unit of information that the position of the zeros can give on the function $f(t)$ is that which corresponds to their regular position at points $z_n = n\pi/T$ ($n = \pm 1, \pm 2, \ldots$). This is the *fundamental lattice* [4.24], [4.25]. It only contains information about the extent $2T$ of the support of $f(t)$. However, the function $f(t)$ whose Fourier transform has such a distribution of zeros is a constant inside this support. Any information about variations of $f(t)$ within the support is encoded by movements of the zeros out of their fundamental lattice positions. Also, their presence is a direct consequence of the finite support of $f(t)$.

The perturbation of each zero from its lattice position encodes a harmonic of $f(t)$, [4.24], [4.25], therefore, the more harmonics $f(t)$ contains (and hence, the greater the information content of $f(t)$ is), the larger is the number of zeros perturbed. This perturbation of the position of the zeros from their fundamental lattice locations is subject to constraints. As seen before, their asymptotic behavior has to be kept such that they lie on a line parallel to the real axis. Movements of zeros too far from the lattice positions will result in a change of the support of $f(t)$. On the other hand, due to *Carlson's theorem* [4.22], no additional zeros can be inserted for $F(z)$: a density larger than the one corresponding to $z_n = n\pi/T$ would make $F(z)$ vanish everywhere. In other words, *no function $F(z)$ satisfying the Paley–Wiener theorem can have a density of zeros larger than the one corresponding to the fundamental lattice, unless it is identically zero everywhere in* **C**.

In this sense, the Hadamard product constitutes a decomposition of the function $F(z)$ into information elements (the binomial $(1 - z/z_n)$), an alternative (although related) to the *Shannon sampling theorem* decomposition for bandlimited functions [4.10], [4.17], [4.18] (Appendix 4.3).

Therefore, all information about $f(t)$ is in the displacement of a certain number of zeros from their fundamental lattice positions. As an example, Fig. 4.2 shows the location of the zeros of the Fourier transform of a Gaussian: $\exp(-t^2)$, truncated at $|t| = 1$.

Appendix 4.2

Dispersion Relations

In this appendix we briefly discuss the dispersion relations that are a consequence of the analyticity of a function in the complex plane. These are based on *Cauchy's integral formula* [4.16]:

$$F(\mathbf{z}) = \frac{1}{(2\pi i)^N} \int_{\mathscr{L}} \frac{F(\mathbf{z}')}{\mathbf{z}' - \mathbf{z}} d\mathbf{z}', \qquad \mathbf{z} = (z_1, \ldots, z_N), \tag{A4.10}$$

relating the values of $F(\mathbf{z})$ at a point of the region of analyticity of $F(\mathbf{z})$ in terms of its values on the boundary $\mathscr{L}$ of that region. The integral Eq. (A4.10) is written in operational form, and hence, it must be understood as:

$$\int_{\mathscr{L}} \frac{(\cdot)}{\mathbf{z} - \mathbf{z}'} d\mathbf{z}' = \int \frac{(\cdot)}{z_1' - z_1} dz_1 \int \frac{(\cdot)}{z_2' - z_2} dz_2' \ldots \int \frac{(\cdot)}{z_N' - z_N} dz_N. \tag{A4.11}$$

For $N = 1$ and by choosing $f(t)$ *causal*, that is to say, $f(t)$ is zero for $t < 0$, as a consequence of Cauchy's integral, the following theorem by Titchmarsh holds [4.32] (we quote it in the form given by Nussenzveig [4.33]):

Titchmarsh Theorem. *If a function $F(z)$, which is square integrable along the real axis, satisfies one of the four conditions cited below, it satisfies all four of them:*

1. *The inverse Fourier transform $f(t)$ of $F(z)$ vanishes for $t < 0$.*
2. *$F(x)$ is for almost all x the limit as $y \to 0$ of an analytic function in the upper half plane $y > 0$, square integrable along any line parallel to the real axis in $y > 0$:*

$$\int_{-\infty}^{\infty} |F(x + iy)|^2 \, dx < C, \qquad y > 0, \tag{A4.12}$$

   C being a positive constant.

3. *The real and imaginary parts of $F(x)$ satisfy:*

$$\text{Re}\, F(x) = \frac{1}{\pi} \mathscr{P} \int_{-\infty}^{\infty} \frac{\text{Im}\, F(x')}{x' - x} dx', \tag{A4.13a}$$

4. *The real and imaginary parts of $F(x)$ satisfy:*

$$\operatorname{Im} F(x) = -\frac{1}{\pi} \mathscr{P} \int_{-\infty}^{\infty} \frac{\operatorname{Re} F(x')}{x' - x}\, dx', \tag{A4.13b}$$

The relations (A4.13a) and (A4.13b) are known as *Hilbert transforms* or *dispersion relations.* $\mathscr{P}$ denotes the *principal value* of the integral. ■

In the time domain, Titchmarsh's theorem is the basis of a number of dispersion relations for several physical quantities. For example, if $f(t)$ represents the dielectric polarizability, the real and imaginary parts of the dielectric permittivity, considered as functions of the temporal frequency ω, are connected by the Kramers–Kronig dispersion relations. In the theory of linear systems [4.17], dispersions relations apply to the transfer function. These relations also hold for the scattering amplitude in quantum scattering theory [4.34], [4.35], and in Fourier spectroscopy for the complex degree of coherence [4.36].

Another consequence of the dispersion relations for $F(x)$ is the possibility of connecting the modulus $|F(x)|$ and the phase $\arg[F(x)]$, [4.36], [4.37], [4.40]:

$$\arg[F(x)] = -\frac{x}{\pi} \mathscr{P} \int_{-\infty}^{\infty} \frac{\log|F(x')|}{x'(x' - x)}\, dx', \tag{A4.14a}$$

$$\log|F(x)| = \frac{x}{\pi} \mathscr{P} \int_{-\infty}^{\infty} \frac{\arg[F(x')]}{x'(x' - x)}\, dx', \tag{A4.14b}$$

Equations (A4.14) are derived from Cauchy's integral for $\log[F(z)]/z$. This function has the same domain of analyticity as $F(z)$, except at those points where $F(z)$ vanishes. If the half plane $y > 0$ is zero free, Eqs. (A4.14) hold. Otherwise, the contribution from these zeros has to be added to the dispersion relations, Eqs. (A4.14) by means of their *Blaschke product* [4.37]–[4.41]. A function $F(x)$ that satisfies Eqs. (A4.14) is known as a *minimum phase function* [4.17].

For $N > 1$ obtaining a zero free half-space is a much more complicated matter, (see, e.g. Ref. 4.27). In addition, the dispersion relations obtainable from Cauchy's integral, applied to a function $F(\mathbf{z})$ of several complex variables, do not yield an explicit expression for either the real or the imaginary part of $F(z)$.

For example, for $N = 2$, Cauchy's integral leads to dispersions relations that, according to Ref. 4.42, are:

$$\operatorname{Re} F(x_1, x_2) = \frac{1}{\pi^2} \mathscr{P} \iint_{-\infty}^{\infty} \frac{\operatorname{Re} F(x_1', x_2')}{(x_1', - x_1)(x_2' - x_2)} \, dx_1' \, dx_2' + \frac{1}{\pi} \mathscr{P} \int_{-\infty}^{\infty} \frac{\operatorname{Im} F(x_1', x_2)}{x_1' - x_1} \, dx_1' + \frac{1}{\pi} \mathscr{P} \int_{-\infty}^{\infty} \frac{\operatorname{Im} F(x_1, x_2')}{x_2' - x_2} \, dx_2' \tag{A4.15a}$$

$$\operatorname{Im} F(x_1, x_2) = \frac{1}{\pi^2} \mathscr{P} \iint_{-\infty}^{\infty} \frac{\operatorname{Im} F(x_1', x_2')}{(x_1' - x_1)(x_2' - x_2)} \, dx_1' \, dx_2' - \frac{1}{\pi} \mathscr{P} \int_{-\infty}^{\infty} \frac{\operatorname{Re} F(x_1', x_2)}{x_1' - x_1} \, dx_1' - \frac{1}{\pi} \mathscr{P} \int_{-\infty}^{\infty} \frac{\operatorname{Re} F(x_1, x_2')}{x_2' - x_2} \, dx_2'. \tag{A4.15b}$$

Equations (A4.15a) and (A4.15b) are the inverse of each other. That is, by Fourier transformation of one, the other is obtained. On the other hand, each of these equations constitute an inhomogeneous singular integral equation [4.43], [4.44] of the form:

$$g(x_1, x_2) = \lambda \mathscr{P} \iint_{-\infty}^{\infty} \frac{g(x_1', x_2')}{(x_1' - x_1)(x_2' - x_2)} \, dx_1' \, dx_2' + h(x_1, x_2), \tag{A4.16}$$

whose solution, easily obtained by means of a Fourier transformation, is [4.42]:

$$g(x_1, x_2) = \frac{1}{1 - \lambda^2 \pi^4} \left[h(x_1, x_2) + \lambda \mathscr{P} \iint_{-\infty}^{\infty} \frac{h(x_1', x_2')}{(x_1' - x_1)(x_2' - x_2)} \, dx_1' \, dx_2' \right]. \tag{A4.17}$$

Equation (A4.17) shows that Eq. (A4.16) has no solution for the eigenvalues $\lambda = \pm 1/\pi^2$. Note, however, that $\lambda = 1/\pi^2$ is the value appearing in the first term of the right-hand side of either Eq. (A4.15a) or (A4.15b). Hence, those two equations cannot be solved for either $\operatorname{Re} F(x_1, x_2)$ or $\operatorname{Im} F(x_1, x_2)$. It similarly occurs for the dispersion relations between $\arg[(F(x_1, x_2)]$ and $\log|F(x_1, x_2)|$. Like $\operatorname{Re} F(x_1, x_2)$ and $\operatorname{Im} F(x_1, x_2)$, those relations constitute a pair of inhomogeneous singular integral equations.

Only in the most trivial case in which $F(x_1, x_2)$ is separable, namely:

$$F(x_1, x_2) = F_1(x_1)F_2(x_2), \tag{A4.18}$$

is it obvious from Eqs. (A4.18) and (A4.15) that, providing $F_1(\mathbf{z}_1)$ and $F_2(\mathbf{z}_2)$ each have a zero-free half plane, the pairs $\arg[F_1(x_1)]$ and $\log|F_2(x_1)|$, and $\arg[F_2(x_2)]$ and $\log|F_2(x_2)|$ satisfy Eqs. (A4.14), then one easily obtains, for instance, for $\arg[F(x_1, x_2)]$:

$$\begin{aligned}\arg[F(x_1, x_2)] = &-\frac{1}{\pi}\mathscr{P}\int_{-\infty}^{\infty}\frac{\log|F_1(x_1')|}{x_1' - x_1}\,dx_1' \\ &-\frac{1}{\pi}\mathscr{P}\int_{-\infty}^{\infty}\frac{\log|F_2(x_2')|}{x_2' - x_2}\,dx_2',\end{aligned} \tag{A4.19}$$

with an analogous expression for $\log|F(x_1, x_2)|$ in terms of $\arg[F(x_1, x_2)]$.

Appendix 4.3

The Whittaker–Shannon Sampling Theorem

Theorem. *Let the Fourier transform $F(x)$ of a function $f(t)$ be zero for $|x| \geqslant x_0$ then $f(t)$ is uniquely determined from its sampling values:*

$$f_m = f\left(m \frac{\pi}{x_0}\right),$$

regularly distributed on the t-axis at intervals at the Nyquist frequency: π/x_0. The representation of $f(t)$ is:

$$f(t) = \sum_{m=-\infty}^{\infty} f_m \frac{\sin(x_0 t - m\pi)}{(x_0 t - m\pi)}. \quad \blacksquare$$

The proof of this theorem is straightforward by repeating periodically $F(x)$ through convolution with the comb function:

$$\tilde{s}(x) = \sum_{m=-\infty}^{\infty} \delta(x - nx_0),$$

whose inverse Fourier transform is the comb of reciprocal sampling:

$$s(t) = \frac{x_0}{2\pi} \sum_{m=-\infty}^{\infty} \delta(t - nT), \qquad T = \frac{2\pi}{x_0},$$

and operating on the Fourier transform of this convolution [4.10], [4.17], [4.18].

5

S-Matrix and Reciprocity

In this chapter some facts concerning the scattering of optical wavefields are presented. These are a consequence of the asymptotic nature of the fields and therefore do not depend on the particular details of the interaction. For this analysis, the *scattering matrix* is introduced. It connects incoming and outgoing fields. One can also define the *transition matrix*, which relates scattered fields with incoming waves. As it will be shown, the *unitarity* of the scattering matrix entails the conservation of energy in the scattering process. On the other hand, certain symmetry relations in this operator underline the *reciprocity* of the scattering process. Therefore, both unitarity and reciprocity are universal properties of the S-matrix, and hence, of the scattering process.

This chapter starts by addressing the S-matrix for electromagnetic fields. A wider discussion is then made concerning scattering situations in which a certain direction is preferential, thus defining two different regions of space, namely, one of backscattering and the other of forward scattering. In this case, the *partitioned* S-*matrix* is introduced in the context of scalar wavefields. Its elements are the *generalized reflection* and *transmission coefficients*. The extension of this theory to electromagnetic fields is then easy. The usefulness of this scheme appears in connection with the reflection and transmission properties of materials, as well as with wavefront restoration based on phase conjugation techniques.

Although the S-matrix has been widely studied in quantum mechanics [5.1], [5.2], it is presented in this chapter, emphasizing its optical applications. The starting point is based on the studies by Saxon and Gerjuoy [5.3], [5.4]; then, from the work by Nieto-Vesperinas and Wolf [5.5], [5.6], the generalized transmission and reflection coefficients of the partitioned S-matrix are addressed.

Some of these concepts have also been introduced in electromagnetic theory [5.7], as well as in waveguide theory [5.8]. All the fields studied in this chapter are assumed to be recorded far from sources, namely, they are approximated by source-free fields. By then taking into account their angular spectrum representation, spectral amplitudes of incoming and outgoing fields are related through the elements of the **S**-matrix. The fact that this relation holds for amplitudes of homogeneous components only is connected with other procedures which introduce the **S**-matrix by means of far field amplitudes. It is left as exercises to prove that, although a scattering matrix could equally well be introduced also for the inhomogeneous component spectral amplitudes, the reciprocity relations are satisfied only for the homogeneous spectral amplitudes. The evanescent spectral amplitudes are irrelevant to conservation of energy, and thus, to the unitarity of the matrix.

5.1. REPRESENTION OF FIELDS OUTSIDE A SCATTERER

Consider a monochromatic field with electric and magnetic vectors $\mathbf{E}^{(i)}(\mathbf{r}, t)$, $\mathbf{H}^{(i)}(\mathbf{r}, t)$, respectively, incident upon a scatterer. We shall assume that these vectors have arbitrary complex amplitudes $\mathbf{E}^{(i)}(\mathbf{r})$, and $\mathbf{H}^{(i)}(\mathbf{r})$, and a time dependence given by the factor $\exp(-i\omega t)$. The scatterer will be supposed to be a linear medium, bounded by a surface S, outside of which there is vacuum (see Fig. 5.1). The spatial parts of the electric and magnetic vectors of the scattered field are: $\mathbf{E}^{(s)}(\mathbf{r}, t)$, and $\mathbf{H}^{(s)}(\mathbf{r}, t)$, respectively. Their time dependence also is: $\exp(-i\omega t)$. The total field is, of course, the sum of the incident and the scattered fields.

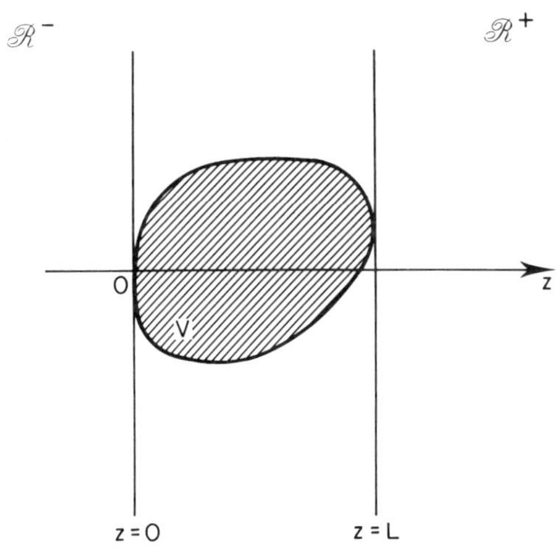

Figure 5.1. Scattering geometry.

Let us choose a Cartesian system of coordinates such that the scatterer is situated within the strip $0 \leqslant z \leqslant L$, and let $\mathscr{R}^+$ and $\mathscr{R}^-$ be the two half-spaces at each side of this strip, respectively (Fig. 5.1).

Since we shall be interested in the fields far away from the scatterer, and under the conditions studied in Chapter 2, we shall omit the contribution of evanescent waves; then the angular spectrum representation of the time independent part of the total field may be written in each half-space (cf. Section 2.2), following Ref. 5.5:

$$\mathbf{E}(\mathbf{r}) = -\frac{ik}{2\pi}\int_{\sigma^{(+)}} \mathbf{C}^{(-)}(\mathbf{n}) \exp(ik\mathbf{n}\cdot\mathbf{r})\, d\Omega + \frac{ik}{2\pi}\int_{\sigma^{(-)}} \mathbf{D}^{(-)}(\mathbf{n}) \exp(ik\mathbf{n}\cdot\mathbf{r})\, d\Omega, \tag{5.1}$$

in $\mathscr{R}^-$.

$$\mathbf{E}(\mathbf{r}) = -\frac{ik}{2\pi}\int_{\sigma^{(-)}} \mathbf{C}^{(+)}(\mathbf{n}) \exp(ik\mathbf{n}\cdot\mathbf{r})\, d\Omega + \frac{ik}{2\pi}\int_{\sigma^{(+)}} \mathbf{D}^{(+)}(\mathbf{n}) \exp(ik\mathbf{n}\cdot\mathbf{r})\, d\Omega, \tag{5.2}$$

in $\mathscr{R}^+$.

In Eqs. (5.1) and (5.2) $\mathbf{n} = (n_x, n_y, n_z)$ is the unit vector of components $n_x = p$, $n_y = q$, and $n_z = m$, where p, q, and m are related by Eqs. (2.9) and are given in terms of the polar and azimuthal angles θ and ϕ by (cf. Eqs. (2.21)):

$$p = \sin\theta\cos\phi, \tag{5.3a}$$

$$q = \sin\theta\sin\phi, \tag{5.3b}$$

$$m = \cos\theta. \tag{5.3c}$$

The symbol $d\Omega$ is the element of solid angle: $d\Omega = d\phi\, d\theta \sin\theta$, and $\sigma^{(+)}$ and $\sigma^{(-)}$ are unit hemispheres of integration, characterized by:

$$\sigma^{(+)}\colon n^2 = 1, \qquad n_z \geqslant 0, \tag{5.4a}$$

$$\sigma^{(-)}\colon n^2 = 1, \qquad n_z < 0. \tag{5.4b}$$

In addition, according to Maxwell's equations, outside the scatterer, and assuming no exterior sources, the time independent parts of the field vectors satisfy:

$$\nabla \times \mathbf{E} = ik\mathbf{H}, \tag{5.5a}$$

$$\nabla \times \mathbf{H} = -ik\mathbf{E}, \tag{5.5b}$$

$$\nabla\cdot\mathbf{E} = 0, \tag{5.5c}$$

$$\nabla\cdot\mathbf{H} = 0. \tag{5.5d}$$

As a consequence of Eq. (5.5c), we have:

$$\mathbf{C}^{(\pm)}(\mathbf{n})\cdot\mathbf{n} = \mathbf{D}^{(\pm)}(\mathbf{n})\cdot\mathbf{n} = 0, \tag{5.6}$$

The factors $\mathbf{C}^{(\pm)}(\mathbf{n})$ and $\mathbf{D}^{(\pm)}(\mathbf{n})$ have the physical significance of complex amplitudes of homogeneous plane waves that propagate in different directions, either towards the scatterer (waves with amplitudes $\mathbf{C}^{(+)}(\mathbf{n})$ and $\mathbf{C}^{(-)}(\mathbf{n})$); or away from it (waves with amplitudes $\mathbf{D}^{(+)}(\mathbf{n})$ and $\mathbf{D}^{(-)}(\mathbf{n})$).

On the other hand, by examining the form of the field far away from the scatterer, one finds, according to the results of Section 2.8, that, as the distance r of the field point from the origin O (Fig. 5.1), increases along any fixed direction specified by the unit vector $\mathbf{u} = (u_x, u_y, u_z)$, one has the following asymptotic form for $\mathbf{E}(\mathbf{r})$ in $\mathscr{R}^+$, given by Eq. (5.1):

$$\mathbf{E}(r\mathbf{u}) \sim \mathbf{C}^{(-)}(-\mathbf{u})\,\frac{\exp(-ikr)}{r} + \mathbf{D}^{(-)}(\mathbf{u})\,\frac{\exp(ikr)}{r}, \tag{5.7}$$

as $kr \to \infty$.

Analogously, for $\mathbf{E}(\mathbf{r})$ in $\mathscr{R}^+$, given by Eq. (5.2), one has:

$$\mathbf{E}(r\mathbf{u}) \sim \mathbf{C}^{(+)}(-\mathbf{u})\,\frac{\exp(-ikr)}{r} + \mathbf{D}^{(+)}(\mathbf{u})\,\frac{\exp(ikr)}{r}, \tag{5.8}$$

as $kr \to \infty$.

The plus and minus signs in Eqs. (5.7) and (5.8) are taken according to whether the field point $\mathbf{r} = r\mathbf{u}$ is in $\mathscr{R}^+$, $u_z > 0$, or in $\mathscr{R}^-$, $u_z < 0$, respectively. These equations express the far field in each of the two half-spaces as the sum of a convergent and a divergent spherical wave with complex amplitudes $\mathbf{C}^{(\pm)}(\mathbf{u})$ and $\mathbf{D}^{(\pm)}(\mathbf{u})$ (see Fig. 5.2). This agrees with the representation of a field which is *incoming at infinity* by the integrals of Eqs. (5.1) and (5.2) which contain the spectral amplitudes $\mathbf{C}^{(\pm)}(\mathbf{n})$, and with the representation of a field which is *outgoing at infinity* by the corresponding integrals containing the spectral amplitudes $\mathbf{D}^{(\pm)}(\mathbf{n})$. (In this connection, it is worth recalling that in the theory of quantum-mechanical potential scattering, the scalar amplitudes $C^{(\pm)}(\mathbf{n})$ represent *incoming states*, whereas the amplitudes $D^{(\pm)}(\mathbf{n})$ represent *outgoing states* [5.1].)

It must be remarked that the term *incident* should not be confused with *incoming*. For instance, the plane homogeneous wave:

$$U^{(i)}(\mathbf{r}) = \exp(ik\mathbf{n}_0\cdot\mathbf{r}), \tag{5.9}$$

which is often taken as an incident wavefield, has asymptotic behavior (cf. Problem 2.3):

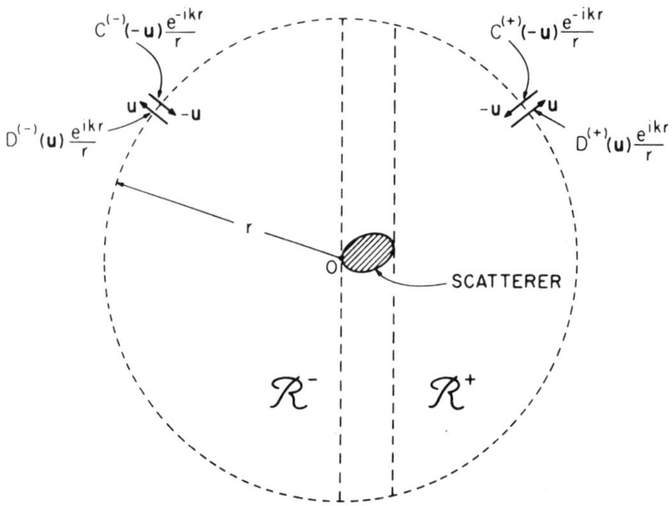

Figure 5.2. The far fields in the half-spaces $\mathcal{R}^+$ and $\mathcal{R}^-$ on either side of the scatterer. (Reproduced with permission from paper by M. Nieto-Vesperinas and E. Wolf, *J. Opt. Soc. Am. A* **3**, 2039–2042 (1986).)

$$\exp(ik\mathbf{n}_0\cdot\mathbf{r}) \sim \frac{2\pi}{ik}\left[\Delta(\mathbf{n}-\mathbf{n}_0)\frac{\exp(ikr)}{r} - \Delta(\mathbf{n}+\mathbf{n}_0)\frac{\exp(-ikr)}{r}\right], \qquad (5.10)$$

as $kr \to \infty$, with the unit vector $\mathbf{n}$ fixed, and Δ being the *spherical* delta function:

$$\Delta(\mathbf{n}-\mathbf{n}_0) = \frac{1}{|\sin\theta|}\delta(\theta-\theta_0)\delta(\phi-\phi_0). \qquad (5.11)$$

Thus, according to Eq. (5.10), a plane wave contains both incoming and outgoing waves, as shown by its behavior at infinity. (Note also that this agrees, for this particular field, with the general result according to which any source-free field is the difference of an outgoing and an incoming field (cf. Problem 2.11).)

5.2. DEFINITION OF THE S-MATRIX

The partition of the space outside the scatterer into two half-spaces $\mathcal{R}^+$ and $\mathcal{R}^-$ is appropriate when the geometry of the problem is such that one has to distinguish between waves that are reflected (backscattered) into $\mathcal{R}^-$, and those waves that are transmitted (forwardly scattered) into $\mathscr{R}^+$. This occurs when there is a preferential direction, such as, for example, phase conjugation, or scattering from rough surfaces or from a plane parallel slab.

In many cases in scattering theory, however [5.3], [5.4], no preferential direction needs to be considered, and hence, as in potential scattering [5.1], [5.2], one deals with an expression for the far field which corresponds to Eqs. (5.7) and (5.8), without distinction between the half-spaces $\mathscr{R}^+$ and $\mathscr{R}^-$. Namely, one then considers the far field as given by:

$$\mathbf{E}(r\mathbf{n}) \sim \mathbf{F}_1(\mathbf{n})\frac{\exp(ikr)}{r} + \mathbf{F}_2(\mathbf{n})\frac{\exp(-ikr)}{r}; \tag{5.12}$$

where, of course, one has that:

$$\mathbf{F}_1(\mathbf{n})\cdot\mathbf{n} = \mathbf{F}_2(\mathbf{n})\cdot\mathbf{n} = 0. \tag{5.13}$$

As a consequence of the linearity of Maxwell's equations (see, e.g. Ref. 5.4), and hence of the wave equation governing the interaction, the outgoing field must be linearly related to the incoming field. Consequently, the amplitudes $\mathbf{F}_1(\mathbf{n})$ and $\mathbf{F}_2(\mathbf{n})$ will be coupled by a relation of the form [5.4]:

$$\mathbf{F}_1(\mathbf{n}) = -\int_\sigma \mathsf{S}(\mathbf{n}, \mathbf{n}')\mathbf{F}_2(-\mathbf{n}')\, d\Omega', \tag{5.14}$$

where the integration is extended over the whole unit sphere σ generated by the unit vector $\mathbf{n}'$.

The operator $\mathsf{S}(\mathbf{n}, \mathbf{n}')$, connecting the outgoing and the incoming parts of $\mathbf{E}(r\mathbf{n})$, is called the *tensor scattering matrix* or the *tensor* S-*matrix*. As will be seen, the choice of signs in Eq. (5.14) is such that the operator $\mathsf{S}(\mathbf{n}, \mathbf{n}')$ reduces to the unit operator when there is no scattering.

Evidently, as a consequence of Eq. (5.13) we have:

$$\mathbf{n}\cdot\mathsf{S}(\mathbf{n}, \mathbf{n}') = \mathsf{S}(\mathbf{n}, \mathbf{n}')\cdot\mathbf{n}. \tag{5.15}$$

5.3. RECIPROCITY AND UNITARITY

Lorentz's reciprocity theorem (Section 1.5) reads:

$$\int_{S_R} \mathbf{n}\cdot[\mathbf{E}'(r\mathbf{n}) \times \mathbf{H}(r\mathbf{n}) - \mathbf{E}(r\mathbf{n}) \times \mathbf{H}'(r\mathbf{n})]\, dS = 0, \tag{5.16}$$

where $dS = r^2\, d\Omega$ and the integration in Eq. (5.16) is extended to a sphere of infinite radius, $\mathbf{n}$ being the local outward normal to this surface.

By using Eq. (5.5a) and the decomposition for both **E** and **E**′ into incoming and outgoing components, one obtains from Eq. (5.16):

$$\int_{\sigma} [\mathbf{F}_1'(\mathbf{n})\cdot\mathbf{F}_2(\mathbf{n}) - \mathbf{F}_1(\mathbf{n})\cdot\mathbf{F}_2'(\mathbf{n})]\, d\Omega = 0. \tag{5.17}$$

Introducing Eq. (5.14) into (5.17), replacing **n** by $-\mathbf{n}'$ in the second term of Eq. (5.17), and transposing the product, one obtains, since $\mathbf{F}_1'(\mathbf{n})$ and $\mathbf{F}_1(\mathbf{n})$ are arbitrary, the following *reciprocity relation* for the **S**-operator [5.4]:

$$\mathsf{S}(\mathbf{n}, \mathbf{n}') = \mathsf{S}^T(-\mathbf{n}', -\mathbf{n}), \tag{5.18}$$

where S^T denotes the transposed tensor, namely: $(\mathsf{S}^T)_{ij} = (\mathsf{S})_{ji}$, $i, j = 1, 2, 3, \ldots$

Since Maxwell's equations are invariant under phase conjugation, one may also write the following expression for the conjugated far field, analogous to Eq. (5.12):

$$\mathbf{E}^*(r\mathbf{n}) \sim \mathbf{F}_2^*(\mathbf{n}) \frac{\exp(ikr)}{r} + \mathbf{F}_1^*(\mathbf{n}) \frac{\exp(-ikr)}{r}. \tag{5.19}$$

Also from the definition of scattering matrix, one has:

$$\mathbf{F}_2^*(\mathbf{n}) = -\int_{\sigma} \mathsf{S}(\mathbf{n}, \mathbf{n}')\mathbf{F}_1^*(-\mathbf{n}')\, d\Omega'. \tag{5.20}$$

The time average flow of energy through a sphere S_R of infinite radius R will be such that, according to Chapter 1:

$$\frac{c}{8\pi} \lim_{R\to\infty} \int_{S_R} (\mathbf{E} \times \mathbf{H}^*)\cdot\mathbf{n}r^2\, d\Omega = 0. \tag{5.21}$$

On using Eq. (5.5a) to write **H** in terms of **E**, and by expressing **E** and **H** according to their far field expressions, Eqs. (5.12) and (5.19), we get from Eq. (5.21):

$$\int_{\sigma} [\mathbf{F}_2^*(\mathbf{n})\cdot\mathbf{F}_2(\mathbf{n}) - \mathbf{F}_1^*(\mathbf{n})\cdot F_1(\mathbf{n})] = 0.$$

This expression, together with Eqs. (5.14) and (5.20), yields, after considering that $\mathbf{F}_1(\mathbf{n})$ and $\mathbf{F}_2(\mathbf{n})$ are arbitrary and transverse, the following two *unitarity relations* for the scattering tensor (Gerjuoy and Saxon [5.3], [5.4]):

$$\int_\sigma \mathbf{S}^{*^T}(\mathbf{n}, \mathbf{n}')\mathbf{S}(\mathbf{n}, \mathbf{n}'')\, d\Omega = \mathbf{I}_T(\mathbf{n}')\Delta(\mathbf{n}' - \mathbf{n}''), \tag{5.22}$$

$$\int_\sigma \mathbf{S}(\mathbf{n}', \mathbf{n})\mathbf{S}^{*^T}(\mathbf{n}'', \mathbf{n})\, d\Omega = \mathbf{I}_T(\mathbf{n}')\Delta(\mathbf{n}' - \mathbf{n}''). \tag{5.23}$$

In Eqs. (5.23) and (5.24), $\Delta(\mathbf{n}' - \mathbf{n}'')$ is the spherical delta function given by Eq. (5.11). $\mathbf{I}_T(\mathbf{n})$ is the transverse part of the unit tensor $\mathbf{I}$. It may be proven (Problem 5.1) that:

$$\mathbf{I}_T(\mathbf{n}) = \mathbf{n} \otimes \mathbf{n} - \mathbf{I}, \tag{5.24}$$

where $\mathbf{n} \otimes \mathbf{n}$ means the *dyadic product* of the unit vector $\mathbf{n}$ by itself; namely, by taking elements:

$$(\mathbf{I}_T(\mathbf{n}))_{ij} = n_i n_j - \delta_{ij}, \qquad (i, j = 1, 2, 3). \tag{5.25}$$

In Ref. 5.4 the reader can find generalizations of Eqs. (5.23) and (5.22) to the case in which there is absorption by the scatterer of the incident electromagnetic energy. The result, which we quote here without proof, is:

$$\int_\sigma \mathbf{S}^{*^T}(\mathbf{n}, \mathbf{n}')\mathbf{S}(\mathbf{n}, \mathbf{n}'')\, d\Omega = \mathbf{I}_T(\mathbf{n}')\Delta(\mathbf{n}' - \mathbf{n}'') - \mathscr{K}(\mathbf{n}', \mathbf{n}''), \tag{5.26}$$

$$\int_\sigma \mathbf{S}(\mathbf{n}', \mathbf{n})\mathbf{S}^{*^T}(\mathbf{n}'', \mathbf{n})\, d\Omega = \mathbf{I}_T(\mathbf{n}')\Delta(\mathbf{n}' - \mathbf{n}'') - \mathscr{K}(\mathbf{n}', \mathbf{n}''). \tag{5.27}$$

$\mathscr{K}(\mathbf{n}', \mathbf{n}'')$ being a Hermitian transverse tensor defined by (cf. Ref. 5.4):

$$\mathscr{P} = \frac{c}{8\pi}\int_\sigma\int_\sigma d\Omega'\, d\Omega''\mathbf{F}_1^*(-\mathbf{n}')\cdot\mathscr{K}(\mathbf{n}', \mathbf{n}'')\cdot\mathbf{F}_1(-\mathbf{n}''), \tag{5.28}$$

and characterizing the energy loss $\mathscr{P}$, due to absorption by the scatterer. $\mathscr{P}$ should be added to the left-hand side of Eq. (5.21) when there is absorption.

5.4. THE S-MATRIX FOR SCALAR WAVEFIELDS

In this section we shall address the important case in which the fields can be described by a scalar function.

The asymptotic expression for the wavefunction $U(\mathbf{r})$ turns out to be given, instead of by Eq. (5.12), by the following expression:

$$U(r\mathbf{n}) \sim F_1(\mathbf{n}) \frac{\exp(ikr)}{r} + F_2(\mathbf{n}) \frac{\exp(-ikr)}{r}. \tag{5.29}$$

The **S**-matrix $\mathscr{S}(\mathbf{n}, \mathbf{n}')$ connects the outgoing amplitudes $F_1(\mathbf{n})$ and the incoming amplitudes $F_2(\mathbf{n})$ as:

$$F_1(\mathbf{n}) = -\int_\sigma \mathscr{S}(\mathbf{n}, \mathbf{n}') F_2(-\mathbf{n}')\, d\Omega', \tag{5.30}$$

where, like in Eq. (5.14), the integration extends over the whole unit sphere σ generated by the unit vector $\mathbf{n}'$. Observe that now the **S**-matrix $\mathscr{S}(\mathbf{n}, \mathbf{n}')$ has only continuous indices, as its components are labeled by the vectors $\mathbf{n}$ and $\mathbf{n}'$.

The *reciprocity* and *unitarity* relations for $\mathscr{S}$ are easily obtained by using the scalar version of Lorentz's reciprocity theorem (see Problem 5.2), and following a procedure analogous to that of Section 5.3. The *reciprocity relation* is:

$$\mathscr{S}(-\mathbf{n}', -\mathbf{n}) = \mathscr{S}(\mathbf{n}, \mathbf{n}'). \tag{5.31}$$

Neglecting absorption losses (if they existed their inclusion would be strictly analogous to Section 5.3), the *unitarity relations* are:

$$\int_\sigma \mathscr{S}^*(\mathbf{n}, \mathbf{n}')\mathscr{S}(\mathbf{n}, \mathbf{n}'')\, d\Omega = \Delta(\mathbf{n}' - \mathbf{n}''), \tag{5.32}$$

$$\int_\sigma \mathscr{S}(\mathbf{n}', \mathbf{n})\mathscr{S}^*(\mathbf{n}'', \mathbf{n})\, d\Omega = \Delta(\mathbf{n}' - \mathbf{n}''). \tag{5.33}$$

5.5. THE PARTITIONED S-MATRIX

Let us consider scalar fields. (The case of vector fields could be accounted for on the basis of the formalism addressed in Section 5.3.) We shall recall expressions for the total field $U(\mathbf{r})$ analogous to Eqs. (5.1), (5.2), (5.7), and (5.8), with reference to Figs. 5.1 and 5.2.

This field is then written as:

$$U(\mathbf{r}) = -\frac{ik}{2\pi}\int_{\sigma^{(+)}} C^{(-)}(\mathbf{n}) \exp(ik\mathbf{n}\cdot\mathbf{r})\, d\Omega + \frac{ik}{2\pi}\int_{\sigma^{(-)}} D^{(-)}(\mathbf{n}) \exp(ik\mathbf{n}\cdot\mathbf{r})\, d\Omega, \tag{5.34}$$

in $\mathscr{R}^-$;

$$U(\mathbf{r}) = -\frac{ik}{2\pi}\int_{\sigma^{(-)}} C^{(+)}(\mathbf{n}) \exp(ik\mathbf{n}\cdot\mathbf{r})\, d\Omega + \frac{ik}{2\pi}\int_{\sigma^{(+)}} D^{(+)}(\mathbf{n}) \exp(ik\mathbf{n}\cdot\mathbf{r})\, d\Omega, \tag{5.35}$$

in $\mathscr{R}^+$.

And asymptotically as $kr \to \infty$ in a fixed direction:

$$U(r\mathbf{u}) \sim C^{(-)}(-\mathbf{u})\frac{\exp(-ikr)}{r} + D^{(-)}(\mathbf{u})\frac{\exp(ikr)}{r}, \tag{5.36}$$

in $\mathscr{R}^-$;

$$U(r\mathbf{u}) \sim C^{(+)}(-\mathbf{u})\frac{\exp(-ikr)}{r} + D^{(+)}(\mathbf{u})\frac{\exp(ikr)}{r}, \tag{5.37}$$

in $\mathscr{R}^+$.

All quantities in Eqs. (5.34)–(5.37) are to be understood as in Section 5.1, except the incoming and outgoing amplitudes $C^{\pm}$ and $D^{\pm}$, which are now scalars. Following Nieto-Vesperinas and Wolf [5.5], the *partitioned* **S**-*matrix* $\mathbf{S}(\mathbf{n}, \mathbf{n}')$ is introduced as:

$$\mathbf{S}(\mathbf{n}, \mathbf{n}') = \begin{pmatrix} S^{(+-)}(\mathbf{n}, \mathbf{n}') & S^{(++)}(\mathbf{n}, \mathbf{n}') \\ S^{(--)}(\mathbf{n}, \mathbf{n}') & S^{(-+)}(\mathbf{n}, \mathbf{n}') \end{pmatrix}, \tag{5.38}$$

and connects the D and C-amplitudes by means of the relation:

$$\mathbf{D}(\mathbf{n}) = -\int \mathbf{S}(\mathbf{n}, \mathbf{n}')\mathbf{C}(\mathbf{n}')\, d\Omega'. \tag{5.39}$$

If there is no scattering, $\mathbf{S}(\mathbf{n}, \mathbf{n}')$ reduces to $\mathbf{I}\Delta(\mathbf{n} - \mathbf{n}')$, where $\mathbf{I}$ is the 2×2 unit matrix.

In Eq. (5.39) **C(n)** and **D(n)** are the column vectors:

$$\mathbf{C}(\mathbf{n}) = \begin{pmatrix} C^{(-)}(\mathbf{n}) \\ C^{(+)}(\mathbf{n}) \end{pmatrix}, \tag{5.40a}$$

$$\mathbf{D}(\mathbf{n}) = \begin{pmatrix} D^{(+)}(\mathbf{n}) \\ D^{(-)}(\mathbf{n}) \end{pmatrix}. \tag{5.40b}$$

Written more explicitly, Eq. (5.39) gives:

$$\begin{aligned} D^{(+)}(\mathbf{n}) = &-\int_{\sigma^{(+)}} S^{(+-)}(\mathbf{n}, \mathbf{n}')C^{(-)}(\mathbf{n}')\, d\Omega' \\ &-\int_{\sigma^{(-)}} S^{(++)}(\mathbf{n}, \mathbf{n}')C^{(+)}(\mathbf{n}')\, d\Omega', \end{aligned} \tag{5.41a}$$

$$\begin{aligned} D^{(-)}(\mathbf{n}) = &-\int_{\sigma^{(+)}} S^{(--)}(\mathbf{n}, \mathbf{n}')C^{(-)}(\mathbf{n}')\, d\Omega' \\ &-\int_{\sigma^{(-)}} S^{(-+)}(\mathbf{n}, \mathbf{n}')C^{(+)}(\mathbf{n}')\, d\Omega'. \end{aligned} \tag{5.41b}$$

From the significance of the functions $C^{(\pm)}$ and $D^{(\pm)}$ as complex amplitudes of waves that propagate either towards the scatterer or away from it, and recalling the definitions, Eqs. (5.4), concerning the domains of integration in Eqs. (5.41), it is clear that each of the four elements of the 2×2 matrix, Eq. (5.38), has the following domain of existence of its arguments:

$$S^{(+-)}(\mathbf{n}, \mathbf{n}'):\ n_z > 0,\ n_z' > 0, \tag{5.42a}$$

$$S^{(++)}(\mathbf{n}, \mathbf{n}'):\ n_z > 0,\ n_z' < 0, \tag{5.42b}$$

$$S^{(--)}(\mathbf{n}, \mathbf{n}'):\ n_z < 0,\ n_z' > 0, \tag{5.42c}$$

$$S^{(-+)}(\mathbf{n}, \mathbf{n}'):\ n_z < 0,\ n_z' < 0. \tag{5.42d}$$

$\mathbf{S}(\mathbf{n}, \mathbf{n}')$ is the partitioned version of the **S**-matrix $\mathscr{S}(\mathbf{n}, \mathbf{n}')$ introduced in Section 5.4, for a scattering geometry in which it is convenient to divide the whole space into the two half-spaces $\mathscr{R}^{(+)}$ and $\mathscr{R}^{(-)}$, as indicated in Figs. 5.1 and 5.2. The elements of both $\mathscr{S}(\mathbf{n}, \mathbf{n}')$ and $\mathbf{S}(\mathbf{n}, \mathbf{n}')$ are labelled by the pairs of unit vectors **n** and **n**′. However, once the vectors **n** and **n**′ are fixed, whereas the elements of $\mathscr{S}(\mathbf{n}, \mathbf{n}')$ are numbers, those elements of $\mathbf{S}(\mathbf{n}, \mathbf{n}')$ are, in turn, 2×2 matrices.

5.6. INCIDENT PLANE WAVE

Let $U^{(i)}(\mathbf{r}) = \exp(ik\mathbf{n}_0 \cdot \mathbf{r})$ be a plane wave of unit amplitude, incident upon the scatterer and propagating along the direction defined by the unit vector $\mathbf{n}_0$.

The scattered field in the far zone is given by:

$$U^{(s)}(r\mathbf{n}) \sim A(\mathbf{n}, \mathbf{n}_0) \frac{\exp(ikr)}{r}, \tag{5.43}$$

as $kr \to \infty$, where $A(\mathbf{n}, \mathbf{n}_0)$ is the scattering amplitude. (It was given under different approaches in Chapter 3.)

The total field, namely, the sum of the incident and the scattered fields, is then:

$$U(r\mathbf{n}) \sim \exp(ik\mathbf{n}_0 \cdot \mathbf{r}) + A(\mathbf{n}, \mathbf{n}_0) \frac{\exp(ikr)}{r}, \tag{5.44}$$

as $kr \to \infty$.

On introducing into Eq. (5.44) the asymptotic form, Eq. (5.10), of the plane wave, Eq. (5.44), reads:

$$U(r\mathbf{n}) \sim F_1(\mathbf{n}) \frac{\exp(ikr)}{r} + F_2(\mathbf{n}) \frac{\exp(-ikr)}{r}, \tag{5.45}$$

with:

$$F_1(\mathbf{n}) = \frac{2\pi}{ik} \Delta(\mathbf{n} - \mathbf{n}_0) + A(\mathbf{n}, \mathbf{n}_0), \tag{5.46a}$$

$$F_2(\mathbf{n}) = -\frac{2\pi}{ik} \Delta(\mathbf{n} + \mathbf{n}_0). \tag{5.46b}$$

Equation (5.46a) may be readily written in terms of the scattering matrix $\mathscr{S}$. Substituting from Eqs. (5.46) into (5.30), we get:

$$A(\mathbf{n}, \mathbf{n}_0) = \frac{2\pi}{ik} [\mathscr{S}(\mathbf{n}, \mathbf{n}_0) - \Delta(\mathbf{n} - \mathbf{n}_0)]. \tag{5.47}$$

Also, on comparing Eqs. (5.47) and (5.46a), we obtain:

$$F_1(\mathbf{n}) = \frac{2\pi}{ik} \mathscr{S}(\mathbf{n}, \mathbf{n}_0). \tag{5.48}$$

Obviously, one sees from Eq. (5.47) that if there is no scattering, i.e., if $A(\mathbf{n}, \mathbf{n}_0)$ is zero, then $\mathscr{S}(\mathbf{n}, \mathbf{n}_0)$ reduces to $\Delta(\mathbf{n} - \mathbf{n}_0)$.

5.7. INCIDENT SOURCE-FREE FIELD

Let the field incident upon the scatterer be source-free with plane wave components propagating in all directions (cf. Section 2.7.1). Its angular spectrum representation is:

$$U^{(i)}(\mathbf{r}) = \int_\sigma a(\mathbf{n}) \exp(ik\mathbf{n}\cdot\mathbf{r})\, d\Omega, \tag{5.49}$$

Expressing the far field as in Eq. (5.43), the total field is asymptotically given by:

$$U(r\mathbf{n}) \sim U^{(i)}(r\mathbf{n}) + A(\mathbf{n}) \frac{\exp(ikr)}{r}, \tag{5.50}$$

as $kr \to \infty$ in a fixed direction $\mathbf{n}$.

On the other hand, from Eq. (5.49), the asymptotic expression for $U^{(i)}(r\mathbf{n})$ is given by (cf. Eq. (2.41)):

$$U^{(i)}(r\mathbf{n}) \sim \frac{2\pi}{ik}\left[a(\mathbf{n}) \frac{\exp(ikr)}{r} - a(-\mathbf{n}) \frac{\exp(-ikr)}{r}\right], \tag{5.51}$$

as $kr \to \infty$.

On substituting Eq. (5.51) into (5.50), $U(r\mathbf{n})$ can be expressed in the form of Eq. (5.45), where now F_1 and F_2 are:

$$F_1(\mathbf{n}) = \frac{2\pi}{ik} a(\mathbf{n}) + A(\mathbf{n}), \tag{5.52a}$$

$$F_2(\mathbf{n}) = -\frac{2\pi}{ik} a(-\mathbf{n}), \tag{5.52b}$$

$A(\mathbf{n})$ being the scattering amplitude.

Equation (5.52a) may be written in terms of the S-matrix $\mathscr{S}$ by introducing Eqs. (5.52) into (5.30). The following relationship between $A(\mathbf{n})$ and $\mathscr{S}$ is then obtained:

$$A(\mathbf{n}) = \frac{2\pi}{ik}\left[\int_\sigma \mathscr{S}(\mathbf{n}, \mathbf{n}')a(\mathbf{n}')\, d\Omega' - a(\mathbf{n})\right]. \tag{5.53}$$

Hence, on comparing Eqs. (5.53) and (5.52a) we obtain:

$$F_1(\mathbf{n}) = \frac{2\pi}{ik} \int_\sigma \mathscr{S}(\mathbf{n}, \mathbf{n}')a(\mathbf{n}')\, d\Omega'. \tag{5.54}$$

Once again, we see from Eq. (5.53) that if there is no scattering, so that $A(\mathbf{n})$ vanishes, the matrix $\mathscr{S}(\mathbf{n}, \mathbf{n}')$ reduces to the delta function $\Delta(\mathbf{n} - \mathbf{n}')$.

5.8. THE GENERALIZED TRANSMISSION AND REFLECTION COEFFICIENTS

In this Section we shall show that the four elements of the partitioned S-matrix, introduced in Eqs. (5.41), have a simple interpretation [5.5].

Suppose that a monochromatic plane wave of unit amplitude and direction of propagation specified by the unit vector $\mathbf{n}_0$ is incident upon the scatterer. We shall write this wave as:

$$U^{(i)}(\mathbf{r}) = \exp(ik\mathbf{n}_0 \cdot \mathbf{r}). \tag{5.55}$$

From Eqs. (5.48), (5.46b), (5.45), (5.37), and (5.36), it follows that:

$$C^{(\pm)}(\mathbf{n}) = -\frac{2\pi}{ik} \Delta(\mathbf{n} - \mathbf{n}_0), \tag{5.56a}$$

$$D^{(\pm)}(\mathbf{n}) = \frac{2\pi}{ik} \mathscr{S}(\mathbf{n}, \mathbf{n}_0). \tag{5.56b}$$

The upper and lower signs of the left-hand side of Eqs. (5.56) are taken according to whether $n_z \geqslant 0$ or $n_z < 0$. On the other hand, the second integral in the right-hand side of Eqs. (5.34) and (5.35) represents the outgoing field after the interaction; this field behaves as a divergent spherical wave at infinity. Hence, on introducing Eq. (5.56b) into Eq. (5.34) and (5.35), one obtains the following expression in terms of the S-matrix for the outgoing field when a plane wave is incident on the scatterer:

$$U^{\text{out}}(\mathbf{r}, \mathbf{n}_0) = \int_{\sigma^{(-)}} \mathscr{S}(\mathbf{n}, \mathbf{n}_0) \exp(ik\mathbf{n} \cdot \mathbf{r})\, d\Omega, \tag{5.57a}$$

when $\mathbf{r}$ belongs to $\mathscr{R}^{(-)}$

$$U^{\text{out}}(\mathbf{r}, \mathbf{n}_0) = \int_{\sigma^{(+)}} \mathscr{S}(\mathbf{n}, \mathbf{n}_0) \exp(ik\mathbf{n} \cdot \mathbf{r})\, d\Omega. \tag{5.57b}$$

when $\mathbf{r}$ belongs to $\mathscr{R}^{(+)}$

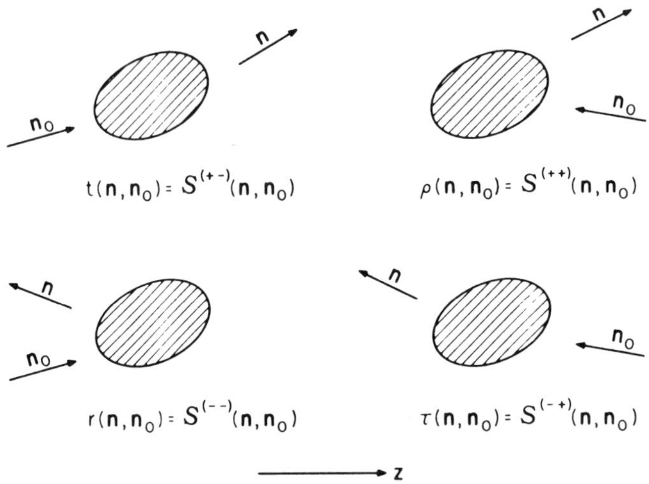

Figure 5.3. Illustration on the significance of the elements of the partitioned S-matrix as generalized transmission and reflection coefficients. (Reproduced with permission from paper by M. Nieto-Vesperinas and E. Wolf, *J. Opt. Soc. Am. A* **3,** 2039–2042 (1986).)

Equations (5.57) represent the outgoing field in each half-space $\mathscr{R}^{(+)}$ and $\mathscr{R}^{(-)}$ in the form of an angular spectrum of plane waves, with complex spectral amplitudes $\mathscr{S}(\mathbf{n}, \mathbf{n}_0)$. When the z-component of $\mathbf{n}$ is positive, $\mathscr{S}(\mathbf{n}, \mathbf{n}_0)$ has the physical meaning of a *generalized transmission coefficient* (which we shall denote by $t(\mathbf{n}, \mathbf{n}_0)$), if $n_z \geqslant 0$, and of a *generalized reflection coefficient*, (denoted by $r(\mathbf{n}, \mathbf{n}_0)$), *if* $n_z \leqslant 0$ (see Fig. 5.3). Recalling Eqs. (5.42a) and (5.42b), we see that these coefficients are two elements of the partitioned **S**-matrix (5.38) (Fig. 5.3), namely:

$$t(\mathbf{n}, \mathbf{n}_0) = S^{(+-)}(\mathbf{n}, \mathbf{n}_0), \qquad n_z > 0, n_{0_z} > 0 \tag{5.58a}$$

$$r(\mathbf{n}, \mathbf{n}_0) = S^{(--)}(\mathbf{n}, \mathbf{n}_0), \qquad n_z < 0, n_{0_z} > 0, \tag{5.58b}$$

Analogously, the quantities:

$$\tau(\mathbf{n}, \mathbf{n}_0) = S^{(-+)}(\mathbf{n}, \mathbf{n}_0), \qquad n_z < 0, n_{0_z} < 0 \tag{5.58c}$$

$$\rho(\mathbf{n}, \mathbf{n}_0) = S^{(++)}(\mathbf{n}, \mathbf{n}_0), \qquad n_z > 0, n_{0_z} < 0, \tag{5.58d}$$

have the physical meaning of a *generalized transmission coefficient* and a *generalized reflection coefficient*, respectively, for incidence from $\mathscr{R}^{(+)}$ (Fig. 5.3).

Therefore, in what follows, the partitioned **S**-matrix will be expressed in terms of these generalized transmission and reflection coefficients as:

$$\mathbf{S}(\mathbf{n}, \mathbf{n}') = \begin{pmatrix} t(\mathbf{n}, \mathbf{n}') & \rho(\mathbf{n}, \mathbf{n}') \\ r(\mathbf{n}, \mathbf{n}') & \tau(\mathbf{n}, \mathbf{n}') \end{pmatrix}. \tag{5.59}$$

In view of the reciprocity (Eq. (5.31)) for $\mathscr{S}$, it is straightforward to derive the following reciprocity relations:

$$t(-\mathbf{n}', -\mathbf{n}) = \tau(\mathbf{n}, \mathbf{n}'), \tag{5.60a}$$

$$\tau(-\mathbf{n}', -\mathbf{n}) = t(\mathbf{n}, \mathbf{n}'), \tag{5.60b}$$

$$\rho(-\mathbf{n}', -\mathbf{n}) = \rho(\mathbf{n}, \mathbf{n}'), \tag{5.60c}$$

$$r(-\mathbf{n}', -\mathbf{n}) = r(\mathbf{n}, \mathbf{n}'), \tag{5.60d}$$

Equations (5.60) could be also derived from the scalar version of Lorentz's reciprocity theorem, applied to the hemispheres $\sigma^{(+)}$ and $\sigma^{(-)}$ (see Problem 5.2).

It is evident that this definition of the generalized transmission and reflection coefficients depends on the choice of the z-axis. Therefore, a transmission coefficient, defined with respect to a particular choice of this axis, could become a reflection coefficient, or vice versa, when the z-direction is chosen differently. However, in many situations of practical interest, a particular direction is preferential. Examples include stratified media, inhomogeneous slabs, rough surfaces and bodies of arbitrary shape in the presence of phase conjugate mirrors.

5.9. THE TRANSITION MATRIX

Also known as the **T**-matrix, the *transition matrix*, $\mathscr{T}$, is defined from the **S**-matrix by subtracting the unit matrix:

$$\mathscr{T}(\mathbf{n}, \mathbf{n}') = \mathscr{S}(\mathbf{n}, \mathbf{n}') - \Delta(\mathbf{n} - \mathbf{n}'). \tag{5.61}$$

For vector fields, the **T**-operator is defined as:

$$\mathsf{T}(\mathbf{n}, \mathbf{n}') = \mathsf{S}(\mathbf{n}, \mathbf{n}') - \mathsf{I}\Delta(\mathbf{n} - \mathbf{n}'). \tag{5.62}$$

The *partitioned* **T**-*matrix* is defined as:

$$\mathbf{T}(\mathbf{n}, \mathbf{n}') = \mathbf{S}(\mathbf{n}, \mathbf{n}') - \mathbf{I}\Delta(\mathbf{n} - \mathbf{n}'), \tag{5.63}$$

I being the 2×2 unit matrix.

While the **S**-matrix transforms *incoming* waves into *outgoing* waves, the **T**-matrix transforms *incoming* waves into *scattered* waves. Namely, the **T**-matrix is zero in the absence of a scatterer. Note that the **T**-matrix also satisfies the *reciprocity relation.*

As an illustration, let us consider the case discussed in Section 5.6, corresponding to an incident plane wave. From Eqs. (5.61) and (5.47), we get:

$$A(\mathbf{n}, \mathbf{n}_0) = \frac{2\pi}{ik} \mathscr{T}(\mathbf{n}, \mathbf{n}_0). \tag{5.64}$$

Obviously, if there is no scattering, $A(\mathbf{n}, \mathbf{n}_0) = 0$, and consequently, $\mathscr{T}(\mathbf{n}, \mathbf{n}_0) = 0$.

In the case of an incident source-free field, one obtains from Eqs. (5.53) and (5.61) the following expression for the scattering amplitude in terms of the **T**-matrix:

$$A(\mathbf{n}) = \frac{2\pi}{ik} \int_{\sigma} \mathscr{T}(\mathbf{n}, \mathbf{n}')a(\mathbf{n}')\, d\Omega'. \tag{5.65}$$

Equation (5.65) illustrates the property of the **T**-matrix mentioned above, namely, it transforms the spectral amplitude, $a(\mathbf{n})$, of the incoming wave into the spectral amplitude $A(\mathbf{n})$ of the scattered wave. When the incident wave is plane, $a(-\mathbf{n})$ becomes $\Delta(\mathbf{n} - \mathbf{n}_0)$ and Eq. (5.65) becomes equal to Eq. (5.64).

5.10. THE GENERALIZED TRANSMISSION AND REFLECTION COEFFICIENTS FOR AN INCIDENT SOURCE-FREE FIELD

Let $U^{(i)}(\mathbf{r})$ be a source-free field, propagating into $z > 0$, incident on the scatterer from $\mathscr{R}^{(-)}$.

Let us refer to Figs. 5.1 and 5.3. The field $U^{(i)}(\mathbf{r})$ may be written both in $\mathscr{R}^{(-)}$ and $\mathscr{R}^{(+)}$ as:

$$U^{(i)}(\mathbf{r}) = \int_{\sigma^{(+)}} a(\mathbf{n}) \exp(ik\mathbf{n}\cdot\mathbf{r})\, d\Omega. \tag{5.66}$$

According to Section 2.8, $U^{(i)}(\mathbf{r})$ behaves as a convergent wave in $\mathscr{R}^{(-)}$, and as a divergent wave in $\mathscr{R}^{(+)}$.

On the other hand, the scattered field may be expressed as:

$$U^{(s)}(\mathbf{r}) = \frac{ik}{2\pi} \int_{\sigma^{(-)}} A^{(-)}(\mathbf{n}) \exp(ik\mathbf{n}\cdot\mathbf{r})\, d\Omega, \tag{5.67}$$

in $\mathscr{R}^{(-)}$.

$$U^{(s)}(\mathbf{r}) = \frac{ik}{2\pi} \int_{\sigma^{(+)}} A^{(+)}(\mathbf{n}) \exp(ik\mathbf{n}\cdot\mathbf{r})\, d\Omega, \tag{5.68}$$

in $\mathscr{R}^{(+)}$.

All integrations are restricted to homogeneous components only.

From Eqs. (5.66)–(5.68), the total field may be written as:

$$U(\mathbf{r}) = \int_{\sigma^{(+)}} a(\mathbf{n}) \exp(ik\mathbf{n}\cdot\mathbf{r})\, d\Omega + \frac{ik}{2\pi} \int_{\sigma^{(-)}} A^{(-)}(\mathbf{n}) \exp(ik\mathbf{n}\cdot\mathbf{r})\, d\Omega, \tag{5.69a}$$

in $\mathscr{R}^{(-)}$.

$$U(\mathbf{r}) = \int_{\sigma^{(+)}} a(\mathbf{n}) \exp(ik\mathbf{n}\cdot\mathbf{r})\, d\Omega + \frac{ik}{2\pi} \int_{\sigma^{(+)}} A^{(+)}(\mathbf{n}) \exp(ik\mathbf{n}\cdot\mathbf{r})\, d\Omega, \tag{5.69b}$$

in $\mathscr{R}^{(+)}$.

On comparing Eqs. (5.69a) and (5.69b) with (5.34) and (5.35), one obtains the following spectral amplitudes:

$$C^{(-)}(\mathbf{n}) = -\frac{2\pi}{ik} a(\mathbf{n}), \tag{5.70a}$$

$$C^{(+)}(\mathbf{n}) = 0, \tag{5.70b}$$

$$D^{(-)}(\mathbf{n}) = A^{(-)}(\mathbf{n}), \tag{5.70c}$$

$$D^{(+)}(\mathbf{n}) = -\frac{2\pi}{ik} a(\mathbf{n}) + A^{(+)}(\mathbf{n}). \tag{5.70d}$$

By using Eqs. (5.39)–(5.41) and (5.59), and substituting $C^{(\pm)}$ and $D^{(\pm)}$ by their corresponding expressions, Eqs. (5.70a)–(5.70d), we obtain:

$$A^{(+)}(\mathbf{n}) = \frac{2\pi}{ik} \left[\int_{\sigma^{(+)}} t(\mathbf{n}, \mathbf{n}') a(\mathbf{n}')\, d\Omega' - a(\mathbf{n}) \right], \tag{5.71}$$

$$\mathrm{A}^{(-)}(\mathbf{n}) = \frac{2\pi}{ik} \int_{\sigma^{(+)}} r(\mathbf{n}, \mathbf{n}') a(\mathbf{n}')\, d\Omega'. \tag{5.72}$$

From Eqs. (5.71) and (5.70d) we get:

$$D^{(+)}(\mathbf{n}) = \frac{2\pi}{ik} \int_{\sigma^{(+)}} t(\mathbf{n}, \mathbf{n}')a(\mathbf{n}')\, d\Omega', \tag{5.73}$$

and from Eqs. (5.72) and (5.70c):

$$D^{(-)}(\mathbf{n}) = \frac{2\pi}{ik} \int_{\sigma^{(+)}} r(\mathbf{n}, \mathbf{n}')a(\mathbf{n}')\, d\Omega'. \tag{5.74}$$

Let us consider now a source-free field $V^{(i)}(\mathbf{r})$, propagating into $z < 0$, incident on the scatterer from $\mathscr{R}^+$. Once again, with reference to Figs. 5.1 and 5.3, the field $V^{(i)}(\mathbf{r})$ may be written in either half-space as:

$$V^{(i)}(\mathbf{r}) = \int_{\sigma^{(-)}} b(\mathbf{n}) \exp(ik\mathbf{n}\cdot\mathbf{r})\, d\Omega. \tag{5.75}$$

Where now, according to Section 3.8, $V^{(i)}(\mathbf{r})$ behaves as a convergent wave in $\mathscr{R}^+$ and as a divergent wave in $\mathscr{R}^-$.

On the other hand, the scattered field may be written as:

$$V^{(s)}(\mathbf{r}) = \frac{ik}{2\pi} \int_{\sigma^{(-)}} B^{(-)}(\mathbf{n}) \exp(ik\mathbf{n}\cdot\mathbf{r})\, d\Omega, \tag{5.76}$$

in $\mathscr{R}^{(-)}$.

$$V^{(s)}(\mathbf{r}) = \frac{ik}{2\pi} \int_{\sigma^{(+)}} B^{(+)}(\mathbf{n}) \exp(ik\mathbf{n}\cdot\mathbf{r})\, d\Omega, \tag{5.77}$$

in $\mathscr{R}^{(+)}$. Where again the integration is restricted to homogeneous components only.

From Eqs. (5.75)–(5.77) the total field may be written as:

$$V(\mathbf{r}) = \int_{\sigma^{(-)}} b(\mathbf{n}) \exp(ik\mathbf{n}\cdot\mathbf{r})\, d\Omega + \frac{ik}{2\pi} \int_{\sigma^{(-)}} B^{(-)}(\mathbf{n}) \exp(ik\mathbf{n}\cdot\mathbf{r})\, d\Omega, \tag{5.78a}$$

in $\mathscr{R}^{(-)}$.

$$V(\mathbf{r}) = \int_{\sigma^{(-)}} b(\mathbf{n}) \exp(ik\mathbf{n}\cdot\mathbf{r})\, d\Omega$$

$$+ \frac{ik}{2\pi} \int_{\sigma^{(+)}} B^{(+)}(\mathbf{n}) \exp(ik\mathbf{n}\cdot\mathbf{r})\, d\Omega, \tag{5.78b}$$

in $\mathscr{R}^{(+)}$.

On comparing Eqs. (5.78a) and (5.78b) with (5.34) and (5.35), one obtains the following expressions for the spectral amplitudes:

$$C^{(-)}(\mathbf{n}) = 0, \tag{5.79a}$$

$$C^{(+)}(\mathbf{n}) = -\frac{2\pi}{ik} b(\mathbf{n}), \tag{5.79b}$$

$$D^{(-)}(\mathbf{n}) = B^{(-)}(\mathbf{n}) + \frac{2\pi}{ik} b(\mathbf{n}), \tag{5.79c}$$

$$D^{(+)}(\mathbf{n}) = B^{(+)}(\mathbf{n}). \tag{5.79d}$$

By means of Eqs. (5.39)–(5.41) and (5.59), and substituting $C^{(\pm)}$ and $D^{(\pm)}$ by their expressions in Eqs. (5.79a) and (5.79b), we obtain:

$$B^{(+)}(\mathbf{n}) = \frac{2\pi}{ik} \int_{\sigma^{(-)}} \rho(\mathbf{n}, \mathbf{n}') b(\mathbf{n}')\, d\Omega', \tag{5.80}$$

$$B^{(-)}(\mathbf{n}) = \frac{2\pi}{ik} \left[\int_{\sigma^{(-)}} \tau(\mathbf{n}, \mathbf{n}') b(\mathbf{n}')\, d\Omega' - b(\mathbf{n}) \right]. \tag{5.81}$$

From Eqs. (5.81) and (5.79c) we get:

$$D^{(-)}(\mathbf{n}) = \frac{2\pi}{ik} \int_{\sigma^{(-)}} \tau(\mathbf{n}, \mathbf{n}') b(\mathbf{n}')\, d\Omega', \tag{5.82}$$

and from Eqs. (5.80) and (5.79d):

$$D^{(+)}(\mathbf{n}) = \frac{2\pi}{ik} \int_{\sigma^{(-)}} \rho(\mathbf{n}, \mathbf{n}') b(\mathbf{n}')\, d\Omega'. \tag{5.83}$$

It should be remarked that, according to Eq. (5.59) and the definition in Eq. (5.61) for the **T**-matrix, the partitioned transition matrix may be expressed as:

$$\mathbf{T}(\mathbf{n}, \mathbf{n}') = \begin{pmatrix} t(\mathbf{n}, \mathbf{n}') - \delta(\mathbf{n} - \mathbf{n}') & \rho(\mathbf{n}, \mathbf{n}') \\ r(\mathbf{n}, \mathbf{n}') & \tau(\mathbf{n}, \mathbf{n}') - \delta(\mathbf{n} - \mathbf{n}') \end{pmatrix}, \tag{5.84}$$

So that, if $\mathbf{T}(\mathbf{n}, \mathbf{n}')$ is expressed as:

$$\mathbf{T}(\mathbf{n}, \mathbf{n}') = \begin{pmatrix} T^{(+-)}(\mathbf{n}, \mathbf{n}') & T^{(++)}(\mathbf{n}, \mathbf{n}') \\ T^{(--)}(\mathbf{n}, \mathbf{n}') & T^{(-+)}(\mathbf{n}, \mathbf{n}') \end{pmatrix}, \tag{5.85}$$

from Eqs. (5.85) and (5.84) the elements of the **T**-matrix are:

$$T^{(+-)}(\mathbf{n}, \mathbf{n}') = t(\mathbf{n}, \mathbf{n}') - \delta(\mathbf{n} - \mathbf{n}'), \tag{5.86a}$$

$$T^{(++)}(\mathbf{n}, \mathbf{n}') = \rho(\mathbf{n}, \mathbf{n}'), \tag{5.86b}$$

$$T^{(--)}(\mathbf{n}, \mathbf{n}') = r(\mathbf{n}, \mathbf{n}'), \tag{5.86c}$$

$$T^{(-+)}(\mathbf{n}, \mathbf{n}') = \tau(\mathbf{n}, \mathbf{n}') - \delta(\mathbf{n} - \mathbf{n}'). \tag{5.86d}$$

It is straightforward to see that these elements satisfy the reciprocity relations analogous to Eqs. (5.60). By means of these definitions, Eqs. (5.71), (5.72), (5.80), and (5.81) may be expressed as:

$$A^{(+)}(\mathbf{n}) = \frac{2\pi}{ik} \int_{\sigma^{(+)}} T^{(+-)}(\mathbf{n}, \mathbf{n}') a(\mathbf{n}') \, d\Omega', \tag{5.87a}$$

$$A^{(-)}(\mathbf{n}) = \frac{2\pi}{ik} \int_{\sigma^{(+)}} T^{(--)}(\mathbf{n}, \mathbf{n}') a(\mathbf{n}') \, d\Omega', \tag{5.87b}$$

$$B^{(+)}(\mathbf{n}) = \frac{2\pi}{ik} \int_{\sigma^{(-)}} T^{(++)}(\mathbf{n}, \mathbf{n}') b(\mathbf{n}') \, d\Omega', \tag{5.87c}$$

$$B^{(-)}(\mathbf{n}) = \frac{2\pi}{ik} \int_{\sigma^{(-)}} T^{(-+)}(\mathbf{n}, \mathbf{n}') b(\mathbf{n}') \, d\Omega'. \tag{5.87d}$$

Equations (5.87) yield directly the scattered amplitudes $A^{(\pm)}(\mathbf{n})$ and $B^{(\pm)}(\mathbf{n})$ in terms of the incoming amplitudes $a(\mathbf{n})$ and $b(\mathbf{n})$ through the transition matrix elements.

5.11. THE GENERALIZED STOKES RELATIONS

Stokes [5.9] considered a plane monochromatic wave, incident on a plane boundary separating two semi-infinite, homogeneous, isotropic dielectric media. Suppose that θ_i and θ_t are the angles of incidence and refraction, respectively, for light propagating from the first medium into the second medium (see Fig. 5.4(a)), the reflection and transmission coefficients being r and t, respectively.

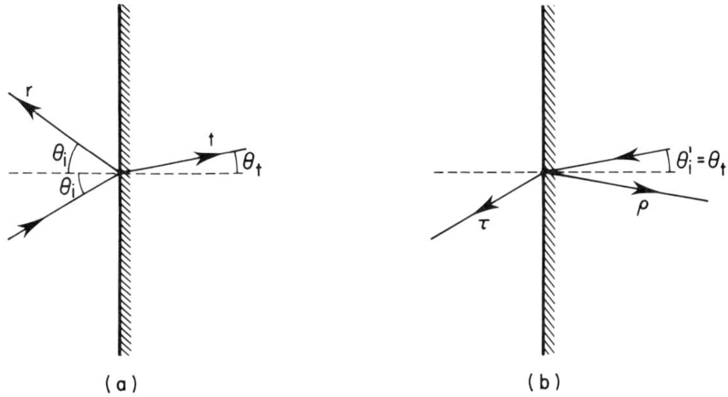

Figure 5.4. Illustration of Stokes' reciprocity relations.

On the other hand, for light incident at an angle $\theta_i' = \theta_t$ from the second medium into the first medium (Fig. 5.4(b)), let ρ and τ be the corresponding reflection and transmission coefficients. The relations derived by Stokes are [5.10], [5.11]:

$$\tau t + r^2 = 1, \tag{5.88a}$$

$$\rho + r = 0. \tag{5.88b}$$

These relations have been later generalized to more general situations including stratified media, and have been used extensively in optics of thin films (see, e.g. Vasiček [5.12]). It will be seen in Chapter 8 that these equations, together with the generalizations presented in this section, are of importance in the study of cancellation of wave distortions by phase conjugation.

In stratified media and thin films, a homogeneous, or a succession of homogeneous dielectrics with mutually parallel planar boundaries is considered. Hence, for an incident plane wave, there is only one plane wave reflected and one plane wave transmitted. In scattering situations where a certain direction (for example, along the z-axis) is preferential, it is useful to broaden the Stokes relations by means of the generalized transmission and reflection coefficients [5.5]. This will be addressed here.

We first separate the contribution from the hemispheres $\sigma^{(-)}$ and $\sigma^{(+)}$ in the unitarity condition (Eq. (5.32)) of $\mathscr{S}(\mathbf{n}, \mathbf{n}')$:

$$\int_{\sigma^{(-)}} \mathscr{S}^*(\mathbf{n}, \mathbf{n}')\mathscr{S}(\mathbf{n}, \mathbf{n}'')\, d\Omega + \int_{\sigma^{(+)}} \mathscr{S}^*(\mathbf{n}, \mathbf{n}')\mathscr{S}(\mathbf{n}, \mathbf{n}'')\, d\Omega = \Delta(\mathbf{n}' - \mathbf{n}''). \tag{5.89}$$

By choosing: $n'_z > 0$ and $n''_z > 0$, and recalling the physical significance of $\mathscr{S}(\mathbf{n}, \mathbf{n}')$ in terms of the generalized reflection and transmission coefficients, as discussed in Section 5.8, we obtain from Eq. (5.89):

$$\int_{\sigma^{(-)}} r^*(\mathbf{n}, \mathbf{n}')r(\mathbf{n}, \mathbf{n}'')\, d\Omega + \int_{\sigma^{(+)}} t^*(\mathbf{n}, \mathbf{n}')t(\mathbf{n}, \mathbf{n}'')\, d\Omega = \Delta(\mathbf{n}' - \mathbf{n}''). \tag{5.90a}$$

Let us next choose the unit vectors $\mathbf{n}'$ and $\mathbf{n}''$ with $n'_z < 0$ and $n''_z > 0$. By using Eqs. (5.58), Eq. (5.89) then gives:

$$\int_{\sigma^{(-)}} \tau^*(\mathbf{n}, \mathbf{n}')r(\mathbf{n}, \mathbf{n}'')\, d\Omega + \int_{\sigma^{(+)}} \rho^*(\mathbf{n}, \mathbf{n}')t(\mathbf{n}, \mathbf{n}'')\, d\Omega = 0. \tag{5.90b}$$

In a similar way, the choice $n'_z > 0$, $n''_z < 0$ gives:

$$\int_{\sigma^{(-)}} r^*(\mathbf{n}, \mathbf{n}')\tau(\mathbf{n}, \mathbf{n}'')\, d\Omega + \int_{\sigma^{(+)}} t^*(\mathbf{n}, \mathbf{n}')\rho(\mathbf{n}, \mathbf{n}'')\, d\Omega = 0, \tag{5.90c}$$

and the choice $n'_z < 0$, $n''_z < 0$ leads to:

$$\int_{\sigma^{(-)}} \tau^*(\mathbf{n}, \mathbf{n}')\tau(\mathbf{n}, \mathbf{n}'')\, d\Omega + \int_{\sigma^{(+)}} \rho^*(\mathbf{n}, \mathbf{n}')\rho(\mathbf{n}, \mathbf{n}'')\, d\Omega = \Delta(\mathbf{n}' - \mathbf{n}''). \tag{5.90d}$$

It is straightforward to verify that the four relations (Eqs. (5.90)) can be synthesized in the following matrix equation which expresses the unitarity of the partitioned S-matrix:

$$\int_{\sigma} \mathbf{S}^{\dagger}(\mathbf{n}, \mathbf{n}')\mathbf{S}(\mathbf{n}, \mathbf{n}'')\, d\Omega = \mathbf{I}\Delta(\mathbf{n}' - \mathbf{n}''). \tag{5.91}$$

$\mathbf{S}^{\dagger}$ being the *adjoint* matrix of $\mathbf{S}$, and $\mathbf{I}$ being the 2×2 unit matrix.

In a similar way, from the second unitarity condition (Eq. (5.33)) of $\mathscr{S}$ one obtains the following four relations:

$$\int_{\sigma^{(-)}} \rho(\mathbf{n}', \mathbf{n})\rho^*(\mathbf{n}'', \mathbf{n})\, d\Omega + \int_{\sigma^{(+)}} t(\mathbf{n}', \mathbf{n})t^*(\mathbf{n}'', \mathbf{n})\, d\Omega = \Delta(\mathbf{n}' - \mathbf{n}''), \tag{5.92a}$$

$$\int_{\sigma^{(-)}} \tau(\mathbf{n}', \mathbf{n})\rho^*(\mathbf{n}'', \mathbf{n})\, d\Omega + \int_{\sigma^{(+)}} r(\mathbf{n}', \mathbf{n})t^*(\mathbf{n}'', \mathbf{n})\, d\Omega = 0, \tag{5.92b}$$

$$\int_{\sigma^{(-)}} \rho(\mathbf{n}', \mathbf{n})\tau^*(\mathbf{n}'', \mathbf{n})\, d\Omega + \int_{\sigma^{(+)}} t(\mathbf{n}', \mathbf{n})r^*(\mathbf{n}'', \mathbf{n})\, d\Omega = 0, \tag{5.92c}$$

$$\int_{\sigma^{(-)}} \tau(\mathbf{n}', \mathbf{n})\tau^*(\mathbf{n}'', \mathbf{n})\, d\Omega + \int_{\sigma^{(+)}} r(\mathbf{n}', \mathbf{n})r^*(\mathbf{n}'', \mathbf{n})\, d\Omega = \Delta(\mathbf{n}' - \mathbf{n}''). \tag{5.92d}$$

As before, it is easy to show that the relations (Eqs. (5.92)) may be condensed in the following expression representing the unitarity condition (Eq. (5.33)) in terms of the partitioned **S**-matrix:

$$\int_{\sigma} \mathbf{S}(\mathbf{n}', \mathbf{n})\mathbf{S}^{\dagger}(\mathbf{n}'', \mathbf{n})\, d\Omega = \mathbf{I}\Delta(\mathbf{n}' - \mathbf{n}''). \tag{5.93}$$

Equations (5.90) and (5.92) can be regarded as generalizations of the Stokes reciprocity relations, Eqs. (5.88). We shall verify in Section 5.12 that they reduce to Eqs. (5.88) in the special case originally considered by Stokes.

Among the eight relations, Eqs. (5.90) and (5.92), only two are actually independent of each other. To see this let us first apply the reciprocity relations Eqs. (5.60a) and (5.60c) to Eq. (5.92a), and take the complex conjugate of the resulting equation. This gives us:

$$\int_{\sigma^{(-)}} \rho^*(-\mathbf{n}, -\mathbf{n}')\rho(-\mathbf{n}, -\mathbf{n}'')\, d\Omega + \int_{\sigma^{(+)}} \tau^*(-\mathbf{n}, -\mathbf{n}')\tau(-\mathbf{n}, -\mathbf{n}'')\, d\Omega = \Delta(\mathbf{n}' - \mathbf{n}''). \tag{5.94}$$

If we now change the variables by letting $\mathbf{n} \to -\mathbf{n}$, $\mathbf{n}' \to -\mathbf{n}'$ and $\mathbf{n}'' \to -\mathbf{n}''$, the relation Eq. (5.94) becomes:

$$\int_{\sigma^{(+)}} \rho^*(\mathbf{n}, \mathbf{n}')\rho(\mathbf{n}, \mathbf{n}'')\, d\Omega + \int_{\sigma^{(-)}} \tau^*(\mathbf{n}, \mathbf{n}')\tau(\mathbf{n}, \mathbf{n}'')\, d\Omega = \Delta(\mathbf{n}' - \mathbf{n}''). \tag{5.95}$$

Which is the relation (5.90d). In a similar fashion, one can show with the help of the reciprocity relations Eqs. (5.60), that Eqs. (5.92b), (5.92c) and (5.92d) are equivalent to (5.90c), (5.90b) and (5.90a), respectively. Hence, if we take the reciprocity relations Eqs. (5.60) into account, the set of Eqs. (5.90) contains the same laws as the set Eq. (5.92). We may therefore consider one of these sets only, e.g. Eqs. (5.90).

Since the two half-spaces $\mathscr{R}^+$ and $\mathscr{R}^-$ play the same role in this theory, it is clear that from any of the generalized Stokes relations derived above one will obtain a valid relation through the transformations:

$$\sigma^{(+)} \leftrightarrow \sigma^{(-)}, \tag{5.96a}$$

$$t \leftrightarrow \tau, \tag{5.96b}$$

$$\rho \leftrightarrow r. \tag{5.96c}$$

Two equations that transform into each other in this way are *dual* of each other [5.5]. Hence, Eqs. (5.90a) and (5.90d) form a dual pair, and also Eqs. (5.90b) and (5.90c). Thus, only Eqs. (5.90a) and (5.90b) constitute independent relations. The others are derived from these by using reciprocity and duality.

5.12. EXAMPLE: STOKES RELATIONS FOR STRATIFIED MEDIA

The general relations, Eqs. (5.90a) and (5.90b), may be illustrated in the case of reflection and transmission of a plane monochromatic electromagnetic wave in a stratified dielectric, then, they reduce to well known relations derived in thin film theory [5.12]. We shall here summarize the results; the reader can find more details in Ref. 5.5.

As illustrated in Fig. 5.5, a stratified dielectric occupies the strip $0 \leqslant z \leqslant L$

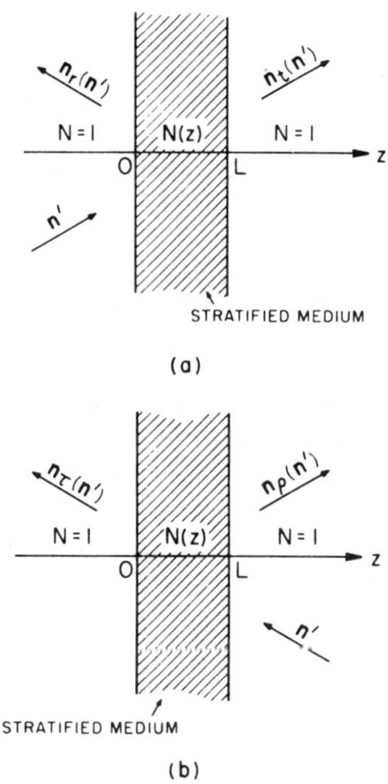

Figure 5.5. Notation for the derivation of the Stokes relations for stratified dielectric media. (Reproduced with permission from paper by M. Nieto-Vesperinas and E. Wolf, *J. Opt. Soc. Am. A* 3, 2039–2042 (1986).)

and has a refractive index: $n = n(z)$. Outside this strip there is free space, $n(z) = 1$. The electromagnetic wave is assumed to be linearly polarized, with the electric vector either perpendicular to the plane of incidence (TE polarization), or in this plane (TM polarization). It is well known [5.13] that the interaction does not change the state of polarization, and therefore the wave can be expressed as a linear combination of those two mutually independent states.

Let the wave be incident from the half-space $z < 0$, with propagation along the unit vector $\mathbf{n}'$ ($n'_z > 0$), and denote by $\mathbf{n}_r(\mathbf{n}')$ and $\mathbf{n}_t(\mathbf{n}')$ the unit vectors specifying the direction of propagation of the reflected and transmitted waves, respectively (Fig. 5.5(a)). Since there is only one reflected and one transmitted plane wave, it is possible to express the generalized reflection and transmission coefficients in the form:

$$r(\mathbf{n}, \mathbf{n}') = \bar{r}(\mathbf{n}')\Delta[\mathbf{n} - \mathbf{n}_r(\mathbf{n}')], \tag{5.97a}$$

$$t(\mathbf{n}, \mathbf{n}') = \bar{t}(\mathbf{n}')\Delta[\mathbf{n} - \mathbf{n}_t(\mathbf{n}')], \tag{5.97b}$$

where $\bar{r}$ and $\bar{t}$ are the usual reflection and transmission coefficients, respectively, for incidence from $z < 0$. The explicit dependence of $\mathbf{n}_r$ and $\mathbf{n}_t$ on $\mathbf{n}'$ is given by the laws of reflection and refraction in stratified media (see, e.g. Sections 1.6.1 and 1.6.3 of Ref. 5.13).

Analogously, we can express:

$$\rho(\mathbf{n}, \mathbf{n}') = \bar{\rho}(\mathbf{n}')\Delta[\mathbf{n} - \mathbf{n}_\rho(\mathbf{n}')], \tag{5.98a}$$

$$\tau(\mathbf{n}, \mathbf{n}') = \bar{\tau}(\mathbf{n}')\Delta[\mathbf{n} - \mathbf{n}_\tau(\mathbf{n}')], \tag{5.98b}$$

where $\bar{\rho}$ and $\bar{\tau}$ are the usual reflection and transmission coefficients, respectively, for incidence from $z > 0$, $\mathbf{n}_\rho$ and $\mathbf{n}_\tau$ being vectors for reflection and transmission given by the laws of reflection and refraction in stratified media.

By using Eqs. (5.97), it is easy to obtain after some algebra the following expression from Eq. (5.90a):

$$\bar{r}^*(\mathbf{n}')\bar{r}(\mathbf{n}') + \bar{t}^*(\mathbf{n}')\bar{t}(\mathbf{n}') = 1. \tag{5.99}$$

Equation (5.99) expresses conservation of energy.

From Eq. (5.60b) one derives the following reciprocity relation:

$$\bar{\tau}(-\mathbf{n}') = \bar{t}(\mathbf{n}'). \tag{5.100}$$

On substituting Eq. (5.100) into (5.99), we obtain:

$$\bar{r}^*(\mathbf{n}')\bar{r}(\mathbf{n}') + \bar{t}^*(\mathbf{n}')\bar{\tau}(-\mathbf{n}') = 1. \tag{5.101}$$

Equation (5.101) is one of the *Stokes relations for stratified dielectric media*, and is well known in thin film theory [5.12].

In a similar way, one can obtain from Eq. (5.90b):

$$\bar{\tau}^*(\mathbf{n}_r'')\bar{r}(\mathbf{n}'') + \bar{\rho}^*(\mathbf{n}_r'')\bar{t}(\mathbf{n}'') = 0. \tag{5.102}$$

If we make use in Eq. (5.102) of the reciprocity relation, Eq. (5.100), and write $\mathbf{n}'$ in place of $\mathbf{n}''$, we get:

$$\bar{t}^*(-\mathbf{n}_r')\bar{r}(\mathbf{n}') + \bar{\rho}^*(\mathbf{n}_r')\bar{t}(\mathbf{n}') = 0. \tag{5.103}$$

Equations (5.102) and (5.103) can be expressed in other forms. For example, since the unit vectors $\mathbf{n}'$ and $-\mathbf{n}_r'$ make the same angle with the z-axis, we have that $\bar{t}(-\mathbf{n}_r') = \bar{t}(\mathbf{n}')$ and $\bar{\rho}(\mathbf{n}_r') = \bar{t}(-\mathbf{n}')$, then, using these relations in Eq. (5.103), we obtain:

$$\bar{t}^*(\mathbf{n}')\bar{r}(\mathbf{n}') + \bar{t}(\mathbf{n}')\bar{\rho}^*(-\mathbf{n}') = 0; \tag{5.104}$$

which is another Stokes relation for stratified media and is well known in the theory of thin films [5.12].

As will be seen in Chapter 8, Eqs. (5.101) and (5.104) are of importance in the theory of distortion correction by phase conjugation.

5.13. RECIPROCITY OF THE IMPULSE RESPONSE FOR SCATTERING FROM INHOMOGENEOUS MEDIA

The reciprocity relations for the generalized reflection and transmission coefficients of the partitioned S-matrix can be employed, as another example of application, in the study on the reciprocity of the impulse response of waves scattered by volumes in which a preferential direction can be introduced, marking two separate regions corresponding to forward and backward scattering, respectively [5.6]; this is for instance the case of an inhomogeneous (i.e., with random refractive index) dielectric slab.

The reciprocity of the impulse response for wave propagation constitutes an important property in optical data processing [5.14], [5.15], on which a number of restoration methods are based. For example, in the presence of layers of turbulent atmosphere [5.16], [5.17], reciprocity is invoked for interchanging the transmitter and the receiver in order to compensate the wavefront distortions involved in atmospheric scattering [5.18], [5.19].

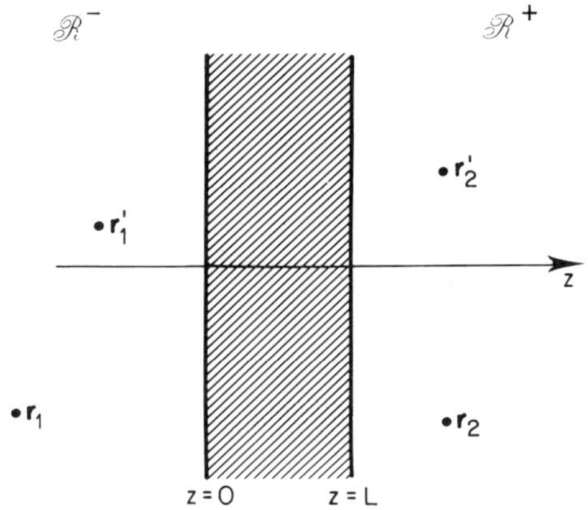

Figure 5.6. Scattering geometry for the impulse response of an inhomogeneous layer.

5.13.1. Definition of the Impulse Response

With reference to Fig. 5.6, let us choose a Cartesian system of coordinates such that the scattering layer is contained in the strip $0 \leqslant z \leqslant L$. If the medium is turbulent, we shall assume that the fluctuation of its refractive index does not change appreciably during the time that the wave spends in traversing this layer.

Let $U(\mathbf{r})$ be the spatial waveform of a monochromatic field in $\mathscr{R}^-$ propagating towards $z > 0$. We shall assume $U(\mathbf{r})$ is considered far enough from the scatterer, so that it does not contain evanescent components. The angular spectrum representation of $U(\mathbf{r})$ at a point $\mathbf{r}_1 = (\mathbf{R}_1, z_1)$, $\mathbf{R}_1 = (x_1, y_1)$ in $\mathscr{R}^-$ is:

$$U(\mathbf{r}_1) = \int_{\sigma^{(+)}} a(\mathbf{n}) \exp(ik\mathbf{n}\cdot\mathbf{r})\, d\Omega. \tag{5.105}$$

Upon interaction of $U(\mathbf{r})$ with the scatterer, the resulting field transmitted at a point $\mathbf{r}_2 = (\mathbf{R}_2, z_2)$, $\mathbf{R}_2 = (x_2, y_2)$ in $\mathscr{R}^+$ will be:

$$U_t(\mathbf{r}_2) = \int_{\sigma^{(+)}} A_t(\mathbf{n}) \exp(ik\mathbf{n}\cdot\mathbf{r})\, d\Omega, \tag{5.106}$$

assuming that it does not contain evanescent components.

Considering the scatterer to be linear, the amplitude $A_t(\mathbf{n})$ of the transmitted field $U_t(\mathbf{r}_2)$ in $\mathscr{R}^+$, will be connected with the amplitude $a(\mathbf{n})$ of the incident field by means of the generalized transmission coefficient (cf. Eq. (5.73)):

$$A_t(\mathbf{n}) = \int_{\sigma^{(+)}} t(\mathbf{n}, \mathbf{n}')a(\mathbf{n}')\, d\Omega'. \tag{5.107}$$

By inverting Eq. (5.105) at a fixed value $z = z_1$, we obtain $a(\mathbf{n})$; then, by introducing this $a(\mathbf{n})$ into Eq. (5.107) one derives an expression for $A_t(\mathbf{n})$, which introduced into Eq. (5.106) yields:

$$U_t(\mathbf{r}_2) = \int h_t(\mathbf{R}_2, z_2; \mathbf{R}_1, z_1)U(\mathbf{r}_1)\, d^2R_1. \tag{5.108}$$

In Eq. (5.108) $h_t(\mathbf{R}_2, z_2; \mathbf{R}_1, z_1)$ is the *point spread function* or *impulse response* [5.6], [5.16], representing the perturbation at the point $\mathbf{r}_2$ of $\mathscr{R}^+$, and originated at a point $\mathbf{r}_1$ of $\mathscr{R}^-$. This function is given by:

$$h_t(\mathbf{R}_2, z_2; \mathbf{R}_1, z_1) = \int_{\sigma^{(+)}} \int_{\sigma^{(+)}} \exp(ik\mathbf{n}\cdot\mathbf{r}_2) \exp(-ik\mathbf{n}'\cdot\mathbf{r}_1)t(\mathbf{n}, \mathbf{n}')\, d\Omega\, d\Omega', \tag{5.109}$$

where $\mathbf{r}_1 = (\mathbf{R}_1, z_1)$, $\mathbf{r}_2 = (\mathbf{R}_2, z_2)$; $z_1 < 0$, $z_2 > 0$.

On the other hand, the field backscattered into $\mathscr{R}^-$ at a point $\mathbf{r}'_1$ will be given by:

$$U_r(\mathbf{r}'_1) = \int_{\sigma^{(-)}} A_r(\mathbf{n}) \exp(ik\mathbf{n}\cdot\mathbf{r})\, d\Omega, \tag{5.110}$$

where the spectral amplitude $A_r(\mathbf{n})$ is connected with $a(\mathbf{n})$ by:

$$A_r(\mathbf{n}) = \int_{\sigma^{(+)}} r(\mathbf{n}, \mathbf{n}')a(\mathbf{n}')\, d\Omega'. \tag{5.111}$$

So that, in a strictly similar manner as before, $U_r(\mathbf{r}'_1)$ is related to $U(\mathbf{r}_1)$ by:

$$U_r(\mathbf{r}'_1) = \int h_r(\mathbf{R}'_1, z'_1; \mathbf{R}_1, z_1)U(\mathbf{r}_1)\, d^2R_1, \tag{5.112}$$

h_r being the impulse response, or perturbation, at a point $\mathbf{r}'_1$ of $\mathscr{R}^-$ originated at a point $\mathbf{r}_1$ of $\mathscr{R}^-$:

$$h_r(\mathbf{R}'_1, z'_1; \mathbf{R}_1, z_1) = \int_{\sigma^{(-)}} \int_{\sigma^{(+)}} \exp(ik\mathbf{n}\cdot\mathbf{r}'_1) \exp(-ik\mathbf{n}'\cdot\mathbf{r}_1)r(\mathbf{n}, \mathbf{n}')\, d\Omega\, d\Omega', \tag{5.113}$$

where $\mathbf{r}'_1 = (\mathbf{R}'_1, z'_1)$, $\mathbf{r}_1 = (\mathbf{R}_1, z_1)$; $z'_1 < 0$, $z_1 < 0$.

Let $V(\mathbf{r})$ be a source-free wavefield in $\mathcal{R}^+$ propagating into $z < 0$. At a point $\mathbf{r}_2 = (\mathbf{R}_2, z_2)$, $\mathbf{R}_2 = (x_2, y_2)$ in $\mathcal{R}^+$ it is written as:

$$V(\mathbf{r}_2) = \int_{\sigma^{(-)}} b(\mathbf{n}) \exp(ik\mathbf{n}\cdot\mathbf{r})\, d\Omega. \tag{5.114}$$

Upon interaction of $V(\mathbf{r})$ with the scatterer, the field transmitted at a point $\mathbf{r}_1$ in $\mathcal{R}^-$ is:

$$V_\tau(\mathbf{r}_1) = \int_{\sigma^{(-)}} B_\tau(\mathbf{n}) \exp(ik\mathbf{n}\cdot\mathbf{r})\, d\Omega, \tag{5.115}$$

where the angular spectrum $B_\tau(\mathbf{n})$ is connected with $b(\mathbf{n})$ by:

$$B_\tau(\mathbf{n}) = \int_{\sigma^{(-)}} \tau(\mathbf{n}, \mathbf{n}')b(\mathbf{n}')\, d\Omega'. \tag{5.116}$$

Also, by a procedure similar to those employed before, one establishes that $V_\tau(\mathbf{r}_1)$ and $V(\mathbf{r}_2)$ are related through:

$$V_\tau(\mathbf{r}_1) = \int h_\tau(\mathbf{R}_1, z_1; \mathbf{R}_2, z_2)V(\mathbf{r}_2)\, d^2R_2, \tag{5.117}$$

h_τ being the impulse response representing the perturbation at the point $\mathbf{r}_1$ of $\mathcal{R}^-$ originating at the point $\mathbf{r}_2$ of $\mathcal{R}^+$. This function is given by:

$$h_\tau(\mathbf{R}_1, z_1; \mathbf{R}_2, z_2) = \int_{\sigma^{(-)}} \int_{\sigma^{(-)}} \exp(ik\mathbf{n}\cdot\mathbf{r}_1) \exp(-ik\mathbf{n}'\cdot\mathbf{r}_2)\tau(\mathbf{n}, \mathbf{n}')\, d\Omega\, d\Omega', \tag{5.118}$$

where $\mathbf{r}_1 = (\mathbf{R}_1, z_1)$, $\mathbf{r}_2 = (\mathbf{R}_2, z_2)$; $z_1 < 0$, $z_2 > 0$.

On the other hand, the field backscattered into $\mathcal{R}^+$ will be given by:

$$V_\rho(\mathbf{r}_2') = \int_{\sigma^{(+)}} B_\rho(\mathbf{n}) \exp(ik\mathbf{n}\cdot\mathbf{r})\, d\Omega, \tag{5.119}$$

where the angular spectrum $B_\rho(\mathbf{n})$ is related to $b(\mathbf{n})$ by:

$$B_\rho(\mathbf{n}) = \int_{\sigma^{(-)}} \rho(\mathbf{n}, \mathbf{n}')b(\mathbf{n}')\, d\Omega'. \tag{5.120}$$

Also, by a procedure identical to the one previously used, it is straightforward to obtain the following relationship between V_ρ and V:

$$V_\rho(\mathbf{r}_2') = \int h_\rho(\mathbf{R}_2', z_2'; \mathbf{R}_2, z_2) V(\mathbf{r}_2)\, d^2R_2, \tag{5.121}$$

h_ρ being the impulse response, or perturbation, at a point $\mathbf{r}_2'$ of $\mathscr{R}^+$ originating at a point $\mathbf{r}_2$ of $\mathscr{R}^+$:

$$h_\rho(\mathbf{R}_2', z_2'; \mathbf{R}_2, z_2) = \int_{\sigma^{(+)}} \int_{\sigma^{(-)}} \exp(ik\mathbf{n}\cdot\mathbf{r}_2') \exp(-ik\mathbf{n}'\cdot\mathbf{r}_2)\rho(\mathbf{n}, \mathbf{n}')\, d\Omega\, d\Omega', \tag{5.122}$$

where $\mathbf{r}_2' = (\mathbf{R}_2, z_2)$, $\mathbf{r}_2 = (\mathbf{R}_2, z_2)$; $z_2 < 0$, $z_2' > 0$.

5.13.2. Reciprocity Relations

From the reciprocity relations for the generalized reflection and transmission coefficients, and by making the changes:

$$\mathbf{n} \rightleftharpoons -\mathbf{n}',$$

in Eqs. (5.109), (5.113), (5.118), and (5.122), one immediately obtains the following reciprocity relations for the impulse responses:

$$h_t(\mathbf{R}_2, z_2; \mathbf{R}_1, z_1) = h_\tau(\mathbf{R}_1, z_1; \mathbf{R}_2, z_2), \tag{5.123a}$$

$$h_r(\mathbf{R}_1', z_1'; \mathbf{R}_1, z_1) = h_r(\mathbf{R}_1, z_1; \mathbf{R}_1', z_1'), \tag{5.123b}$$

$$h_\rho(\mathbf{R}_2', z_2'; \mathbf{R}_2, z_2) = h_\rho(\mathbf{R}_2, z_2; \mathbf{R}_2', z_2'). \tag{5.123c}$$

Equations (5.123) pertain to the interchange between a point of observation and a point source, either in backscattering or forward scattering. It is important to realize that these relations apply only to fields *without evanescent components* (see Problems 5.2 and 5.3). Within the conditions studied in Chapter 2, this will be a good approximation at points many wavelengths away from the boundary planes $z = 0$ and $z = L$.

PROBLEMS

5.1. Prove that the transversal part $\mathbf{I}_T(\mathbf{n})$ of the unit tensor $\mathbf{I}(\mathbf{n})$ (whose elements are $\mathbf{I}_{ij} = \delta_{ij}$, where δ_{ij} is Kronecker's delta) is:

$$\mathbf{I}_T(\mathbf{n}) = \mathbf{n} \otimes \mathbf{n} - \mathbf{I}$$

(where $\otimes$ denotes a dyadic product and **n** is a unit vector).
So that the elements are:

$$[\mathbf{I}_T(\mathbf{n})]_{ij} = \mathbf{n}_i \mathbf{n}_j - \delta_{ij}.$$

Prove also that if **B** is a vector perpendicular to **n**, then:

$$\mathbf{B} \cdot \mathbf{I}_T(\mathbf{n}) = \mathbf{I}_T(\mathbf{n}) \cdot \mathbf{B} = \mathbf{B}.$$

Thus implying that $\mathbf{I}_T(\mathbf{n})$ is a unit tensor for such a vector **B**.

5.2. Consider Lorentz's reciprocity theorem for scalar fields:

$$\mathscr{J} = \int_\sigma \mathbf{J} \cdot \mathbf{n}\, ds = 0, \tag{P.1}$$

where **J** is given by:

$$\mathbf{J} = U_2 \nabla U_1 - U_1 \nabla U_2, \tag{P.2}$$

σ being the unit sphere described by the unit vector **n**. U_1 and U_2 are solutions of the Helmholtz equation:

$$\nabla^2 U + k^2 U = F(\mathbf{r}) U, \tag{P.3}$$

$F(\mathbf{r})$ being a scattering potential.
This theorem holds irrespective of whether $F(\mathbf{r})$ is absorbing or not, namely, of whether $F(\mathbf{r})$ is complex or real.
If the integral $\mathscr{J}$ is decomposed as:

$$\mathscr{J} = \int_{\sigma^{(+)}} \mathbf{J}^{(+)} \cdot \mathbf{n}\, ds + \int_{\sigma^{(-)}} \mathbf{J}^{(-)} \cdot \mathbf{n}\, ds, \tag{P.4}$$

with $\mathbf{J}^{(\pm)}$ defined in $\mathscr{R}^\pm$ as:

$$\mathbf{J}^{(\pm)} = U_2^{(\pm)} \nabla U_1^{(\pm)} - U_1^{(\pm)} \nabla U_2^{(\pm)}. \tag{P.5}$$

By substituting into Eq. (P.5) the angular spectrum representation for

$U_1^{(\pm)}$ and $U_2^{(\pm)}$, given by Eqs. (5.34) and (5.35), and making use of Eq. (P.4), show that Eq. (P.1) leads, after using Eqs. (5.39) and (5.59), to the reciprocity relations for the generalized reflection and transmission coefficients given by Eqs. (5.60).

5.3. Prove that, if the definition in Eq. (5.39) for the **S**-matrix is extended to spectral amplitudes also corresponding to inhomogeneous waves, these do not contribute to the reciprocity relations for the generalized reflection and transmission coefficients [5.6].

5.4. Prove Eqs. (5.123) for the impulse response.

5.5. Consider the case of an incident plane wave $\exp(ik\mathbf{n}_0 \cdot \mathbf{r})$ on a potential. Making use of Eq. (5.47), show that the two unitarity conditions, Eqs. (5.32) and (5.33), lead to the *generalized optical theorems* [5.4]:

$$\frac{1}{2i}[A(\mathbf{n}', \mathbf{n}'') - A^*(\mathbf{n}'', \mathbf{n}')] = \frac{k}{4\pi}\int_\sigma A^*(\mathbf{n}, \mathbf{n}')\mathrm{A}(\mathbf{n}, \mathbf{n}'')\, d\Omega, \qquad \text{(P.6a)}$$

$$\frac{1}{2i}[A(\mathbf{n}', \mathbf{n}'') - A^*(\mathbf{n}'', \mathbf{n}')] = \frac{k}{4\pi}\int_\sigma A^*(\mathbf{n}'', \mathbf{n})A(\mathbf{n}', \mathbf{n})\, d\Omega. \qquad \text{(P.6b)}$$

These expressions reduce to the *optical theorem* [5.20], or *cross section theorem* [5.3], [5.4]:

$$\Sigma \equiv \int_\sigma |A(\mathbf{n}, \mathbf{n}_0)|^2\, d\Omega = \frac{4\pi}{k}\,\mathrm{Im}\, A(\mathbf{n}_0, \mathbf{n}_0), \qquad \text{(P.7)}$$

when $\mathbf{n}' = \mathbf{n}'' = \mathbf{n}_0$.

Equation (P.7) expresses the scattering cross section Σ in terms of the destructive interference between the incident and the scattered field in the forward direction.

5.6. Write the optical theorems for the forward scattered and backscattered amplitudes in the half-spaces $\mathscr{R}^+$ and $\mathscr{R}^-$. Generalize them to the case in which there is an incident arbitrary source-free field.

5.7. Using the expression of $\Delta(\mathbf{n} - \mathbf{n}')$ in spherical coordinates, prove that:

$$\int_\sigma \Delta(\mathbf{n} - \mathbf{n}')\Delta(\mathbf{n} - \mathbf{n}'')\, d\Omega = \Delta(\mathbf{n}' - \mathbf{n}''),$$

where the integration is extended to the whole sphere $\sigma = 4\pi$ described by the unit vector $\mathbf{n}$.

REFERENCES

5.1. P. Roman, *Advanced Quantum Theory*, Addison-Wesley, Reading, Massachusetts, 1965.

5.2. L. I. Schiff, *Quantum Mechanics*, McGraw-Hill, New York, 1968.

5.3. E. Gerjuoy and D. S. Saxon, *Phys. Rev.* **94**, 1445 (1954).

5.4. D. S. Saxon, *Phys. Rev.* **100**, 1771 (1955).

5.5. M. Nieto-Vesperinas and E. Wolf, *J. Opt. Soc. Am. A* **3**, 2038 (1986).

5.6. M. Nieto-Vesperinas, *J. Opt. Soc. Am. A* **5**, 360 (1988).

5.7. D. M. Kerns, *Plane Wave Scattering Matrix Theory of Antennas and Antenna–Antenna Interactions*, U.S. Department of Commerce, National Bureau of Standards, Monograph N. 162, Washington, 1981.

5.8. R. Mittra and T. M. Habashy, *J. Opt. Soc. Am. A* **1**, 1103 (1984).

5.9. G. G. Stokes, *Cambridge Dublin Math. J.* **4**, 1 (1849). Reprinted in *Mathematical and Physical Papers of G. G. Stokes*, Cambridge University Press, Cambridge, 1883, Vol. II, pp. 89–103.

5.10. F. A. Jenkins and H. E. White, *Fundamentals of Optics*, McGraw-Hill, New York, 1976, 4th edition.

5.11. E. Hecht and A. Zajac, *Optics*, Addison-Wesley, Reading, Massachusetts, 1974.

5.12. A. Vasiček, *Optics of Thin Films*, North-Holland, Amsterdam, 1960.

5.13. M. Born and E. Wolf, *Principles of Optics*, Pergamon Press, Oxford, 1980, 6th edition.

5.14. H. J. Butterweck, *J. Opt. Soc. Am.* **67**, 60 (1977).

5.15. H. J. Butterweck, *Arch. Elektron. Übertragungstech.* **31**, 335 (1977).

5.16. J. H. Shapiro, *J. Opt. Soc. Am.* **61**, 492 (1971).

5.17. J. H. Shapiro, in *Laser Beam Propagation in the Atmosphere*, J. W. Strohbehn, ed., *Topics in Applied Physics*, Vol. 25, Springer-Verlag, Berlin, 1978, Chapter 6.

5.18. D. L. Fried and H. T. Yura, *J. Opt. Soc. Am.* **62**, 600 (1972).

5.19. W. T. Cathey, C. L. Hayes, W. C. Davis and V. F. Pizzuro, *Appl. Opt.* **9**, 701 (1970).

5.20. R. G. Newton, *Scattering Theory of Waves and Particles*, McGraw-Hill, New York, 1966.

6

Elements of the Theory of Diffraction

Diffraction has been usually understood as the deviation in the propagation of wavefields from rectilinear paths due to their interaction with obstacles such as apertures on screens, edges, and object boundaries. This phenomenon shares with scattering a common formal description through the basic integral theorems addressed in Chapter 1. In fact, diffraction may be considered as a particular case of scattering at boundary surfaces, hence, its mathematical formulation makes use of surface integrals, and thus, requires surface boundary conditions.

The initial investigations of this effect are associated with the names of Grimaldi, Huygens, Fresnel, and Young. (The reader can find historical introductions in Refs. 6.1–6.3.) The first attempt to put Fresnel's wave theory on a mathematical basis was done by G. Kirchhoff in 1882. However, as will be discussed later in this chapter, Kirchhoff's method presents some formal inconsistencies, which require careful interpretation. These issues have given rise to a vast amount of literature, which is worthwhile outlining here. On the other hand, however, Kirchhoff's procedure constitutes a very simple (and at the same time very accurate), solution in many instances.

In this chapter we shall discuss first the scalar theory. Its use with light, and electromagnetic waves in general, is only justified when the size of the diffracting aperture, or body, is large compared with the wavelength, and the outgoing fields are considered far from the diffracting centers. This will be further discussed later when we make a comparison with the results derived from the vector formulation.

Some aspects of diffraction have already been seen in previous chapters in connection with the angular spectrum of propagating wavefields. We shall return to some of these questions in this chapter, emphasizing their relationship with the analysis of diffraction patterns. Also, in Chapter 2 we studied the inherent spreading of light beams, due to their finite waist, which in fact constitutes another example of the diffraction phenomenon of a wavefield by a tapered aperture.

6.1. THE SCALAR THEORY OF DIFFRACTION

Let $U(\mathbf{r})$ be the spatial part of a wavefield, assumed to have harmonic time dependence: $\exp(-i\omega t)$. The problem of diffraction consists of finding $U(\mathbf{r})$, upon interaction of the incident field $U^{(i)}(\mathbf{r})$ with the object boundary, from knowledge of the surface boundary values of either $U(\mathbf{r})$ or its normal derivative $\partial U/\partial \mathbf{n}$, $\mathbf{n}$ being the local outward normal to the boundary surface. (We shall return later to the question on which boundary values are actually appropriate.)

The region enclosed by the boundary surface is assumed to be free of sources; therefore $U(\mathbf{r})$ satisfies in that region (which we shall denote by $\tilde{V}$) the homogeneous Helmholtz equation:

$$\nabla^2 U + k^2 U = 0. \tag{6.1}$$

The solution to Eq. (6.1) in $\tilde{V}$ was seen in Section 1.6.3 (cf. Eq. (1.56)) to be expressible in terms of its boundary values at the limiting surface S as:

$$U(\mathbf{r}) = \frac{1}{4\pi}\int_S \left[\frac{\partial U}{\partial \mathbf{n}} G - U\frac{\partial G}{\partial \mathbf{n}}\right] dS, \tag{6.2}$$

$G(\mathbf{r}, \mathbf{r}')$ being the outgoing spherical wave $\exp(ik|\mathbf{r} - \mathbf{r}'|)/|\mathbf{r} - \mathbf{r}'|$, and the integration being done in $\mathbf{r}'$. Eq. (6.2) is known as the *integral theorem of Helmholtz and Kirchhoff.* It gives the wavefield $U(\mathbf{r})$ at a generic point $\mathbf{r}$ of the volume $\tilde{V}$ in terms of the values of $U(\mathbf{r}')$ and $\partial U(\mathbf{r}')/\partial \mathbf{n}$ at the limiting surface S of $\tilde{V}$. This geometry is usually adopted in problems of diffraction by screens (see Fig. 6.1(a)): one portion of S is the diffracting screen S_C, and the other portion is a large hemisphere S_R, centered either in the screen or at some point behind it, and whose radius is eventually made infinite. In this geometry, all sources creating the incident field $U^{(i)}$ are assumed to be located at the left of the screen.

Another geometry used is the one depicted in Fig. 6.1(b): The region $\tilde{V}$ is limited by two boundary surfaces: the surface S enclosing the volume V of a diffracting body, and the surface S_R of a sphere whose radius is made infinity. The sources that create the incident field are either infinitely far away or inside the volume V. According to Section 1.6.3 (cf. Eq. (1.52)) the resulting field is given by:

$$U(\mathbf{r}) = U^{(i)}(\mathbf{r}) - \frac{1}{4\pi}\int_S \left[\frac{\partial U}{\partial \mathbf{n}} G - U\frac{\partial G}{\partial \mathbf{n}}\right] dS. \tag{6.3}$$

Observe that the second term of Eq. (6.3) constitutes the diffracting field

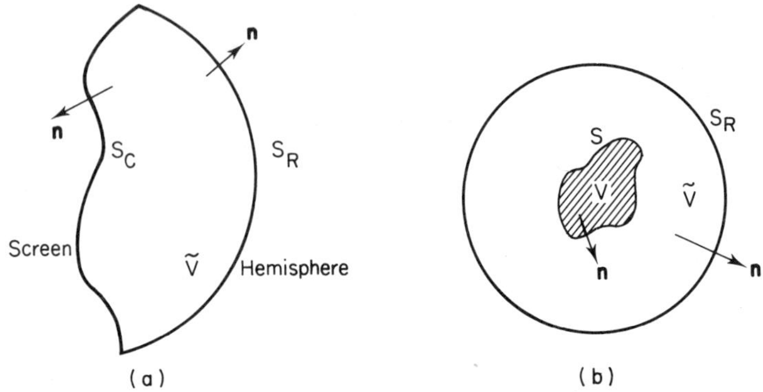

Figure 6.1. (a) Illustration of the geometry corresponding to Eq. (6.2). (b) Illustration of the geometry for Eq. (6.3).

which, apart from the geometry of the surface of integration, is similar to that expressed by Eq. (6.2).

When Sommerfeld's radiation condition:

$$\lim_{r \to \infty} r\left(\frac{\partial U}{\partial r} - ikU\right) = 0, \tag{6.4}$$

is imposed on $U(\mathbf{r})$, the integral in Eq. (6.2) over the hemisphere S_R of Fig. 6.1(a) becomes zero (see Problem 6.1); thus one is left with an integral over the diffracting screen S_C only. On the other hand, the local normal $\mathbf{n}$ to the screen S_R of Fig. 6.1(a) points outward from the volume $\tilde{V}$, whereas according to Fig. 6.1(b) the normal $\mathbf{n}$ in the surface S of the diffracting volume V in Eq. (6.3) points inward from the volume $\tilde{V}$, thus changing the sign of the integral in Eq. (6.3) with respect to the integral in Eq. (6.2). When these considerations are taken into account, one sees that Eqs. (6.2) and (6.3) are in fact equivalent, and only differ in the geometry of the surface of integration over which the boundary conditions are given. Therefore, we shall deal from now on with Eq. (6.2) only. It constitutes the mathematical formulation of Huygens' principle [6.1]–[6.3] on the basis of Fresnel's wave theory.

6.2. UNIQUENESS OF THE SOLUTION. BOUNDARY CONDITIONS

Equation (6.2) expresses $U(\mathbf{r})$ in terms of U and $\partial U/\partial \mathbf{n}$ on the boundary surface S. In mathematical terms, these are *Cauchy's boundary conditions* (see, e.g. Section 6.2 of Ref. 6.4). However, the Helmholtz equation is an elliptic partial differential equation [6.5], and hence, giving the boundary values of both U and $\partial U/\partial \mathbf{n}$ overspecify the requirements to obtain a unique solution. Let us next discuss this point.

Let us see whether the solution of the Helmholtz equation (6.1) is unique when one specifies either U on the surface S only (Dirichlet's boundary conditions), or its normal derivative $\partial U/\partial \mathbf{n}$ on S only (Neumann's boundary conditions). The argument used in this case is similar to the one employed in potential theory in electrostatics (see, e.g. Ref. 6.6).

If the solution were not unique, there would exist two different wavefields U_1 and U_2, satisfying the same boundary conditions on S, then the difference:

$$\psi = U_1 - U_2, \tag{6.5}$$

would obviously satisfy the Laplace equation:

$$\nabla^2 \psi = 0, \tag{6.6}$$

in $\tilde{V}$. In addition, either $\psi = 0$ or $\partial\psi/\partial \mathbf{n} = 0$ on S, according to whether U_1 and U_2 hold Dirichlet's or Neumann's boundary conditions.

Now, by applying Green's first identity to ψ, we obtain:

$$\int_{\tilde{V}} (\psi \nabla^2 \psi + \nabla\psi \cdot \nabla\psi)\, dv = \int_S \psi \frac{\partial \psi}{\partial \mathbf{n}}\, dS. \tag{6.7}$$

However, from Eq. (6.6), and since either ψ or $\partial\psi/\partial \mathbf{n}$ is assumed to be zero on S, we get:

$$\int_{\tilde{V}} \nabla\psi \cdot \nabla\psi\, dv = 0. \tag{6.8}$$

Equation (6.8) implies that $\nabla\psi = 0$. Hence, ψ is a constant in $\tilde{V}$. Due to the well known properties of harmonic functions [6.4], ψ is zero in $\tilde{V}$ if $\psi = 0$ on S, and ψ is a constant in $\tilde{V}$ if $\partial\psi/\partial \mathbf{n} = 0$ on S. Therefore, U_1 and U_2 are identical for Dirichlet's boundary conditions, or differ by a constant for Neumann's boundary conditions. As a consequence, giving either U or $\partial U/\partial \mathbf{n}$ on S is enough to determine U in $\tilde{V}$ uniquely. The imposition of arbitrary values of both U and $\partial U/\partial \mathbf{n}$ simultaneously *overspecifies* the problem and can, in fact, constitute a pair of incompatible boundary values. When arbitrary U and $\partial U/\partial \mathbf{n}$ are prescribed on the surface S, they are, in general, inconsistent with the limiting values of U and $\partial U/\partial \mathbf{n}$ as the field point **r** in $\tilde{V}$ tends to S, a fact first noted by Poincaré [6.7] (see also Baker and Copson [6.8]). In this respect, the uniqueness of the diffraction problem is analogous to that of electrostatic potentials, as mentioned before, as stems from the elliptic nature of the governing partial differential equations. Besides, in accordance with the Cauchy–Kowalewski theorem, quoted in Section 4.4, the function $U(\mathbf{r})$ can be extended as an analytic function for complex values of **r** in $\tilde{V}$.

6.3. THE KIRCHHOFF APPROXIMATION

Let us consider a wavefield $U^{(i)}(\mathbf{r})$ incident on an aperture $\mathscr{A}$ punched in an infinite opaque screen S_C, assumed to be in the plane $z = 0$ of a Cartesian system of coordinates (see Fig. 6.2). The field diffracted by the aperture $U(\mathbf{r})$, evaluated at a point P, is obtained from Eq. (6.2) with a surface of integration formed by the screen S_C and a sphere S_R of radius R that will be made to tend to infinity and be centered at P. As quoted before, due to Sommerfeld's radiation condition, Eq. (6.4), the surface integral over S_R vanishes. Therefore, Eq. (6.2) reduces to:

$$U(\mathbf{r}) = \frac{1}{4\pi} \int_{S_C} \left[\frac{\partial U}{\partial \mathbf{n}} G - U \frac{\partial G}{\partial \mathbf{n}} \right] dS. \tag{6.9}$$

The *Kirchhoff approximation*, sometimes also referred to as the *physical optics approximation* (see also Section 7.5), assumes the boundary values of the

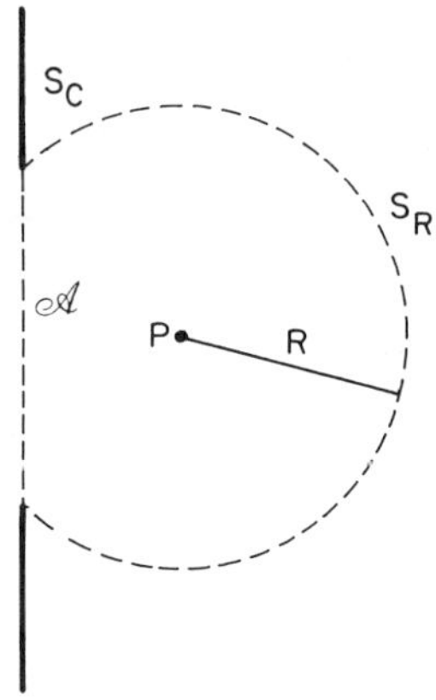

Figure 6.2. Notation for the Kirchhoff approximation of diffraction by an aperture $\mathscr{A}$ on a screen S_C.

field U and its normal derivative $\partial U/\partial \mathbf{n}$ on the screen S_C as if they were determined from geometrical optics. Namely, in a plane immediately behind the screen, these boundary values are considered equal to the incident field $U^{(i)}$ and its normal derivative $\partial U^{(i)}/\partial \mathbf{n}$, respectively inside the aperture, being zero outside the aperture due to the geometric shadow of the rest of the screen.

On the other hand, in a plane immediately in front of the screen, the Kirchhoff approximation assumes those boundary values to be equal to the corresponding unperturbed values of the incident field. Observe that these assumptions imply the existence of *saltus* both in U and $\partial U^{(i)}/\partial \mathbf{n}$ across the screen and outside the aperture $\mathscr{A}$.

In summary, from Eq. (6.9) this approximation gives the following *Kirchhoff diffraction formula*:

$$U(\mathbf{r}) = \frac{1}{4\pi} \int_{\mathscr{A}} \left[\frac{\partial U^{(i)}}{\partial \mathbf{n}} G - U^{(i)} \frac{\partial G}{\partial \mathbf{n}} \right] dS. \tag{6.10}$$

As pointed out in Section 6.2, the requirements of both U and $\partial U^{(i)}/\partial \mathbf{n}$ on S_C overspecify the problem, and hence constitute incompatible boundary conditions. This argument has been used in many criticisms of the Kirchhoff approximation (see, e.g. Refs. 6.1, 6.6, and 6.8). It is often argued that since U is analytic in $\tilde{V}$, if U is zero in a portion of the boundary S_C, it must be zero elsewhere. However, from the analysis of Chapter 4, it is easy to understand that U cannot be analytic everywhere on S_C, since the vicinities of the rim of the aperture constitute scattering centers of the field. (We shall return later to this point in connection with Young's interpretation.) Therefore, it is better to consider the screen as a real physical surface, rather than as a mathematically fictitious surface at which the boundary conditions are to be recovered by letting the field point tend towards the screen.

An attempt in this direction was done by Kottler [6.9], [6.10]; he replaced the problem by that with *saltus conditions across the screen*. These discontinuities being in accordance with the saltus in the boundary values of the field and its normal derivative across the screen, as mentioned above. These jumps also constitute the requirements for the screen to be *black*; namely, it does not transmit nor reflect any portion of energy of the incident wave.

On the other hand, in Kottler's theory, U is supposed to satisfy the homogeneous Helmholtz equation (6.1) everywhere in $\tilde{V}$, except on the surface that constitutes the screen. Under these assumptions, it can be shown that Kirchhoff's diffraction formula (Eq. (6.10)) is the rigorous solution to the problem of diffraction by the screen [6.8].

Another point of view for solving the apparent inconsistency of Kirchhoff's approach was adopted by Marchand and Wolf in connection with the role played by the rim of the aperture. This bears a close relationship with Young's earlier ideas of diffraction. We shall address this work later.

It should be mentioned that Born [6.11] suggested considering the Kirchhoff approximation as the first term of a Neumann series for Eq. (6.9), in analogy with the first Born approximation in scattering problems; this interpretation can also be found in Section 9.3 of Ref. 6.4; however, according to investigations by Franz [6.12] and Schelkunoff [6.13] (see also Refs. 6.14 and 6.15), no proof of convergence of this iterative series was obtained.

6.4. THE RAYLEIGH–SOMMERFELD DIFFRACTION INTEGRALS

There exists a certain freedom in choosing the Green function associated with the diffraction problem for a wavefield U that satisfies Helmholtz's Eq. (6.1) in a domain $\tilde{V}$, with prescribed boundary conditions in the limiting surface S of $\tilde{V}$ (see, e.g. Ref. 6.4). So far, in Eq. (6.2) $G(\mathbf{r}, \mathbf{r}')$ has been considered to be the outgoing spherical wave $\exp(ik|\mathbf{r} - \mathbf{r}'|)/|\mathbf{r} - \mathbf{r}'|$ which satisfies:

$$(\nabla^2 + k^2)G(\mathbf{r}, \mathbf{r}') = -4\pi\delta(\mathbf{r} - \mathbf{r}'). \tag{6.11}$$

However, the outgoing spherical wave is only one of the possible solutions to Eq. (6.11). Another function:

$$G_F(\mathbf{r}, \mathbf{r}') = \frac{\exp(ik|\mathbf{r} - \mathbf{r}'|)}{|\mathbf{r} - \mathbf{r}'|} + F(\mathbf{r}, \mathbf{r}') \tag{6.12}$$

is valid and fulfils Eq. (6.11). The function $F(\mathbf{r}, \mathbf{r}')$ should satisfy:

$$(\nabla^2 + k^2)F(\mathbf{r}, \mathbf{r}') = 0, \tag{6.13}$$

and it can be chosen such that one of the two terms of the integrand of Eq. (6.2) vanishes. This freedom in the choice of $F(\mathbf{r}, \mathbf{r}')$ is the basis of the well known method of images, extensively used in electrostatics (see, e.g. Ref. 6.6), and was adopted by Sommerfeld (p. 199 of Ref. 6.1) and earlier by Rayleigh [6.16].

A convenient choice for the function $F(\mathbf{r}, \mathbf{r}')$ is:

$$F(\mathbf{r}, \mathbf{r}') = \pm\frac{\exp(ik|\mathbf{r}^s - \mathbf{r}'|)}{|\mathbf{r}^s - \mathbf{r}'|}, \tag{6.14}$$

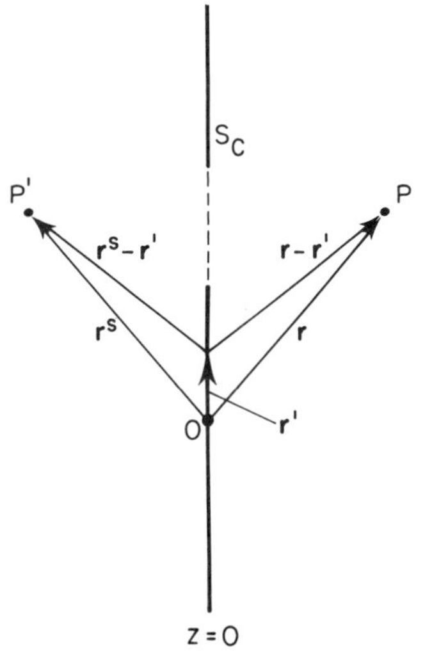

Figure 6.3. Notation for the Rayleigh–Sommerfeld integrals.

where $\mathbf{r}^s$ is the position vector of the mirror image P' of the observation point P with respect to the plane of the screen (Fig. 6.3).

On introducing Eq. (6.14) into (6.12), we then get the following expression for the alternative Green functions:

$$G_{I,II}(\mathbf{r}, \mathbf{r}') = \frac{\exp(ik|\mathbf{r} - \mathbf{r}'|)}{|\mathbf{r} - \mathbf{r}'|} \mp \frac{\exp(ik|\mathbf{r}^s - \mathbf{r}'|)}{|\mathbf{r}^s - \mathbf{r}'|}. \tag{6.15}$$

In Eq. (6.15) the subindices I and II are taken according to whether the sign "minus" or "plus" is considered in the right-hand side.

It is not difficult to verify that $G_I(\mathbf{r}, \mathbf{r}')$ and $G_{II}(\mathbf{r}, \mathbf{r}')$ satisfy the following relations for $\mathbf{r}'$ in the screen:

$$G_I|_{S_C} = 0, \tag{6.16a}$$

$$\left.\frac{\partial G_I}{\partial \mathbf{n}}\right|_{S_C} \equiv -\left.\frac{\partial G_I}{\partial z}\right|_{S_C} = -2\left.\frac{\partial}{\partial z}\frac{\exp(ik|\mathbf{r} - \mathbf{r}'|)}{|\mathbf{r} - \mathbf{r}'|}\right|_{S_C} \equiv 2\left.\frac{\partial G}{\partial \mathbf{n}}\right|_{S_C}, \tag{6.16b}$$

and:

$$G_{II}|_{S_C} = 2\frac{\exp(ik|\mathbf{r} - \mathbf{r}'|)}{|\mathbf{r} - \mathbf{r}'|} \equiv 2G|_{S_C}, \tag{6.17a}$$

$$\left.\frac{\partial G_{II}}{\partial \mathbf{n}}\right|_{S_C} \equiv -\left.\frac{\partial G_{II}}{\partial z}\right|_{S_C} = 0. \tag{6.17b}$$

By substituting the functions G_I and G_{II} in the Helmholtz–Kirchhoff integral, Eq. (6.9), and making use of Eqs. (6.16) and (6.17) we obtain:

$$U_I(\mathbf{r}) = -\frac{1}{2\pi}\int_{S_C} U(\mathbf{r}')\frac{\partial G}{\partial \mathbf{n}}\,dS, \tag{6.18}$$

when $G_I(\mathbf{r}, \mathbf{r}')$ is chosen. Equation (6.18) is the *first Rayleigh–Sommerfeld integral formula*, and corresponds to Dirichlet boundary conditions.

On the other hand, when $G_{II}(\mathbf{r}, \mathbf{r}')$ is chosen, one gets from Eq. (6.9):

$$U_{II}(\mathbf{r}) = \frac{1}{2\pi}\int_{S_C} \frac{\partial U}{\partial \mathbf{n}} G(\mathbf{r}, \mathbf{r}')\,dS. \tag{6.19}$$

Equation (6.19) is the *second Rayleigh–Sommerfeld integral formula*, and corresponds to Neumann boundary conditions.

Since the screen S_C has been assumed to be in the plane $z = 0$ (Fig. 6.3), U_I and U_{II} may also be expressed as:

$$U_I(\mathbf{r}) = \frac{1}{2\pi}\iint_{-\infty}^{\infty} U(x', y', 0)\left[\frac{\partial}{\partial z'} G(\mathbf{r}, \mathbf{r}')\right]_{z'=0} dx'\,dy', \tag{6.20}$$

$$U_{II}(\mathbf{r}) = -\frac{1}{2\pi}\iint_{-\infty}^{\infty} \left[\frac{\partial U(x', y', z')}{\partial z'}\right]_{z'=0} [G(\mathbf{r}, \mathbf{r}')]_{z'=0}\,dx'\,dy'. \tag{6.21}$$

On comparing Eq. (6.9) with Eqs. (6.18) and (6.19), one can see immediately that the Helmholtz–Kirchhoff solution (Eq. (6.9)) is:

$$U = \tfrac{1}{2}[U_I + U_{II}]. \tag{6.22}$$

The substitution in Eq. (6.20) or (6.21) of the boundary values for $U(x, y, 0)$ and $\partial U/\partial z|_{z=0}$ by those corresponding to the incident field $U^{(i)}$, according to the Kirchhoff approximation, although mathematically consistent with respect to the analysis of Section 6.2 and 6.3, still represents an approximation which is only valid when the dimensions of the aperture, or diffracting body, are significantly larger than the wavelength. This will be addressed later.

6.5. THE RAYLEIGH–SOMMERFELD INTEGRALS AND THE ANGULAR SPECTRUM REPRESENTATION

In applications of the diffraction integrals to specific problems, it is often necessary to make additional approximations as regards the field observation point, (e.g. the Fresnel or Fraunhofer approximations to be discussed later). However, as mentioned in Chapter 2, the angular spectrum representation permits an exact determination at any plane $z = constant$ of the field diffracted at any other plane, for example: $z = 0$. This procedure of formulating diffraction problems has been discussed in Refs. 2.1–2.17 of Chapter 2. An extensive discussion can be also found in Ref. 6.17. Applications to image formation and holography can be seen in Refs. 6.18–6.20. A formal study of this technique in the context of functional analysis is given in Refs. 6.21 and 6.22, also in connection with self-imaging [6.23], [6.24].

We have seen in Section 2.3 that the angular spectrum representation allows one to determine a wavefield $U(\mathbf{r})$, governed by the homogeneous Helmholtz equation (6.1) in a half-space, from its boundary value $U(x, y, 0) = f(x, y)$ at a certain plane $z = 0$. We shall next prove that this is equivalent to the first Rayleigh–Sommerfeld integral of Eq. (6.20) if the plane $z = 0$ contains the diffracting screen.

We shall also show that when one uses the Neumann boundary condition $\partial U/\partial z|_{z=0} = g(x, y)$ in $z = 0$, the determination of the diffracted field by means of the angular spectrum representation coincides with the second Rayleigh–Sommerfeld integral, Eq. (6.21).

The following analysis is based on work by Bouwkamp [6.14], Osterberg [6.25], and Sherman [6.26].

Let the screen be at the plane $z = 0$, and the Dirichlet boundary condition at this plane be:

$$U(x, y, z) = f(x, y). \tag{6.23}$$

According to Section 2.3, Eq. (2.18), the field diffracted in $z > 0$ is given by:

$$U(\mathbf{r}) = \iint_{-\infty}^{\infty} f(x', y')\mathscr{K}(x - x', y - y', z)\, dx'\, dy', \tag{6.24}$$

where:

$$\mathscr{K}(x - x', y - y', z) = \left(\frac{k}{2\pi}\right)^2 \iint_{-\infty}^{\infty} \exp[ik(p(x - x') + q(y - y') + mz)]\, dp\, dq, \tag{6.25}$$

p, q, and m satisfying Eq. (2.9).

Next, let us consider the plane wave expansion, Eq. (2.51), of the outgoing spherical wave in $z > 0$:

$$G(\mathbf{r}, \mathbf{r}') = \frac{ik}{2\pi} \iint_{-\infty}^{\infty} \frac{1}{m} \exp[ik(p(x - x') + q(y - y') + m(z - z'))]\, dp\, dq, \tag{6.26}$$

r being a point in $z > 0$ and **r**′ being a point in the plane of the screen whose z-coordinate will eventually be made $z = 0$.

The derivative of Eq. (6.26) with respect to z' is:

$$\left.\frac{\partial G(\mathbf{r}, \mathbf{r}')}{\partial z'}\right|_{z'=0} = \frac{k^2}{2\pi} \iint_{-\infty}^{\infty} \exp[ik(p(x - x') + q(y - y') + mz)]\, dp\, dq, \qquad (6.27)$$

On comparing Eqs. (6.27) and (6.25) one obtains at once:

$$\mathscr{K}(x - x', y - y', z) = \frac{1}{2\pi}\left[\frac{\partial G(\mathbf{r}, \mathbf{r}')}{\partial z'}\right]_{z'=0}. \qquad (6.28)$$

By introducing Eq. (6.28) into (6.24), and taking Eq. (6.23) into account, we obtain the first Rayleigh–Sommerfeld integral, Eq. (6.20).

Let us consider now the Neumann boundary condition in $z = 0$:

$$\left[\frac{\partial U(x, y, z)}{\partial z}\right]_{z=0} = g(x, y). \qquad (6.29)$$

The angular spectrum $a(p, q)$ of $U(x, y, z)$ can easily be related to $g(x, y)$. Let us take the derivative with respect to z in (2.12), ($z > 0$):

$$\frac{\partial U(\mathbf{r})}{\partial z} = \iint_{-\infty}^{\infty} ikma(p, q) \exp[ik(px + qy + mz)]\, dp\, dq. \qquad (6.30)$$

Hence, in $z = 0$ we obtain:

$$\left.\frac{\partial U(\mathbf{r})}{\partial z}\right|_{z=0} = \iint_{-\infty}^{\infty} ikma(p, q) \exp[ik(px + qy)]\, dp\, dq. \qquad (6.31)$$

On using Eq. (6.29), the inversion of (6.31) gives:

$$a(p, q) = \frac{k}{i(2\pi)^2 m} \iint_{-\infty}^{\infty} g(x, y) \exp[ik(px + qy)]\, dx\, dy. \qquad (6.32)$$

By introducing the expression (6.32) of $a(p, q)$ into Eq. (2.12), we obtain the wavefield $U(\mathbf{r})$ in $z > 0$ in terms of the normal derivative $g(x, y)$ at $z = 0$:

$$U(\mathbf{r}) = \iint_{-\infty}^{\infty} g(x', y')\mathscr{H}(x - x', y - y', z)\, dx'\, dy'; \qquad (6.33)$$

where the impulse response $\mathscr{H}$ is:

$$\mathscr{H}(x - x', y - y', z) = -\frac{ik}{(2\pi)^2} \iint_{-\infty}^{\infty} \frac{1}{m} \exp[ik(p(x - x') + q(y - y') + mz)]\, dp\, dq. \quad (6.34)$$

On comparing Eqs. (6.34) and (6.26), we see that:

$$\mathscr{H}(x - x', y - y', z) = -\frac{1}{2\pi} [G(\mathbf{r}, \mathbf{r}')]_{z'=0}. \quad (6.35)$$

By introducing Eq. (6.35) into (6.33), and using Eq. (6.29), we obtain the second Rayleigh–Sommerfeld integral (6.21).

6.6. RECIPROCITY. DIFFRACTION FOR SMALL ANGLES

As seen in Chapter 5, reciprocity is a fundamental property of scattering. Diffraction, which can be regarded as a particular case of that phenomenon, should therefore satisfy this property. We are next going to analyze whether the basic equations, corresponding to the formulations discussed so far in this chapter, indeed satisfy reciprocity.

Let us assume that the screen S_C, which contains the aperture $\mathscr{A}$, is illuminated by a spherical wave arising from a point source P_0 as indicated in Fig. 6.4. The field at any point P behind the screen, and at distance s from a generic point of the aperture $\mathscr{A}$, represents the *impulse response* of this system in passing from the region $\mathscr{R}^-$ into the region $\mathscr{R}^+$, in the sense discussed in Section 5.13. Hence, if the Kirchhoff approximation is used, the field at the screen is:

$$U(\mathbf{s}) \equiv U^{(i)}(\mathbf{s}) = \frac{\exp(iks)}{s}. \quad (6.36)$$

Letting $\mathbf{n}$ be the unit normal, pointing towards $\mathscr{R}^-$, one has:

$$\frac{\partial U^{(i)}}{\partial \mathbf{n}} = \cos(\mathbf{n}, \mathbf{s})\left(ik - \frac{1}{s}\right)\frac{\exp(iks)}{s}. \quad (6.37)$$

On the other hand, we have:

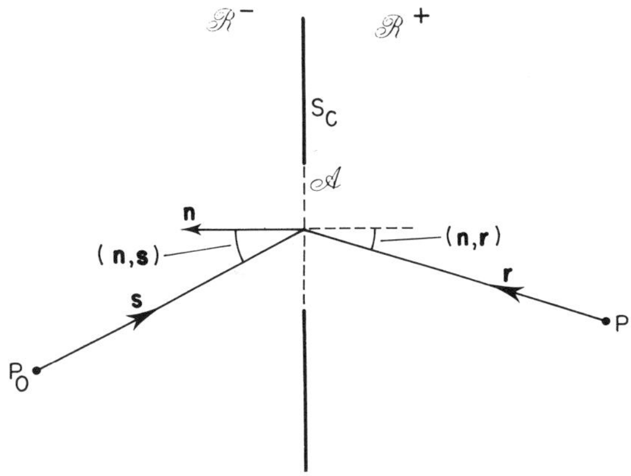

Figure 6.4. Illustration of the notation for diffraction of a wave created at a point source at P_0, incident on an aperture.

$$\frac{\partial G}{\partial \mathbf{n}} = \cos(\mathbf{n}, \mathbf{r})\left(ik - \frac{1}{r}\right)\frac{\exp(ikr)}{r}, \tag{6.38}$$

r being the distance from the observation point P in $\mathscr{R}^+$ to a generic point of the aperture $\mathscr{A}$. By assuming that the distances of both P_0 and P to the aperture are much larger than the wavelength λ, the second term inside the parentheses of Eqs. (6.37) and (6.38) can be neglected compared with the first term ik. Therefore, these two expressions become:

$$\frac{\partial U^{(i)}}{\partial \mathbf{n}} \sim ik \cos(\mathbf{n}, \mathbf{s}) \frac{\exp(iks)}{s}; \tag{6.39}$$

and:

$$\frac{\partial G}{\partial \mathbf{n}} \sim ik \cos(\mathbf{n}, \mathbf{r}) \frac{\exp(ikr)}{r}. \tag{6.40}$$

By substituting Eqs. (6.36), (6.39), (6.40), and the expression for G into Eq. (6.10), we get:

$$U(P) = \frac{ik}{4\pi}\int_{\mathscr{A}} \frac{\exp[ik(s + r)]}{sr}[\cos(\mathbf{n}, \mathbf{s}) - \cos(\mathbf{n}, \mathbf{r})]\, dS. \tag{6.41}$$

Equation (6.41) is symmetrical with respect to $\mathbf{r}$ and $\mathbf{s}$. Hence, the impulse response at P, due to a point source P_0, is the same as the corresponding one at P_0 due to a point source P. This impulse response thus satisfies reciprocity as it should in agreement with Section 5.13. This result of diffraction theory is usually quoted as the *reciprocity theorem of Helmholtz*.

On the other hand, under the same conditions as before, and with the Kirchhoff boundary conditions in the screen S_C, the wavefield according to the first Rayleigh–Sommerfeld integral in Eq. (6.20) is:

$$U_I(P) = -\frac{ik}{2\pi}\int_{\mathscr{A}} \frac{\exp[ik(s+r)]}{sr}\cos(\mathbf{n},\mathbf{r})\,dS, \tag{6.42}$$

which is not reciprocal by interchanging P and P_0, since it is not symmetrical with respect to $\mathbf{r}$ and $\mathbf{s}$. In fact, interchanging $\mathbf{r}$ and $\mathbf{s}$ in Eq. (6.42) leads to:

$$\frac{ik}{2\pi}\int_{\mathscr{A}} \frac{\exp[ik(s+r)]}{sr}\cos(\mathbf{n},\mathbf{s})\,dS = U_{II}(P_0), \tag{6.43}$$

which is the wavefield at P_0 according to the second Rayleigh–Sommerfeld integral. An analogous result would be obtained by starting from the second Rayleigh–Sommerfeld integral.

We shall see, however, that since the numerical difference between the diffraction patterns obtained from the Kirchhoff integral, Eq. (6.10), and the Kirchhoff approximation in either Rayleigh–Sommerfeld integral is very small when the Kirchhoff approximation is valid, the distinction between the three, originating from the aforementioned mathematical inconsistencies, is of not all that practical relevance. It is worth emphasizing, however, that when the "consistent" Rayleigh–Sommerfeld integrals are used in conjunction with the boundary conditions at the screen given by the Kirchhoff approximation, they violate the reciprocity property.

In particular, in the case often found in many practical instances of application of the Kirchhoff approximation, in which the distance from the points P and P_0 to the screen are large compared with the linear dimensions of the aperture, and the angle made by the line P_0P with P_00 and $0P$ is small, 0 being a point of the aperture (Fig. 6.5), one can obviously approximate:

$$\begin{gathered}\cos(\mathbf{n},\mathbf{s}) = \cos\delta; \qquad \cos(\mathbf{n},\mathbf{r}) = -\cos\delta; \\ \cos(\mathbf{n},\mathbf{s}) - \cos(\mathbf{n},\mathbf{r}) \sim 2\cos\delta.\end{gathered} \tag{6.44}$$

Under such conditions, it is easy to see that the three integrals, Eqs. (6.41)–(6.43) coincide, and are equal to:

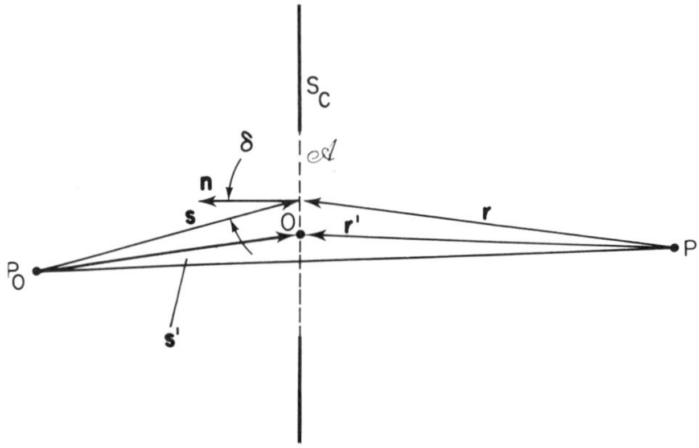

Figure 6.5. Diffraction at low angles.

$$U(P) = \frac{ik}{2\pi} \cos \delta \int_{\mathscr{A}} \frac{\exp[ik(s + r)]}{sr} \, dS. \tag{6.45}$$

In this case, the factor $1/rs$ may be replaced by $1/r's'$ (r' and s' being the distances from the origin O to P and P_0, respectively (Fig. 6.5)), and Eq. (6.45) may be expressed as:

$$U(P) = \frac{ik}{2\pi} \frac{\cos \delta}{s'r'} \int_{\mathscr{A}} \exp[ik(s + r)] \, dS. \tag{6.46}$$

Equation (6.46) is often the starting point of the so-called *Fresnel* and *Fraunhofer approximations* for diffraction of light from a point source by an aperture (cf. Section 8.3.3 of Ref. 6.2).

Observe also that when the dimensions of the aperture are large compared with the wavelength λ, the diffraction pattern is restricted to low angles of observation; then, Eq. (6.46) is also valid provided that the point source is far from the screen.

6.7. THE FRESNEL APPROXIMATION

Starting from the Helmholtz–Kirchhoff integral, Eq. (6.9), we shall derive next some approximations that are usually made to the diffraction formulas in connection with the position of the field point P. At the moment, we shall not make assumptions about the field $U(\mathbf{r})$ in the plane of the screen S_C, except that it vanishes outside an aperture $\mathscr{A}$ of S_C.

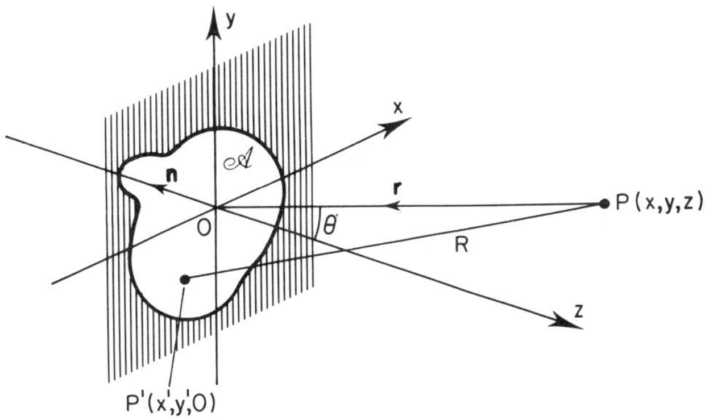

Figure 6.6. Illustration of the geometry of diffraction.

Let us assume that the distance r from an origin of Cartesian coordinates, situated in the aperture $\mathscr{A}$, to the field observation point P is much larger than the maximum dimension of this aperture. In this framework, we choose the plane $z = 0$ containing the screen. In addition, we shall consider the coordinates x and y of the point P being much smaller than the z coordinate; in other words, the point P is close to the z-axis (Fig. 6.6); then:

$$\frac{\partial G(\mathbf{r}, \mathbf{r}')}{\partial \mathbf{n}} \sim -\frac{\partial G(\mathbf{r}, \mathbf{r}')}{\partial z'} = -\frac{\partial G(\mathbf{r}, \mathbf{r}')}{\partial R}\frac{\partial R}{\partial z'};$$
$$R = \sqrt{(x - x')^2 + (y - y')^2 + z^2}. \tag{6.47}$$

It is easy to see that Eq. (6.47) yields:

$$\frac{\partial G(\mathbf{r}, \mathbf{r}')}{\partial \mathbf{n}} = \frac{\exp(ikR)}{R}\left(ik - \frac{1}{R}\right)\cos\theta \sim ik\cos\theta\frac{\exp(ikR)}{R}; \qquad \theta = (\mathbf{n}, \mathbf{r}). \tag{6.48}$$

By introducing Eq. (6.48) into (6.9) we obtain:

$$U(\mathbf{r}) = \frac{1}{4\pi}\int_{\mathscr{A}}\left[\frac{\partial U}{\partial \mathbf{n}} - ik\cos\theta U(\mathbf{r}')\right]\frac{\exp(ik|\mathbf{r} - \mathbf{r}'|)}{|\mathbf{r} - \mathbf{r}'|}\,dS. \tag{6.49}$$

However, since $r \ll r'$, the distance $|\mathbf{r} - \mathbf{r}'|$ is expanded as:

$$|\mathbf{r} - \mathbf{r}'| = r\left[1 + \frac{1}{2r^2}(r'^2 - 2\mathbf{r}\cdot\mathbf{r}') - \frac{1}{8r^4}(r'^2 - 2\mathbf{r}\cdot\mathbf{r}')^2 + \cdots\right]. \tag{6.50a}$$

The Fresnel approximation consists of neglecting the higher order terms in the expansion, Eq. (6.50a), so that:

$$|\mathbf{r} - \mathbf{r}'| \sim r\left[1 + \frac{1}{2r^2}(r'^2 - 2\mathbf{r}\cdot\mathbf{r}')\right]. \tag{6.50b}$$

It is evident that the approximation, Eq. (6.50b), requires that the third term of Eq. (6.50a), and those of higher order, do not contribute to the integral in Eq. (6.49). A sufficient condition for this is obviously:

$$\frac{k}{8r^4}(r'^2 - 2\mathbf{r}\cdot\mathbf{r}')^2 \ll 1, \qquad \text{for } \mathbf{r}' \text{ in } \mathscr{A}. \tag{6.51}$$

By substituting Eq. (6.50b) into (6.49) and replacing $|\mathbf{r} - \mathbf{r}'|$ by r in the denominator (note that this denominator is much less sensitive than the phase to this replacement), we obtain the following expression for the diffracted field in the Fresnel approximation:

$$U(\mathbf{r}) = \frac{1}{4\pi}\frac{\exp(ikr)}{r}\int_{\mathscr{A}}\left[\frac{\partial U}{\partial \mathbf{n}} - ik\cos\theta U(\mathbf{r}')\right]\exp\left(ik\frac{r'^2}{2r}\right)\exp\left[-i\frac{k}{r}(\mathbf{r}\cdot\mathbf{r}')\right]dS. \tag{6.52}$$

By writing $\mathbf{r}$ and $\mathbf{r}'$ in terms of their Cartesian components, making $\cos\theta \sim 1$, replacing r by its z-component in the denominators, and approximating: $r \sim z[1 + \frac{1}{2}(x^2 + y^2)/z^2]$ the following current expression for the Fresnel approximation is obtained:

$$\begin{aligned} U(\mathbf{r}) \sim {} & \frac{1}{4\pi}\frac{\exp(ikz)}{z}\exp\left[-i\frac{k}{2z}(x^2 + y^2)\right] \\ & \times \int_{\mathscr{A}}\left(-\left[\frac{\partial U(x', y', z')}{\partial z'}\right]_{z'=0} - ikU(x', y', 0)\right) \\ & \times \exp\left[\frac{ik}{2z}(x'^2 + y'^2)\right]\exp\left[i\frac{k}{z}(xx' + yy')\right]dx'\,dy'; \end{aligned} \tag{6.53}$$

alternatively, Eq. (6.53) may be written as:

$$\begin{aligned} U(\mathbf{r}) \sim {} & \frac{1}{4\pi}\frac{\exp(ikz)}{z}\int_{\mathscr{A}}\left(\left[\frac{\partial U(x', y', z')}{\partial \mathbf{n}}\right]_{z'=0} - ikU(x', y', 0)\right) \\ & \times \exp\left[\frac{ik}{2z}[(x - x')^2 + (y - y')^2]\right]dx'\,dy'. \end{aligned} \tag{6.54}$$

Equation (6.54) expresses the diffracted field under the Fresnel approximation as a convolution of the boundary values of U, and its gradient, with a quadratic phase factor which represents an approximation to the spherical wave in the diffraction integral.

Note that if the screen were to be illuminated by a point source, and one made use of the Kirchhoff approximation, an additional spherical wave factor would be obtained. By making assumptions about the position of the point source, similar to those made for the observation point, this factor can be approximated like the one involved in Eqs. (6.49)–(6.51).

By analogy with Eq. (6.54), the Fresnel approximation for the first and second Rayleigh–Sommerfeld integrals, respectively, is:

$$U_I(\mathbf{r}) \sim \frac{-ik}{2\pi}\frac{\exp(ikz)}{z}\int_{\mathscr{A}} U(x', y', 0)\exp\left[\frac{ik}{2z}[(x-x')^2+(y-y')^2]\right]dx'\,dy', \tag{6.55}$$

and:

$$U_{II}(\mathbf{r}) \sim \frac{-1}{2\pi}\frac{\exp(ikz)}{z}\int_{\mathscr{A}}\left[\frac{\partial U(x', y', z')}{\partial z'}\right]_{z'=0} \times \exp\left[\frac{ik}{2z}[(x-x')^2+(y-y')^2]\right]dx'\,dy'. \tag{6.56}$$

6.8. THE FRAUNHOFER APPROXIMATION

If, in addition to the condition Eq. (6.51), we assume that the linear dimensions of the aperture are so small compared with the distance of the observation point that, in Eq. (6.50b), one has, after multiplying by k:

$$\frac{k}{2r}r'^2 \ll 1; \qquad \text{for } \mathbf{r}' \text{ in } \mathscr{A}; \tag{6.57a}$$

namely:

$$\frac{k}{2}(x'^2+y'^2)_{\max} \ll r, \tag{6.57b}$$

where the subindex max stands for the maximum value of the linear dimensions of the aperture.

Equation (6.57b) permits the following approximation to the diffraction integral in Eq. (6.52):

$$U(\mathbf{r}) \sim \frac{1}{4\pi}\frac{\exp(ikr)}{r}\int_{\mathscr{A}}\left[\frac{\partial U(\mathbf{r}')}{\partial \mathbf{n}} - ik\cos\theta U(\mathbf{r}')\right]\exp\left(-i\frac{k}{r}\mathbf{r}\cdot\mathbf{r}'\right)dS. \tag{6.58}$$

Equation (6.58) constitutes the Fraunhofer approximation to the diffracted field. It renders this field in terms of the two-dimensional Fourier transform of its boundary value (and that of its gradient) at the aperture $\mathscr{A}$. Therefore, apart from the spherical wave envelope $\exp(ikr)/r$, Eq. (6.58) is a function of the angular variables x/r and y/r, which characterize the direction of the position vector of the point of observation P. We shall prove in the next section that the Fraunhofer field is, as a matter of fact, the far field and coincides with the asymptotic expression of the diffracted field when this is represented in terms of its angular spectrum.

The Fraunhofer approximation is sometimes expressed directly from Eq. (6.53) by neglecting the quadratic factor $\exp[(ik/2z)(x'^2 + y'^2)]$ as:

$$U(\mathbf{r}) \sim \frac{1}{4\pi}\frac{\exp(ikz)}{z}\exp\left[-i\frac{k}{2z}(x^2+y^2)\right] \times \int_{\mathscr{A}}\left(-\left[\frac{\partial U(x',y',z')}{\partial z'}\right]_{z'=0} - ikU(x',y',0)\right) \times \exp\left[i\frac{k}{z}(xx'+yy')\right]dx'\,dy'; \tag{6.59}$$

and, by analogy with Eq. (6.59), the Fraunhofer approximation for the first and second Rayleigh–Sommerfeld integrals is, after neglecting the diffraction angle θ:

$$U_I(\mathbf{r}) \sim \frac{-ik}{2\pi}\frac{\exp(ikz)}{z}\int_{\mathscr{A}} U(x',y',0)\exp\left[\frac{ik}{z}(xx'+yy')\right]dx'\,dy', \tag{6.60}$$

and:

$$U_{II}(\mathbf{r}) \sim \frac{-1}{2\pi}\frac{\exp(ikz)}{z}\int_{\mathscr{A}}\left[\frac{\partial U(x',y',z')}{\partial z'}\right]_{z'=0}\exp\left[\frac{ik}{z}(xx'+yy')\right]dx'\,dy'. \tag{6.61}$$

6.9. COMPARISON WITH THE ANGULAR SPECTRUM REPRESENTATION

6.9.1. Fresnel Approximation

We have seen in Section 2.11 that the representation of an optical beam in

$z > 0$ by its angular spectrum $a(p, q)$ is:

$$U(\mathbf{r}) = \exp(ikz) \iint_{-\infty}^{\infty} a(p, q) \exp\left[i\frac{k}{2}(p^2 + q^2)z\right] \exp[ik(px + qy)]\, dp\, dq, \tag{6.62}$$

where the angular spectrum $a(p, q)$ is given in terms of the boundary value $U(x, y, 0) = f(x, y)$ at $z = 0$ as (cf. Eq. (2.16)):

$$a(p, q) = \left(\frac{k}{2\pi}\right)^2 \iint_{-\infty}^{\infty} f(x, y) \exp[-ik(pz + qy)]\, dx\, dy. \tag{6.63}$$

Next, we shall outline the proof of the equivalence between the Fresnel approximation for the first Rayleigh–Sommerfeld integral and the expression (6.62) for a beam. (Original details may be found in Chapter 5 of the book by Mandel and Wolf [6.27].)

By introducing Eq. (6.63) into (6.62) we obtain:

$$U(\mathbf{r}) = \exp(ikz)\left(\frac{k}{2\pi}\right)^2 \iint_{-\infty}^{\infty} dx'\, dy' f(x', y')$$
$$\times \iint_{-\infty}^{\infty} dp\, dq \exp\left[i\frac{k}{2}z(p^2 + q^2)\right] \exp(ik[p(x - x') + q(y - y')]). \tag{6.64}$$

The second integral in Eq. (6.64) is the two-dimensional Fourier transform of $\exp[i(kz/2)(p^2 + q^2)]$, evaluated at $x - x'$, $y - y'$; namely:

$$\left(\frac{k}{2\pi}\right)^2 \iint_{-\infty}^{\infty} dp\, dq \exp\left[i\frac{k}{2}z(p^2 + q^2)\right] \exp(ik[p(x - x') + q(y - y')])$$
$$= -\frac{ik}{2\pi z} \exp\left[i\frac{k}{2z}[(x - x')^2 + (y - y')^2]\right]. \tag{6.65}$$

Therefore, $U(\mathbf{r})$ becomes:

$$U(\mathbf{r}) = -\frac{ik}{2\pi}\frac{\exp(ikz)}{z} \iint_{-\infty}^{\infty} f(x', y') \exp\left[i\frac{k}{2z}[(x - x')^2 + (y - y')^2]\right] dx'\, dy'. \tag{6.66}$$

On comparing Eqs. (6.66) and (6.55), and recalling that $U(x, y, 0) = f(x, y)$, we see that both expressions are identical. Therefore, the Fresnel approximation is applicable to beam propagation, and, as such, characterizes fields diffracted under low angles, namely, in the *paraxial* regime.

6.9.2. Fraunhofer Approximation

Let us address the Fraunhofer approximation to the first Rayleigh–Sommerfeld integral, giving the diffracted field in terms of its boundary value at the screen; viz., from Eq. (6.58):

$$U(\mathbf{r}) = -\frac{ik}{2\pi}\cos\theta\,\frac{\exp(ikr)}{r}\int_{\mathscr{A}}[U(\mathbf{r}')]_{z=0}\exp\left(-i\frac{k}{r}\mathbf{r}\cdot\mathbf{r}'\right)dS. \quad (6.67)$$

Comparing the integral in Eq. (6.67) with the expression (6.63) for the angular spectrum of $U(\mathbf{r})$, it is evident that:

$$\int_{\mathscr{A}}[U(\mathbf{r}')]_{z=0}\exp\left(-i\frac{k}{r}\mathbf{r}\cdot\mathbf{r}'\right)dS = \left(\frac{2\pi}{k}\right)^2 a(n_x, n_y); \quad (6.68)$$

where $n_x = x/r$, $n_y = y/r$. Thus, from Eqs. (6.67) and (6.68), one derives that the Fraunhofer approximation for the diffracted field may be written as:

$$U(\mathbf{r}) \sim -\frac{2\pi i}{k}\cos\theta a(n_x, n_y)\frac{\exp(ikr)}{r}. \quad (6.69)$$

Equation (6.69) coincides with the asymptotic expression (2.36a) for the diffracted field in terms of its angular spectrum $a(n_x, n_y)$ in a fixed direction as $kr \to \infty$.

6.10. EXAMPLE: SCALAR THEORY OF DIFFRACTION BY A CIRCULAR APERTURE

Let us consider a plane wave of unit amplitude, propagating along the direction characterized by the unit vector $\mathbf{n}_0$:

$$U^{(i)}(\mathbf{r}) = \exp(ik\mathbf{n}_0\cdot\mathbf{r}); \quad (6.70)$$

incident on a circular aperture $\mathscr{A}$ of radius a on a screen S_C in the plane $z = 0$.

We shall calculate the field diffracted by this aperture, according to the integral in Eq. (6.58) in the Kirchhoff and Fraunhofer approximations. This

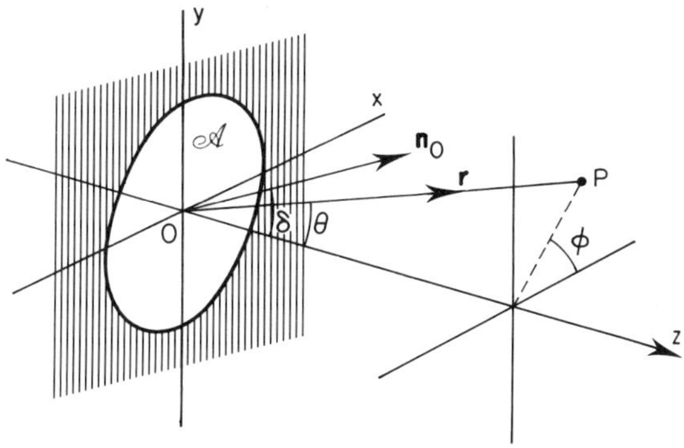

Figure 6.7. Illustration of the geometry of diffraction. The polar and azimuthal angles, θ and ϕ, characterize the unit vector $\hat{\mathbf{r}}$ of the point of observation P.

amounts to replacing the boundary values of U and its normal derivative inside the aperture by those corresponding to the incident wave, being both zero outside.

Let us choose coordinates as indicated in Fig. 6.7. The unit vector $\mathbf{n}_0$ has components:

$$\mathbf{n}_0 = (\sin \delta, 0, \cos \delta). \tag{6.71}$$

The angle $\theta = (\mathbf{n}, \mathbf{r})$ is equal to the angle made by the unit vector $\hat{\mathbf{r}}$ along the direction of the point of observation P with the z-axis in the region $z > 0$. The unit vector $\hat{\mathbf{r}}$ has components:

$$\hat{\mathbf{r}} = (\sin \theta \cos \phi, \sin \theta \sin \phi, \cos \theta). \tag{6.72}$$

Hence, Eq. (6.58) becomes:

$$U(P) = \frac{1}{4\pi} \frac{\exp(ikr)}{r} \int_{\mathscr{A}} dx'\, dy' \left[-\frac{\partial}{\partial z'} [\exp(ik\mathbf{n}_0 \cdot \mathbf{r}') - ik \cos \theta \exp(ik\mathbf{n}_0 \cdot \mathbf{r}')] \times \exp(-ik\hat{\mathbf{r}} \cdot \mathbf{r}') \right]. \tag{6.73}$$

The integration in Eq. (6.73) is extended over the circular aperture $\mathscr{A}$. Taking into account that:

$$\frac{\partial}{\partial z'} [\exp(ik\mathbf{n}_0 \cdot \mathbf{r}')] = ik \cos \delta \exp(ik\mathbf{n}_0 \cdot \mathbf{r}'), \tag{6.74}$$

Eq. (6.73) becomes, after changing the variables of integration x' and y' into polar coordinates: $r' = \sqrt{x'^2 + y'^2}$ and $\beta = \text{angle}(\mathbf{r}', \mathbf{v})$, $\mathbf{v} = (\hat{\mathbf{r}} - \mathbf{n}_0) \times \hat{\mathbf{z}}$:

$$U(P) = -\frac{ik}{4\pi}\frac{\exp(ikr)}{r}(\cos\delta + \cos\theta)\int_0^a dr'r'\int_0^{2\pi} d\beta \exp(-ikvr'\cos\beta); \tag{6.75}$$

where we have defined:

$$v = |(\hat{\mathbf{r}} - \mathbf{n}_0) \times \hat{\mathbf{z}}|, \tag{6.76}$$

$\hat{\mathbf{z}}$ being the unit vector in the z-direction. Performing the β-integration, we obtain:

$$U(P) = -ik\frac{\exp(ikr)}{r}\frac{(\cos\delta + \cos\theta)}{2}\int_0^a dr'r'J_0(kvr'). \tag{6.77}$$

Recalling that the Bessel functions satisfy:

$$\int_0^a xJ_0(x)\,dx = aJ_1(a), \tag{6.78}$$

we finally obtain:

$$U(P) = -ika^2\frac{\exp(ikr)}{r}\frac{(\cos\delta + \cos\theta)}{2}\frac{J_1(kav)}{kav}. \tag{6.79}$$

In the special case in which the incident wave propagates along the z-axis, the value of $U^{(i)}$ is unity at any point of the aperture $\mathscr{A}$, $\delta = 0$, and v becomes: $v = \sin\theta$. Equation (6.79) then reduces to the familiar expression previously given in several textbooks (see, e.g. Eq. (4.17) of Ref. 6.3).

The distribution of diffracted intensity, associated with the wavefield given by Eq. (6.79) is:

$$I(P) = \frac{k^2a^4}{r^2}\left(\frac{\cos\delta + \cos\theta}{2}\right)^2\left[\frac{J_1(kav)}{kav}\right]^2. \tag{6.80}$$

The factor $[J_1(kav)/(kav)]^2$ has the first zero at $kav = 3.83$, namely, at $v = 0.61\lambda/a$. Therefore, the radius of the central peak of the diffraction pattern represented by Eq. (6.80) is: $v = 1.22\lambda/a$, and, thus, it is inversely proportional to the radius a of the aperture. When $ka \gg 1$, this peak is very high and narrow: the diffracted field is mainly concentrated very closely about the z-direction of incidence. This is the regime in which the Kirchhoff approximation is exact. Note that in this case for an incident plane wave of unit amplitude, propagating in the z-direction, the angular factor $(\cos\delta + \cos\theta)/2 = (1 + \cos\theta)/2$ becomes practically equal to unity.

If we had calculated the diffracted intensity starting from the first Rayleigh–Sommerfeld integral, Eq. (6.60), we would have obtained, instead of Eq. (6.80), the following result:

$$I_I(P) = \frac{k^2 a^4}{r^2}\cos^2\theta\left[\frac{J_1(kav)}{kav}\right]^2; \tag{6.81}$$

which differs from Eq. (6.80) by the angular factor $\cos^2\theta$. However, both expressions coincide when $ka \gg 1$, and hence, $\cos^2\theta \sim 1$.

When $ka \leqslant 1$, neither the Kirchhoff approximation nor the scalar theory of diffraction constitutes an adequate description of diffraction. Therefore, within the domain in which these two approximations are appropriate, it is not relevant whether one uses the Kirchhoff diffraction formula or either Rayleigh–Sommerfeld integral.

6.11. THEORY OF THE BOUNDARY DIFFRACTION WAVE

6.11.1. The Vector Potential

When the Kirchhoff approximation is used, the integral of Helmholtz and Kirchhoff represents the diffracted field as a superposition of spherical waves, each being originated at a point source in the screen and inside the aperture. Moreover, all these waves interfere coherently to produce the diffracted wavefield. This interpretation agrees with Huygens' principle and matches with Fresnel's ideas.

An alternative interpretation of diffraction was posed by Young in 1802. He considered this phenomenon as due to the interference between the light passing directly through the aperture following the laws of geometrical optics, and the light originated by scattering of the incident wave at the rim of the aperture. However, this point of view did not find success because of its rather imprecise formulation. Nevertheless, later, Maggi [6.28], Sommerfeld [6.29], and Rubinowicz [6.30] (see also Ref. 6.31), showed the basic correctness of Young's ideas. These studies were improved and further developed by Miyamoto and Wolf [6.32] first, and later by Marchand and Wolf [6.15], [6.33]. In this section an outline of these formulations [6.32], [6.33] is presented.

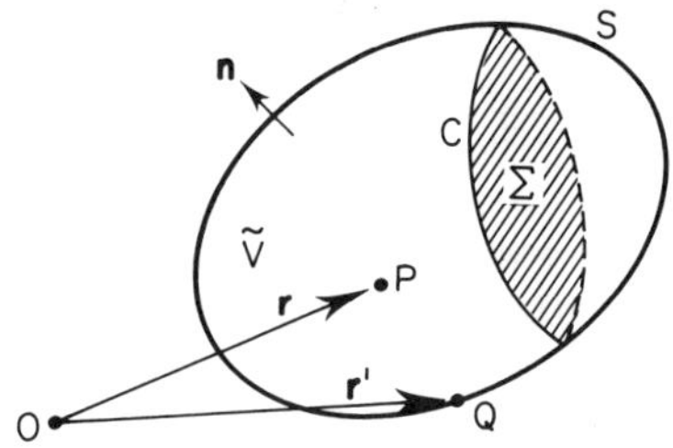

Figure 6.8. Illustration of the notation.

Let us address the theorem of Helmholtz and Kirchhoff, Eq. (6.2), which gives the values of a scalar wavefield $U(P)$ at a point P of position vector $\mathbf{r}$ inside the volume $\tilde{V}$ enclosed by a surface S, in terms of the boundary values of U and $\partial U/\partial \mathbf{n}$ (Fig. 6.8). By taking the divergence of the integrand of Eq. (6.2) with respect to $\mathbf{r}'$, one has, after using the homogeneous Helmholtz equation satisfied by both U and G, the following expression:

$$\nabla' \cdot \mathbf{V}(\mathbf{r}, \mathbf{r}') = 0, \tag{6.82}$$

where:

$$\mathbf{V}(\mathbf{r}, \mathbf{r}') = \frac{1}{4\pi} [G(\mathbf{r}, \mathbf{r}')\nabla' U(\mathbf{r}') - U(\mathbf{r}')\nabla' G(\mathbf{r}, \mathbf{r}')]. \tag{6.83}$$

Equation (6.82) implies that $\mathbf{V}$ can be expressed in terms of a *vector potential* $\mathbf{W}$ (which was originally introduced in the works by Miyamoto and Wolf [6.32] and Marchand and Wolf [6.33]):

$$\mathbf{V}(\mathbf{r}, \mathbf{r}') = \nabla' \times \mathbf{W}(\mathbf{r}, \mathbf{r}') \tag{6.84}$$

(apart, of course, from the gradient of some arbitrary scalar function that can be added to $\mathbf{W}$).

By taking Eqs. (6.83) and (6.84) into account, the Helmholtz–Kirchhoff integral in Eq. (6.2) becomes:

$$U(\mathbf{r}) = \int_S \nabla' \times \mathbf{W}(\mathbf{r}, \mathbf{r}')\, dS. \tag{6.85}$$

The vector potential $\mathbf{W}$, considered as a function of $\mathbf{r}'$, must have singularities on S, and in particular, when $\mathbf{r} = \mathbf{r}'$. Otherwise, by applying Stokes' theorem to each of the two surfaces and their common boundary section C, obtained by cutting the volume $\tilde{V}$ by a plane that intersects it through the section Σ, one would conclude that $U(\mathbf{r}) = 0$ for any point $\mathbf{r}$ inside $\tilde{V}$ (see Fig. 6.8). Therefore, by assuming that these singularities occur at discrete points of S, Q_j, $(j = 1, \ldots, N)$, and surrounding these points by small circumferences Γ_j of radii σ_j that will eventually be made to tend to zero, one obtains, by using Stokes' theorem in Eq. (6.85):

$$U(\mathbf{r}) = \sum_{j=1}^{N} \lim_{\sigma_j \to 0} \int_{\Gamma_j} \mathbf{W} \cdot \mathbf{l} \, dl, \tag{6.86}$$

where $\mathbf{l}$ is a unit vector along the tangent to Γ_j and dl is an arc element of Γ_j.

Hence, according to Eq. (6.86), the diffracted wavefield $U(\mathbf{r})$ may be expressed by the superposition of line integrals arising from singularity points of the vector potential $\mathbf{W}$.

Let us now specialize more the above case. Let us consider the diffraction of an incident wavefield $U^{(i)}(\mathbf{r})$ by an aperture $\mathscr{A}$ on S (Fig. 6.9). We assume

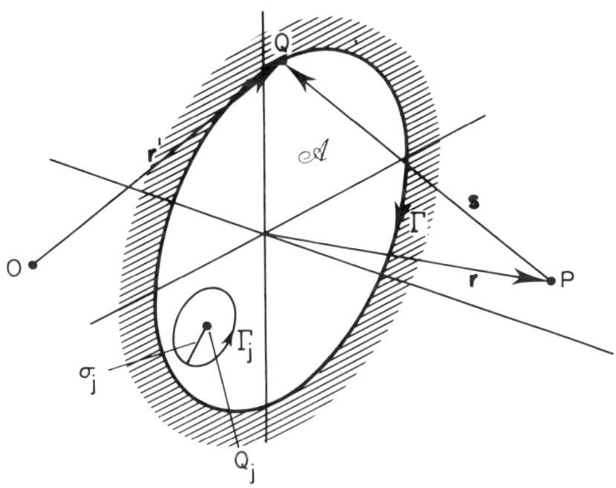

Figure 6.9. Illustration of the notation.

that the linear dimensions of this aperture are large enough, compared with the wavelength, to permit the use of the Kirchhoff approximation. Therefore, in all the analysis that follows, we shall assume that the field U and its normal derivative $\partial U/\partial \mathbf{n}$ are equal to the corresponding values of the incident field inside the aperture $\mathscr{A}$, and they are zero otherwise in the plane of the aperture, then the diffracted field is given by Eq. (6.10).

By an argument identical to the one used above it can be easily shown (the interested reader can consult Refs. 6.32 and 6.33 for details) that $U(\mathbf{r})$ can be expressed in terms of the vector potential $\mathbf{W}^{(i)}$ in the aperture, associated with the incident field, as:

$$U(\mathbf{r}) = \int_{\mathscr{A}} \nabla' \times \mathbf{W}^{(i)}(\mathbf{r}, \mathbf{r}')\, dS. \tag{6.87}$$

Let us surround the singularities of $\mathbf{W}^{(i)}$ in the aperture $\mathscr{A}$ by small circumferences Γ_i of radii σ_i, $(i = 1, \ldots, M)$, Γ being the boundary of the aperture $\mathscr{A}$. We apply Stokes' theorem to the area of the aperture excluding the small circles Γ_i, then, in the limit when the radii σ_i tend to zero we obtain:

$$U(\mathbf{r}) = U^{(B)}(\mathbf{r}) + \sum_{i=1}^{M} F_i(\mathbf{r}), \tag{6.88}$$

where:

$$U^{(B)}(\mathbf{r}) = \int_{\Gamma} \mathbf{W}^{(i)}(\mathbf{r}, \mathbf{r}')\cdot\mathbf{l}\, dl, \tag{6.89a}$$

$$F_i(\mathbf{r}) = \lim_{\sigma_i \to 0} \int_{\Gamma_i} \mathbf{W}^{(i)}(\mathbf{r}, \mathbf{r}')\cdot\mathbf{l}\, dl. \tag{6.89b}$$

The integral along the border Γ of the aperture is taken clockwise, whereas those along the circumferences Γ_i are considered anticlockwise when viewed from the observation point P. $U^{(B)}$ is the *boundary wave* due to scattering in the rim of the aperture.

Examples of this formalism to the diffraction of a plane and a spherical wave have been given by Miyamoto and Wolf [6.32], and by Marchand and Wolf [6.33]. When Eqs. (6.88) and (6.89) are applied to these cases, the contribution from the singularities in the aperture produces the incident wave according to geometrical optics, namely, inside the domain of the aperture, and zero otherwise. Also, an expression for the boundary wave $U^{(B)}$ originating at the rim of the aperture, is obtained, so that Eq. (6.88) becomes equal to the *Maggi–Rubinowicz representation* of the Kirchhoff diffraction integral in Eq. (6.10), and underlines the basic ideas of Young (see also Ref. 6.8 and Section 8.9 of Ref. 6.2 for a detailed discussion on the Maggi–Rubinowicz formulation).

6.11.2. A Mathematically Consistent Interpretation of the Kirchhoff Diffraction Integral

The aforementioned inconsistency of the Kirchhoff integral, Eq. (6.10), is only apparent. It was pointed out by the authors of Ref. 6.34 that it arises from the attempt to recover boundary conditions that are mutually contradictory when the observation point approaches the plane of the aperture. As a matter of fact, the Kirchhoff solution is so accurate in all cases dealing with sufficiently large apertures, that somehow it should constitute a true solution. Marchand and Wolf [6.15] posed the question in a different way, namely: ignore the justification of (6.10) on the grounds of Green's theorem, and ask instead: what problem does Eq. (6.10) actually solve? In other words: *to what boundary value does the Kirchhoff integral represent an exact solution?*

The answer to that question was given in Ref. 6.15 by means of the theory of the boundary wave. According to Marchand and Wolf [6.15], by taking the limit of the field $U(\mathbf{r})$ given by the Maggi–Rubinowicz expression, obtained from Eqs. (6.88) and (6.89), when the point of observation $\mathbf{r}$ tends to a point in the plane of the aperture, one retrieves the boundary value of $U(\mathbf{r})$ in that plane, which, associated with the diffracted field (Eq. (6.10)) constitutes a mathematically consistent solution to the homogeneous Helmholtz equation in the half-space $z > 0$. (Of course, $U(\mathbf{r})$ also satisfies Sommerfeld's radiation condition, Eq. (6.4).)

In spite of this consistency, obtained by taking the limit of the field as indicated above, the uniqueness of the general solution to this boundary problem is still an open question. A uniqueness theorem is proved by the authors of Ref. 6.15 in the particular case of a plane wave incident on a circular aperture.

6.12. COMPARISON WITH EXPERIMENT

Considering all the studies that the justification of Kirchhoff's integral has produced, it is clear that, after all, the best assessment about its validity is a direct comparison with the experiment. We have remarked in Sections 6.6 and 6.10 that this approximation gives exactly the same result as the "consistent" Rayleigh–Sommerfeld integrals for apertures whose linear dimensions are large compared with the wavelength, and at observation points not too close to the aperture.

Marchand and Wolf [6.15] carried out a test using the experimental results obtained by Ehrlich *et al.* [6.34] on diffraction of microwaves of wavelength $\lambda = 3.2$ cm by circular apertures in conducting screens. The

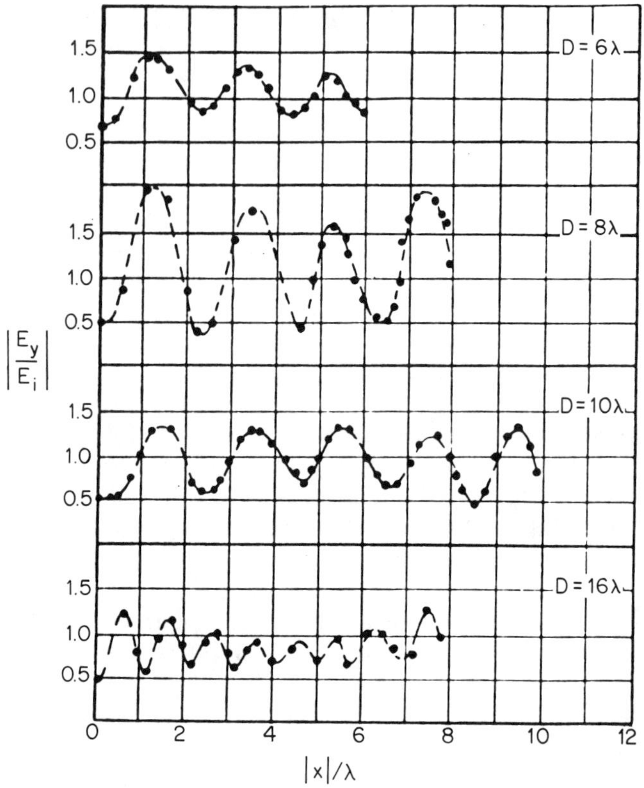

Figure 6.10a. Measured principal component of the electric field in the plane of a circular aperture of diameter D in a conducting screen, as a function of the radial distance x (measured in wavelengths). The incident field is a plane wave of unit amplitude at normal incidence upon the aperture. (After M. J. Ehrlich, S. Silver, and G. Held, *J. Appl. Phys.* **26**, 336 (1955).)

comparison with the experiment was made both with the Kirchhoff integral of Eq. (6.10) and with the Rayleigh–Sommerfeld formulas with boundary conditions given by the Kirchhoff approximation:

$$U_I(\mathbf{r}) = -\frac{1}{2\pi}\int_{\mathscr{A}} U^{(i)}(\mathbf{r}')\frac{\partial G(\mathbf{r}, \mathbf{r}')}{\partial \mathbf{n}}\, dS, \tag{6.90a}$$

$$U_{II}(\mathbf{r}) = \frac{1}{2\pi}\int_{\mathscr{A}} \frac{\partial U^{(i)}(\mathbf{r}')}{\partial \mathbf{n}} G(\mathbf{r}, \mathbf{r}')\, dS. \tag{6.90b}$$

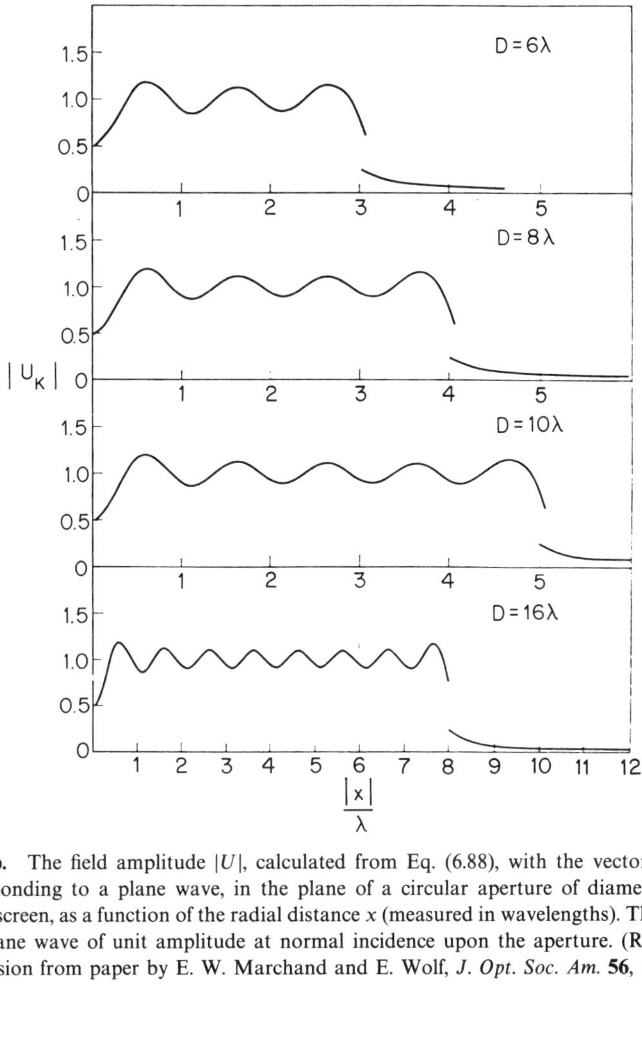

Figure 6.10b. The field amplitude $|U|$, calculated from Eq. (6.88), with the vector potential $\mathbf{W}^{(i)}$ corresponding to a plane wave, in the plane of a circular aperture of diameter D in a conducting screen, as a function of the radial distance x (measured in wavelengths). The incident field is a plane wave of unit amplitude at normal incidence upon the aperture. (Reproduced with permission from paper by E. W. Marchand and E. Wolf, *J. Opt. Soc. Am.* **56**, 1716–1717 (1966).)

Figure 6.10a shows the experimental values of the magnitude of the electric field in the plane of the aperture [6.34]. The corresponding curves obtained from Eq. (6.88) (circular aperture in a black screen), with $U^{(i)}$ being a plane wave, is shown in Fig. 6.10 (Ref. 6.15). The values obtained from Eqs. (6.90a) and (6.90b) are shown in Fig. 6.11. From these figures one concludes that Eq. (6.88) agrees much better with the experiment near the center of the aperture than does U_{II}. On the other hand, U_I gives incorrect values in all cases and U_{II} gives the best results away from the center.

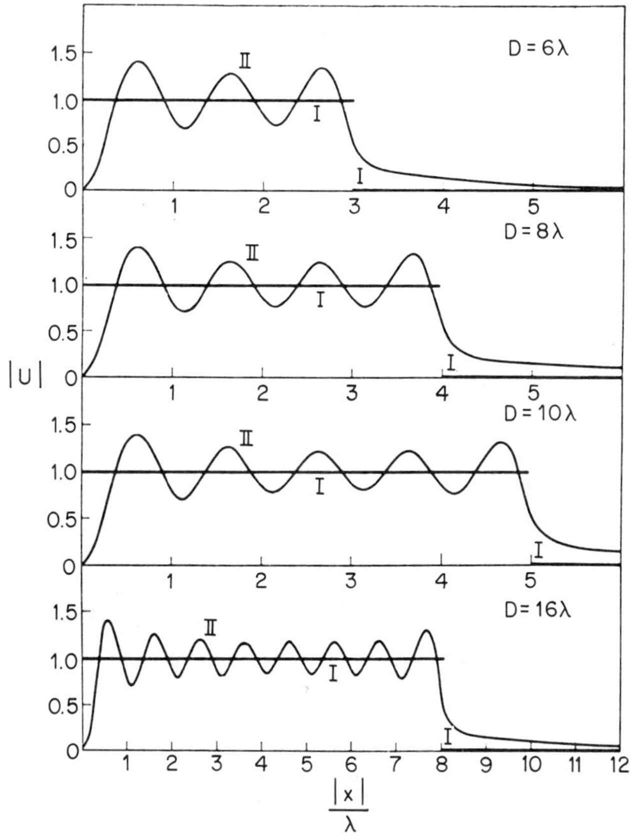

Figure 6.11. Same as Fig. 6.10b for the field amplitudes $|U_I|$ (denoted by I) and U_{II} (denoted by II), from the Rayleigh–Sommerfeld integrals, Eqs. (6.90a) and (6.90b). (Reproduced with permission from paper by E. W. Marchand and E. Wolf, *J. Opt. Soc. Am.* **56**, 1716–1717 (1966).)

It should be mentioned that similar comparisons with the same experimental data, but with different calculation procedures for the theoretical curves, have been done by Stamnes (see Section 4.4 of Ref. 6.35).

The comparison of the Kirchhoff integral, Eqs. (6.10) and (6.88), with the experiment at points situated in the axial direction, and near the aperture, is shown in Figs. 6.12a and 6.12b.

We conclude that, although there are qualitative similarities (mainly concerning the oscillation shown in the results), there are still significant

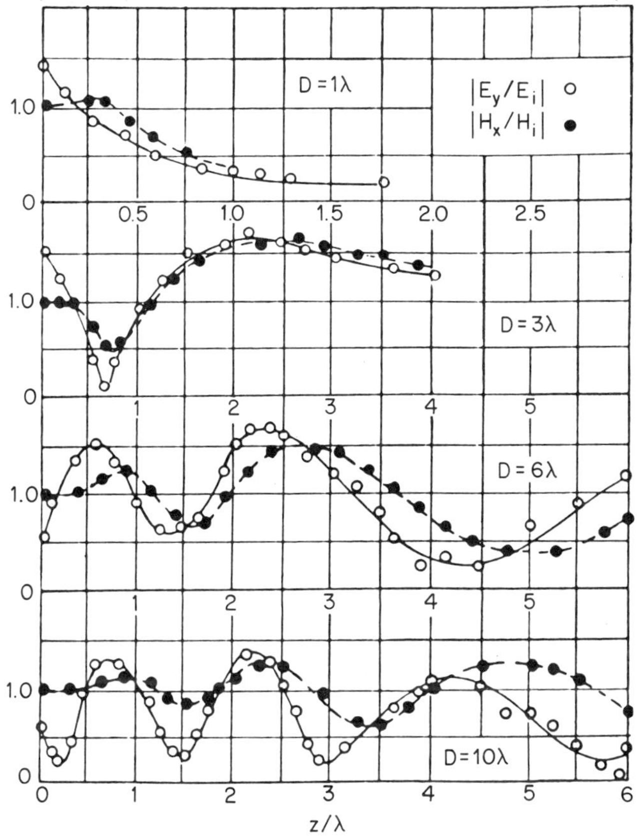

Figure 6.12a. Measured axial distribution of the principal tangential component of the electric and magnetic fields **E** and **H** for a circular aperture of diameter D, as a function of the distance z (measured in wavelengths) from the aperture. The incident field is a plane wave of unit amplitude at normal incidence upon the aperture. (After M. J. Ehrlich, S. Silver, and G. Held, *J. Appl. Phys.* **26**, 336 (1955).)

quantitative differences between the theory and the experiment. We must remember, however, that the theory assumes the screen to be black, rather than conducting as the experiment does. Also very important is the fact that the size of the aperture is *not* much larger than the wavelength, and the measurements are not made far from the aperture and at low angles of diffraction; thus the Kirchhoff approximation and the scalar theory should not be expected to be very accurate.

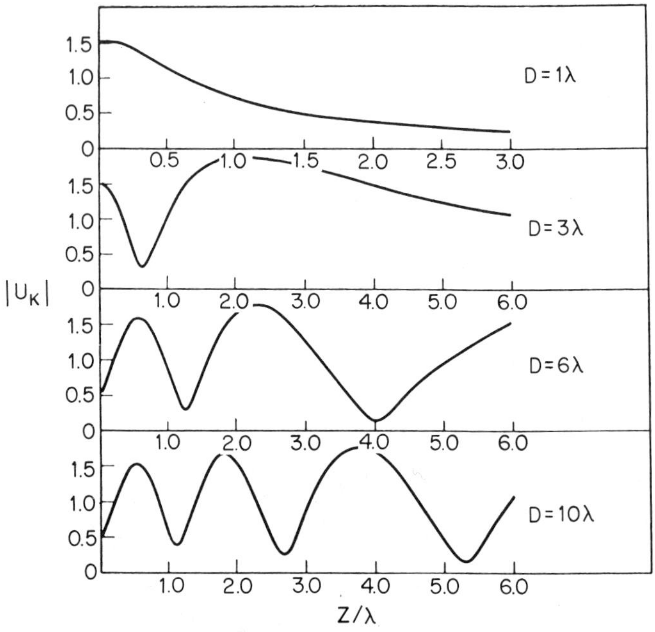

Figure 6.12b. The axial distribution of the field amplitude $|U|$ calculated from Eq. (6.88) (with the vector potential corresponding to an incident plane wave), for a circular aperture of diameter D, as a function of the distance z (measured in wavelengths) from the aperture. (Reproduced with permission from paper by E. W. Marchand and E. Wolf, *J. Opt. Soc. Am.* **56**, 1716–1717 (1966).)

Other comparisons in terms of the boundary wave, between Eqs. (6.88), (6.90a) and (6.90b), for a plane wave normally incident on a circular aperture, evaluated both at the plane of the aperture and in the far zone (Fraunhofer approximation), have been made in Ref. 6.33(b). There it is shown that the difference between the Kirchhoff integral and the Rayleigh–Sommerfeld formulas is given by the boundary wave (see also Problem 6.9). However, as was discussed in Section 6.10, the three formulas predict the same result for large apertures compared with the wavelength, and at low angles of diffraction.

An extension of the boundary wave theory to apertures whose linear dimensions are not large compared with the wavelength is done in Ref. 6.33(c) in terms of multiple scattering of the incident wave on the boundary of the aperture.

6.13. DEBYE APPROXIMATION. SYMMETRIES OF FOCUSED WAVEFIELDS AND PHASE ANOMALY NEAR THE FOCUS

Let us consider the angular spectrum representation for a wavefield satisfying the homogeneous Helmholtz equation in the half-space $z > -f$, and behaving asymptotically as a divergent spherical wave in $z > 0$ (cf. Section 2.2):

$$U(\mathbf{r}) = \iint_{-\infty}^{\infty} a(p, q) \exp[ik(px + qy + mz)]\, dp\, dq. \tag{6.91}$$

The *Debye approximation*, used in problems on focused fields (Debye [6.36]; see also Ref. 6.2), consists of removing all plane wave components of Eq. (6.91) that propagate with directional cosines (p, q, m) larger than those within a solid angle Ω.

Thus, Eq. (6.91) is written in the Debye approximation as:

$$U(\mathbf{r}) = \iint_{\Omega} a(p, q) \exp[ik(px + qy + mz)]\, dp\, dq, \tag{6.92}$$

where:

$$m = \sqrt{1 - p^2 - q^2}, \qquad p^2 + q^2 < \sin^2 \Theta_m. \tag{6.93}$$

The polar angle Θ_m characterizes the maximum divergence of the bundle of plane wave components of $U(\mathbf{r})$ away from the z-axis, contained in Ω.

The Debye integral, Eq. (6.92), is used for expressing a wavefield $U(\mathbf{r})$, converging towards a focal point at the origin $z = 0$, from an aperture in the plane $z = -f$ (see Fig. 6.13).

The integral in Eq. (6.92) represents the field $U(\mathbf{r})$ in the half-space $z \geqslant -f$ behind the aperture. The angle Θ_m defines the semiangular aperture of the system, and hence, all plane wave components of $U(\mathbf{r})$ propagate with wavevectors included in the cone whose apex is at the focal point $z = 0$, and whose boundary consists of straight lines from $z = 0$ through the aperture edge, forming a solid angle Ω. In the problem originally treated by Debye, $a(p, q)$ was taken as:

$$a(p, q) = -\frac{ik}{2\pi}\frac{1}{m}, \tag{6.94}$$

so that $U(\mathbf{r})$ is a bandlimited representation for a spherical wave, convergent towards $z = 0$, for $z < 0$, and divergent from $z = 0$, for $z > 0$ (see Eqs. (2.36) in Chapter 2). However, it is well known that the angular spectrum representation of a spherical wave by homogeneous components only

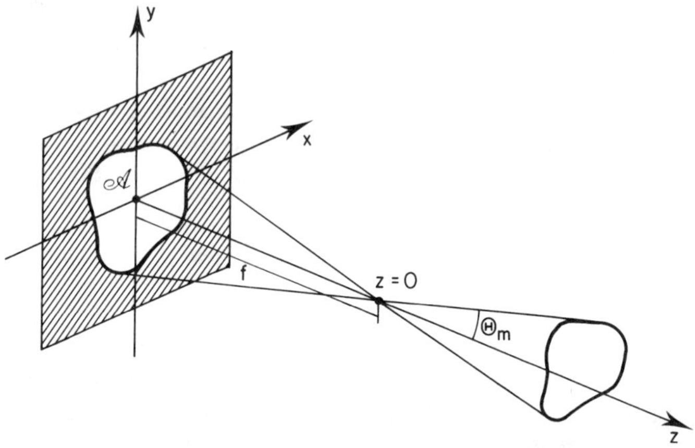

Figure 6.13. Geometry for a focused wavefield.

encounters difficulties (cf. Eqs. (2.88)–(2.90) of Example 2.2), and this poses a problem with the Debye approximation on the axis of a focusing system. An analysis of this anomalous behavior of the Debye integral (Eq. (6.92)) near the axis has been done by Sherman and Chew [6.37] by extending the asymptotic expression of wavefields, as discussed in Section 2.12, to finite aperture fields. The problems related to the study of focused fields have been extensively analyzed in works by Li [6.38], [6.39], Li and Wolf [6.40]–[6.42], and Stamnes (see his detailed study in [6.35], and references cited therein).

It is worth noting, as has been done in Ref. 6.35, the reciprocal analogy between the Kirchhoff approximation and the Debye approximation, as regards the wavefield incident on the aperture of the focusing system, and its angular spectrum. Whereas in the Kirchhoff approximation the field is made zero outside the aperture, in the Debye approximation the angular spectrum of that field is considered to be zero outside a solid angle Ω. Of course, in the focal region both approximations give different results.

Rather than restricting the analysis to uniform spherical waves in the aperture, one can consider more general wavefields for which the angular spectrum $a(p, q)$ is of the form:

$$a(p, q) = -\frac{ik}{2\pi} A(p, q), \tag{6.95}$$

where $A(p, q)$ is assumed to be a real function of p and q.

Introducing Eq. (6.95) into (6.92), it is easy to prove by taking the complex conjugate in the resulting expression for $U(\mathbf{r})$, that the focused field $U(\mathbf{r})$ satisfies the following symmetry relation:

$$U(-\mathbf{r}) = -U^*(\mathbf{r}). \tag{6.96}$$

By writing $U(\mathbf{r})$ as: $U(\mathbf{r}) = |U(\mathbf{r})| \exp[i\phi(\mathbf{r})]$, Eq. (6.96) implies the following relationships for the intensity and the phase about the focal point $z = 0$:

$$|U(-\mathbf{r})|^2 = |U(\mathbf{r})|^2, \tag{6.97a}$$

$$\phi(-\mathbf{r}) = -\phi(\mathbf{r}) - \pi \qquad (\text{mod } 2\pi). \tag{6.97b}$$

Equations (6.96) and (6.97) were originally obtained by Collett and Wolf [6.43]. Observe that these symmetry relations can be equally obtained from the asymptotic expressions of $U(\mathbf{r})$ and $U(-\mathbf{r})$ (cf. Eqs. (2.36) of Section 2.8), by taking Eq. (6.95) into account. Therefore, Eqs. (6.97) hold both near the focus and throughout the space, except close to the aperture at $z = -f$, providing that the aperture is so far from the focus that it may be considered in the far zone, namely, $kf \gg 1$. The additive term $-\pi$ in Eq. (6.97b) represents the *phase anomaly* near the focus (a detailed study on this phase anomaly is done in Section 8.8.4 of Ref. 6.2).

6.14. VECTOR THEORY OF DIFFRACTION

The electromagnetic formulation of diffraction seeks the solution of the homogeneous vector equations for the electric and magnetic vector of the light field in a source-free volume $\tilde{V}$, bounded by a surface S:

$$\nabla \times \nabla \times \mathbf{E} - k^2\mathbf{E} = 0, \tag{6.98a}$$

$$\nabla \times \nabla \times \mathbf{H} - k^2\mathbf{H} = 0, \tag{6.98b}$$

subject to boundary conditions on the limiting surface S, and to the vector form of Sommerfeld's radiation condition (cf. Problem 3.2).

Once again, the geometry can be either that of Fig. 6.1(a), or that of Fig. 6.1(b). The solution for the electric and magnetic vectors in a volume $\tilde{V}$, given the boundary values on its limiting surface S in the geometry of Fig. 6.1(a) is given by Eqs. (1.34) and (1.39), and by the corresponding equation for the magnetic field, with the source terms $\mathbf{F}_e$ and $\mathbf{F}_h$, Eqs. (1.10), involving the polarization $\mathbf{P}$, magnetization $\mathbf{M}$ and electric currents, equal to zero. On the other hand, the solution associated with the geometry of Fig. 6.1(b) is given by Eqs. (1.36) and (1.38). Therefore, the electric vector of the diffracted field is, according to Eqs. (1.34) and (1.39), given by:

$$\mathbf{E}(\mathbf{r}) = -\frac{1}{4\pi}\int_{S^-} [(\mathbf{n} \times \nabla \times \mathbf{E}(\mathbf{r}'))\cdot\mathscr{G}(\mathbf{r}, \mathbf{r}') + (\mathbf{n} \times \mathbf{E}(\mathbf{r}'))\cdot(\nabla \times \mathscr{G}(\mathbf{r}, \mathbf{r}'))]\, dS. \tag{6.99}$$

Since $\tilde{V}$ is nonmagnetic in this case, we have $\mathbf{M} = 0$, then, by using the second Maxwell's equation (1.1b), Eq. (6.99) may be written as:

$$\mathbf{E}(\mathbf{r}) = -\frac{1}{4\pi}\int_{S^-} [ik(\mathbf{n} \times \mathbf{H}(\mathbf{r}'))\cdot\mathscr{G}(\mathbf{r}, \mathbf{r}') + (\mathbf{n} \times \mathbf{E}(\mathbf{r}'))\cdot(\nabla \times \mathscr{G}(\mathbf{r}, \mathbf{r}'))]\, dS. \tag{6.100}$$

With the corresponding equation for the magnetic vector:

$$\mathbf{H}(\mathbf{r}) = -\frac{1}{4\pi}\int_{S^-} [(\mathbf{n} \times \mathbf{H}(\mathbf{r}'))\cdot(\nabla \times \mathscr{G}(\mathbf{r}, \mathbf{r}')) - ik(\mathbf{n} \times \mathbf{E}(\mathbf{r}'))\cdot\mathscr{G}(\mathbf{r}, \mathbf{r}')]\, dS. \tag{6.101}$$

Equations (6.100) and (6.101) represent the vector formulation of the theorem of Helmholtz and Kirchhoff.

Alternative forms of Eqs. (6.100) and (6.101), often used, are the so-called Stratton–Chu integrals [6.44] (see Problem 1.3):

$$\mathbf{E}(\mathbf{r}) = -\frac{1}{4\pi}\int_{S^-} [ik(\mathbf{n} \times \mathbf{H}(\mathbf{r}'))G(\mathbf{r}, \mathbf{r}') + (\mathbf{n} \times \mathbf{E}(\mathbf{r}')) \times \nabla G(\mathbf{r}, \mathbf{r}') + (\mathbf{n}\cdot\mathbf{E}(\mathbf{r}'))\nabla G(\mathbf{r}, \mathbf{r}')]\, dS, \tag{6.102}$$

and:

$$\mathbf{H}(\mathbf{r}) = \frac{1}{4\pi}\int_{S^-} [ik(\mathbf{n} \times \mathbf{E}(\mathbf{r}'))G(\mathbf{r}, \mathbf{r}') - (\mathbf{n} \times \mathbf{H}(\mathbf{r}')) \times \nabla G(\mathbf{r}, \mathbf{r}') - (\mathbf{n}\cdot\mathbf{H}(\mathbf{r}'))\nabla G(\mathbf{r}, \mathbf{r}')]\, dS. \tag{6.103}$$

6.15. UNIQUENESS OF THE ELECTROMAGNETIC SOLUTION. BOUNDARY CONDITIONS

Equations (6.100) and (6.101) require knowledge of the boundary values of the tangential components of the electric and the magnetic vectors. However, we shall see that, in fact, these data overspecify the problem.

Let us prove that the boundary value of only the tangential component of the electric vector is enough to determine the diffracted field uniquely. Similarly, with the tangential component of the magnetic vector.

If the solution to Eq. (6.98a) were not unique, there would exist two different solutions $\mathbf{E}_1$ and $\mathbf{E}_2$, with the same tangential component, namely, the difference: $\mathbf{U} = \mathbf{E}_1 - \mathbf{E}_2$ would satisfy the equation:

$$\nabla \times \nabla \times \mathbf{U} = 0, \tag{6.104}$$

with the boundary condition: $\mathbf{n} \times \mathbf{U} = 0$ on S^-.

Now, by applying the vector form of Green's first identity to $\mathbf{U}$, we obtain:

$$\int_{S^-} [\mathbf{U} \times \nabla \times \mathbf{U}] \cdot \mathbf{n}\, dS = \int_{\tilde{V}} [\mathbf{U} \cdot \nabla \times \nabla \times \mathbf{U} - |\nabla \times \mathbf{U}|^2]\, dv. \tag{6.105}$$

The left-hand side of Eq. (6.105) may be written as:

$$\int_{S^-} [\mathbf{n} \times \mathbf{U}] \cdot \nabla \times \mathbf{U}\, dS = 0. \tag{6.106}$$

In equating to zero the surface integral in Eq. (6.106) we have made use of the fact that: $\mathbf{n} \times \mathbf{U} = 0$ on S^-.

Using Eqs. (6.104) and (6.106), we get from Eq. (6.105):

$$\int_{\tilde{V}} |\nabla \times \mathbf{U}|^2\, dv = 0. \tag{6.107}$$

Equation (6.107) implies that $\nabla \times \mathbf{U} = 0$. Hence, $\mathbf{U}$ is either zero or the gradient of a scalar potential in $\tilde{V}$. As a result, $\mathbf{E}_1$ and $\mathbf{E}_2$ are identical apart from the gradient of a scalar potential. A similar argument could be used to prove the uniqueness of the solution from the boundary value of the tangential component of the magnetic vector $\mathbf{H}$.

In analogy with the scalar case, there is a mathematical "inconsistency" when both boundary conditions are simultaneously used with the Kirchhoff approximation, and one tries to recover them from the diffracted field at the limit in which the point of observation tends to the boundary surface. For an aperture in a screen, the Kirchhoff approximation, analogously to the scalar case, amounts to replacing the tangential component of the field in the aperture by that corresponding to the incident field, being zero in the rest of the screen. Thus, in this approximation $\mathbf{E}$ and $\mathbf{H}$ are substituted in

Eqs. (6.100) and (6.101) (or in their equivalent, Eqs. (6.102) and (6.103)), by the corresponding vectors $\mathbf{E}^{(i)}$ and $\mathbf{H}^{(i)}$ of the incident field; and the S^--integration is restricted to the area occupied by the aperture $\mathscr{A}$. Once again, the geometry of Fig. 6.2 has been used, and the integration over the spherical surface of infinite radius vanishes if the field satisfies Sommerfeld's radiation condition.

A mathematically consistent formulation of the resulting diffraction integral, limited to the aperture $\mathscr{A}$, has been given by Kottler [6.45], who, once again, interpreted the solution associated to a *saltus* problem on a black screen (see also a discussion in Ref. 6.14).

6.16. THREE VECTOR FORMULATIONS OF DIFFRACTION

Based on the discussion in Section 6.15, it is possible to write the diffracted field by choosing a Green function such that the term of the integrand containing the boundary value on the screen of either the tangential component of **E**, or the tangential component of **H**, does not contribute to the integral. This has a resemblance to the choice of the term containing the boundary value of either the wavefield, or its normal derivative, in the scalar theory. This sets the possibility of writing the solution as:

$$\mathbf{E}(\mathbf{r}) = -\frac{ik}{2\pi}\int_{S^-} [\mathbf{n} \times \mathbf{H}(\mathbf{r}')]\cdot\mathscr{G}(\mathbf{r}, \mathbf{r}')\, dS, \tag{6.108a}$$

$$\mathbf{H}(\mathbf{r}) = -\frac{1}{2\pi}\int_{S^-} [\mathbf{n} \times \mathbf{H}(\mathbf{r}')]\cdot\nabla \times \mathscr{G}(\mathbf{r}, \mathbf{r}')\, dS, \tag{6.108b}$$

when the tangential component of **H** is prescribed on the boundary surface; or as:

$$\mathbf{E}(\mathbf{r}) = -\frac{1}{2\pi}\int_{S^-} [\mathbf{n} \times \mathbf{E}(\mathbf{r}')]\cdot\nabla \times \mathscr{G}(\mathbf{r}, \mathbf{r}')\, dS, \tag{6.109a}$$

$$\mathbf{H}(\mathbf{r}) = \frac{ik}{2\pi}\int_{S^-} [\mathbf{n} \times \mathbf{E}(\mathbf{r}')]\cdot\mathscr{G}(\mathbf{r}, \mathbf{r}')\, dS, \tag{6.109b}$$

when the tangential component of **E** is given on the boundary surface.

Equations (6.100) and (6.101), and (6.108) and (6.109), constitute each of three vector formulations of the diffraction problem, established by Kottler, Luneburg, Severin, and Vasseur (the reader can find a summary of these theories in Refs. 6.14 and 6.46). When these expressions are used in conjunction with the Kirchhoff approximation, the tangential components of **E** and **H** are replaced by those corresponding to the incident field, $\mathbf{E}^{(i)}$ and $\mathbf{H}^{(i)}$, and the S^- integration is restricted to the aperture $\mathscr{A}$. With this approximation, those three formulations are known as the *(e, m)-theory*, Eqs. (6.100) and (6.101); the *e-theory*, Eqs. (6.108); and the *m-theory*, Eqs. (6.109). (See also Refs. 6.14 and 6.46, and p. 221 of Ref. 6.47.)

6.17. THE FRAUNHOFER APPROXIMATION FOR THE VECTOR SOLUTION. COMPARISON WITH THE SCALAR RESULT

The Fresnel and Fraunhofer approximations are done in the electromagnetic theory in a way identical to that used in the scalar formulation. By using the Kirchhoff approximation, and the expression in Eq. (1.27) for the dyadic Green function $\mathscr{G}(\mathbf{r}, \mathbf{r}')$, and the expansion (Eq. (6.50a)) for the exponent $|\mathbf{r} - \mathbf{r}'|$ of $G(\mathbf{r}, \mathbf{r}')$, one obtains, for instance from Eq. (6.109a), by a procedure identical to that of Section 6.8, the following expression for the electric vector of the diffracted field in the Fraunhofer approximation:

$$\mathbf{E}(\mathbf{r}) = \frac{1}{2\pi i}\frac{\exp(ikr)}{r} k\hat{\mathbf{r}} \times \int_{\mathscr{A}} \mathbf{n} \times \mathbf{E}^{(i)}(\mathbf{r}') \exp(-ik\hat{\mathbf{r}}\cdot\mathbf{r}')\, dS, \qquad (6.110)$$

where $\hat{\mathbf{r}}$ is the unit vector in the direction of observation $\mathbf{r}$, and $\mathbf{n}$ is the local unit outward normal to the plane of the aperture $\mathscr{A}$.

As an example of the application of Eq. (6.110), let us determine the intensity corresponding to a plane wave of unit amplitude, diffracted by a circular aperture of radius a.

Making again reference to Fig. 6.7, the electric vector of the incident field is assumed to be linearly polarized, and contained in the xz-plane, and its direction of propagation, characterized by the unit vector $\mathbf{n}_0$, is given by Eq. (6.71). In other words:

$$\mathbf{E}^{(i)}(\mathbf{r}) = \mathbf{e} \exp(ik\mathbf{n}_0\cdot\mathbf{r}), \qquad (6.111)$$

where $\mathbf{e}$ is a unit vector of components:

$$\mathbf{e} = (e_1, e_2, e_3) = (\cos\delta, 0, -\sin\delta). \qquad (6.112)$$

The plane defined by the normal $\mathbf{n} = \hat{\mathbf{z}}$ and the incident wavevector $\mathbf{n}_0$ is the *plane of incidence*. The choice of Eq. (6.111), in which the incident electric vector is contained in this plane, is referred to as *p-polarization*.

The tangential component of $\mathbf{E}^{(i)}$ in the plane $z = 0$ containing the aperture is then:

$$[\mathbf{n} \times \mathbf{E}^{(i)}]_{z=0} = (0, \cos\delta, 0)\exp(ikx\sin\delta). \tag{6.113}$$

By substituting Eqs. (6.113) and (6.111) into (6.110), introducing v given by Eq. (6.76), and performing the integrals of Eq. (6.110) as in the scalar case (Section 6.10), we obtain the electric vector of the diffracted field:

$$\mathbf{E}(\mathbf{r}) = -ika^2\cos\delta\,\frac{\exp(ikr)}{r}\,(\hat{\mathbf{r}} \times \hat{\mathbf{y}})\,\frac{J_1(kav)}{kav}, \tag{6.114}$$

$\hat{\mathbf{y}}$ being the unit vector in the y-direction.

Expressing $\hat{\mathbf{r}}$ by the angles θ and ϕ: $\hat{\mathbf{r}} = (\sin\theta\cos\phi, \sin\theta\sin\phi, \cos\theta)$, the diffracted intensity is:

$$I(\theta, \phi) = \frac{k^2a^4}{r^2}\cos^2\delta(\cos^2\theta + \cos^2\phi\sin^2\theta)\left[\frac{J_1(kav)}{kav}\right]^2. \tag{6.115}$$

On comparing the result, Eq. (6.115), with the scalar solution, Eq. (6.80) or (6.81), we observe a difference once again due to the angular factor. Unlike the scalar case, the electromagnetic theory gives a dependence on ϕ due to the polarization of the field.

For unpolarized incident waves, one has to add the mean values of the diffracted intensities corresponding to the two independent directions of polarization of the incident field, namely: with $\mathbf{e}$ in the xz-plane given by Eq. (6.112), and with $\mathbf{e}$ in the xy-plane, given by: $\mathbf{e} = (\cos\delta, \sin\delta, 0)$. This second polarization gives a resulting diffracted intensity which is identical to that of Eq. (6.115), apart from the angular factor which is now: $\sin^2\delta(\cos^2\theta + \sin^2\phi\sin^2\theta)$. Therefore, the result corresponding to unpolarized light is obtained by adding the averages of both intensities with respect to the fluctuating angle δ. Since $\langle\cos^2\delta\rangle = \langle\sin^2\delta\rangle = \frac{1}{2}$, this operation yields:

$$I^{\text{unpol}}(\theta, \phi) = \frac{k^2a^4}{2r^2}(\cos^2\theta + 1)\left[\frac{J_1(kav)}{kav}\right]^2, \tag{6.116}$$

and has no dependence on ϕ, as expected.

When the incident field is *s-polarized*, namely, it has the electric vector perpendicular to the plane of incidence, $\mathbf{e} = (0, 1, 0)$, the diffracted intensity given by the vector theory, Eq. (6.116), coincides with the scalar result, Eq. (6.81), in the plane of incidence (see Problem 6.11). Further details are discussed in Chapter 7.

When the radius a of the aperture is large, such that $ka \gg 1$, and the incidence is nearly normal to the plane of the aperture ($\delta \sim 0$), Eqs. (6.115), (6.116) and the scalar results, Eqs. (6.80) and (6.81) coincide. As a matter of fact, only in this case is the scalar Kirchhoff approximation exact, although numerical calculations by Stratton and Chu [6.44] show that the vector Kirchhoff solution, Eq. (6.115), is close to the exact one, calculated by Morse and Rubinstein [6.48], even for $ka \sim 1$. (A brief discussion may also be found in Section 6.12 of Ref. 6.6.)

6.18. DIFFRACTION PROBLEMS AND THE EXTINCTION THEOREM

Let $\mathbf{E}^{(i)}$ and $\mathbf{H}^{(i)}$ be the electric and magnetic vector, respectively, of a field incident on a plane screen, which we shall consider infinitely thin and perfectly conducting. Let $\mathscr{A}$ be an aperture in that screen. The extinction theorem, applied to the opaque portion $\mathscr{P}$ of the screen reads for the electric vector (cf. Eqs. (1.35) and (1.38) for a perfect conductor):

$$\mathbf{E}^{(i)}(\mathbf{r}) + \frac{ik}{c}\int_{\mathscr{P}} \mathbf{J}(\mathbf{r}')\cdot\mathscr{G}(\mathbf{r}, \mathbf{r}')\, dS = 0. \tag{6.117}$$

The integral of Eq. (6.117) is extended to the metallic portion $\mathscr{P}$ of the screen, and $\mathbf{J}$ is the induced electric current density. The point $\mathbf{r}$, at which the magnitudes of Eq. (6.117) are evaluated, is also situated in $\mathscr{P}$.

Since Eq. (6.117) constitutes a nonlocal boundary condition, from which the current density $\mathbf{J}$ may be found, this equation is sufficient to specify the problem uniquely. Once $\mathbf{J}$ is determined, the diffracted field, at a point $\mathbf{r}$ behind the screen, is obtained from Eqs. (1.36) and (1.38); namely:

$$\mathbf{E}(\mathbf{r}) = \frac{ik}{c}\int_{\mathscr{P}} \mathbf{J}(\mathbf{r}')\mathscr{G}(\mathbf{r}, \mathbf{r}')\, dS. \tag{6.118}$$

Equation (6.118) is equivalent to Eq. (6.108a) with $\mathbf{J} = -(c/2\pi)\mathbf{n} \times \mathbf{H}$, where $\mathbf{H}$ in the integrand of Eq. (6.118) is the field due to the induced electric current $\mathbf{J}$. However, while in the applications of Eq. (6.108a) the integration is extended to the aperture $\mathscr{A}$ (cf. Eq. (6.110)), the integral in Eq. (6.118) covers the opaque region $\mathscr{P}$. Also, $\mathbf{H}$ in Eq. (6.108a) represents the total field: incident plus the one generated by the induced electric current. These considerations lead to the vector formulation of *Babinet's principle*, establishing the equivalence of the diffraction patterns from complementary screens (more details can be found in Refs. 6.2 and 6.6).

Equations (6.117) and (6.118) are especially convenient in the case of two-dimensional diffraction problems for *s*-polarized incident waves. For *p*-polarized incident waves, it is more convenient to use, instead of Eqs. (6.117) and (6.118), the corresponding equations for the magnetic vector. In problems of diffraction by apertures and half-planes, this rigorous method was introduced by Copson [6.49], Levine, and Schwinger [6.50] and others (see, e.g. Section 11.4.3 in Ref. 6.2). The present day availability of fast computers has stimulated the use of this method for numerical calculations of diffraction. Its applications are discussed in Chapter 7 in the study of rough surface scattering and diffraction from gratings.

PROBLEMS

6.1. Prove that if the scalar wavefield U satisfies Sommerfeld's radiation condition, Eq. (6.4), then Eq. (6.2) reduces to Eq. (6.9) (see Fig. 6.2).

6.2. Find the Fresnel transform that gives the wavefield $U(\mathbf{r})$ in the plane $z > 0$ in terms of its normal derivative at $z = 0$. This expression replaces Eq. (6.66) when the angular spectrum is given by Eq. (6.32), rather than by Eq. (6.63).

6.3. Assuming that the field at a square aperture of width a on a screen is:

$$U(x, y) = \text{Rect}\left(\frac{x}{a}\right) \text{Rect}\left(\frac{y}{a}\right),$$

where $\text{Rect}(x/a) = 1$ if $|x| \leqslant a/2$, and $\text{Rect}(x/a) = 0$ if $|x| > a/2$, show that the Kirchhoff diffraction pattern in the Fresnel approximation, Eq. (6.55), is:

$$I(x, y) = \tfrac{1}{4}([C(u_2) - C(u_1)]^2 + [S(u_2) - S(u_1)]^2) \times ([C(v_2) - C(v_1)]^2 + [S(v_2) - S(v_1)]^2),$$

where:

$$u_1 = -\sqrt{\frac{k}{\pi z}}\left(\frac{a}{2} + x\right), \qquad u_2 = \sqrt{\frac{k}{\pi z}}\left(\frac{a}{2} - x\right),$$

$$v_1 = -\sqrt{\frac{k}{\pi z}}\left(\frac{a}{2} + y\right), \qquad v_2 = \sqrt{\frac{k}{\pi z}}\left(\frac{a}{2} - y\right),$$

and $C(\beta)$ and $S(\beta)$ are the *Fresnel integrals*:

$$C(\beta) = \int_0^\beta \cos\left(\frac{\pi\xi^2}{2}\right) d\xi, \qquad S(\beta) = \int_0^\beta \sin\left(\frac{\pi\xi^2}{2}\right) d\xi.$$

6.4. Prove that the power spectral density of $U(x, y, 0)$ (namely, the square modulus of the Fourier transform of $U(x, y, 0)$), is equal to that of its Fresnel transform $U(x, y, z)$, given by Eq. (6.55). Discuss the physical meaning of this result.

6.5. Find the Kirchhoff diffraction pattern in the scalar Fraunhofer approximation for a field that has on the screen the grating structure:

$$U(x, y) = 1 + \frac{h}{2}\cos\frac{2\pi}{a}x.$$

Discuss the position of the diffraction peaks.

6.6. Prove that the scalar Kirchhoff diffraction pattern from a semi-infinite plane, $x < 0$ at $z = 0$, in the Fresnel approximation is given by the intensity:

$$I(x, y) = \tfrac{1}{2}[(C(\beta) + \tfrac{1}{2})^2 + (S(\beta) + \tfrac{1}{2})^2],$$

where $\beta = \sqrt{k/2zx}$, and the observation point P is: $(x, 0, z)$; $C(\beta)$ and $S(\beta)$ are the Fresnel integrals of Problem 6.3.

6.7. Find the Kirchhoff diffraction pattern in the far zone of a linearly polarized plane wave by a circular aperture of radius a by using the diffraction integral, Eq. (6.100). Compare with the result of Eq. (6.115).

6.8. Prove that one can associate with the Rayleigh–Sommerfeld integrals the corresponding vector potentials:

$$W_{I,II}(\mathbf{r}, \mathbf{r}') = W(\mathbf{r}, \mathbf{r}') \mp W(\mathbf{r}^s, \mathbf{r}'),$$

where W is the vector potential of Kirchhoff's theory, introduced in Eq. (6.84), $\mathbf{r}'$ is a point on the screen, and $\mathbf{r}$ and $\mathbf{r}^s$ are mirror images of each other with respect to the plane of the screen.

6.9. By using the result of Problem 6.8, express the diffracted field as the sum of an incident field and a boundary wave, as in Eq. (6.88). Prove that the difference between the two Rayleigh–Sommerfeld solutions is given by the contribution of the boundary wave, namely:

$$U_I - U_{II} = U_I^{(B)} - U_{II}^{(B)}.$$

(Marchand and Wolf [6.33(b)].)

6.10. Using the expression of Problem 1.1 for the dyadic Green function, show that Eqs. (6.108) and (6.109) may be expressed respectively as:

$$\mathbf{E}(\mathbf{r}) = -\frac{2}{ik} \nabla \times \nabla \times \mathbf{\Pi}_e(\mathbf{r}),$$

$$\mathbf{H}(\mathbf{r}) = 2\nabla \times \mathbf{\Pi}_e(\mathbf{r});$$

and:

$$\mathbf{E}(\mathbf{r}) = 2\nabla \times \mathbf{\Pi}_m(\mathbf{r}),$$

$$\mathbf{H}(\mathbf{r}) = \frac{2}{ik} \nabla \times \nabla \times \mathbf{\Pi}_m(\mathbf{r});$$

(Karczewski and Wolf [6.46]), where $\mathbf{\Pi}_e$ and $\mathbf{\Pi}_m$ are the electric and magnetic Hertz vectors, respectively, and are given by:

$$\mathbf{\Pi}_e(\mathbf{r}) = \frac{1}{4\pi} \int_{S^-} \mathbf{n} \times \mathbf{H}(\mathbf{r}') G(\mathbf{r}, \mathbf{r}') \, dS,$$

$$\mathbf{\Pi}_m(\mathbf{r}) = \frac{1}{4\pi} \int_{S^-} \mathbf{n} \times \mathbf{E}(\mathbf{r}') G(\mathbf{r}, \mathbf{r}') \, dS.$$

6.11. Prove that the diffracted intensity in the plane of incidence for an s-polarized plane wave from an aperture is the same in the scalar and vector theories.

6.12. Show that in the two-dimensional case in which the fields depend only on the x and z variables of the plane of incidence, the nonlocal boundary condition, Eq. (6.117), for the diffraction of an s-polarized wave of electric vector E_y becomes:

$$E_y^{(i)} - \frac{\pi k}{c} \int_{\mathscr{P}} J_y(x') H_0^{(1)}\left(k\sqrt{(x - x')^2 + (z - z')^2}\right) dS = 0,$$

where $H_0^{(1)}$ is the Hankel function of the first kind and zero order.

For a plane screen $z = z'$, the boundary condition becomes:

$$\mathbf{E}_y^{(i)} - \frac{\pi k}{c} \int_{\mathscr{P}} J_y(x') H_0^{(1)}(k|x - x'|)\, dx' = 0,$$

and the diffracted field is expressed by:

$$\mathbf{E}_y^{(s)} = -\frac{\pi k}{c} \int_{\mathscr{P}} J_y(x') H_0^{(1)}(k|x - x'|)\, dx'.$$

Considering the diffraction of a plane wave with $E_z^{(i)} = \exp(ikz)$, incident from $z < 0$ on a slit of width a in the plane $z = 0$; the integral can be written as:

$$\int_{\mathscr{P}} (\cdot)\, dx' = \int_{-\infty}^{\infty} (\cdot)\, dx' - \int_{-a}^{a} (\cdot)\, dx'.$$

Using an expansion of $H_0^{(1)}(k|x - x'|)$ into plane waves, analogous to Eq. (2.51) for the three-dimensional Green function, find an approximate expression for the Fourier transform of $J_z(x)$. Use this result to obtain the angular spectrum of the complex amplitude of the diffracted electric vector $E(x, z)$ in the half-space $z > 0$.

6.13. Prove by using the results of Problems 1.1 and 1.2, that the diffraction integral, Eq. (6.100), for the electric vector in a domain $\tilde{V}$ consisting of vacuum, bounded by a surface S^-, may be expressed as:

$$\mathbf{E}(\mathbf{r}) = \frac{1}{4\pi k^2} \nabla \times \nabla \times \int_{S^-} \left[\frac{\partial \mathbf{E}(\mathbf{r}')}{\partial \mathbf{n}} G(\mathbf{r}, \mathbf{r}') - \mathbf{E}(\mathbf{r}') \frac{\partial G(\mathbf{r}, \mathbf{r}')}{\partial \mathbf{n}} \right] dS,$$

$\mathbf{n}$ being the local outward normal to the surface S^-.

REFERENCES

6.1. A. Sommerfeld, *Optics* (Vol. 4 of *Lectures on Theoretical Physics*), Academic Press, New York, 1967, 4th edition.

6.2. M. Born and E. Wolf, *Principles of Optics*, Pergamon Press, Oxford, 1980, 6th edition.

6.3. J. W. Goodman, *Introduction to Fourier Optics*, McGraw-Hill, New York, 1968.

6.4. P. M. Morse and H. Feshbach, *Methods of Theoretical Physics*, McGraw-Hill, New York, 1953.

6.5. P. R. Garabedian, *Partial Differential Equations*, J. Wiley, New York, 1964.

6.6. J. D. Jackson, *Classical Electrodynamics*, J. Wiley, New York, 1975, 2nd edition.

6.7. H. Poincaré, *Theorie Mathematique de la Lumiere*, Vol. II, Georges Carré, Paris, 1892.

6.8. B. B. Baker and E. T. Copson, *The Mathematical Theory of Huygens Principle*, Clarendon Press, Oxford, 1950, 2nd edition.

6.9. F. Kottler, *Ann. Phys. (Leipzig)* **70**, 405 (1923).

6.10. F. Kottler, in *Progress in Optics*, Vol. 4, E. Wolf, ed., North-Holland, Amsterdam, 1965.

6.11. M. Born, *Optik*, Springer-Verlag, Berlin, 1933, reprinted 1965.

6.12. W. Franz, *Z. Phys.* **125**, 563 (1949).

6.13. S. A. Schelkunoff, *Commun. Pure Appl. Math.* **4**, 43 (1951).

6.14. C. J. Bouwkamp, *Rep. Prog. Phys.* **17**, 35 (1954).

6.15. E. W. Marchand and E. Wolf, *J. Opt. Soc. Am.* **56**, 1712 (1966).

6.16. Lord Rayleigh, *Phil. Mag.* **43**, 259 (1897), see also his *Scientific Papers*, Vol. 4, Cambridge University Press, Cambridge, 1903, p. 283.

6.17. G. C. Sherman, *J. Opt. Soc. Am.* **57**, 1490 (1967).

6.18. A. Walther, *J. Opt. Soc. Am.* **57**, 639 (1967); *Am. J. Phys.* **35**, 808 (1967).

6.19. G. C. Sherman, *J. Opt. Soc. Am.* **57**, 1160 (1967).

6.20. E. Wolf and J. R. Shewell, *J. Math. Phys.* **11**, 2254 (1970).

6.21. G. C. Sherman and H. J. Bremerman, *J. Opt. Soc. Am.* **59**, 146 (1969).

6.22. W. D. Montgomery, *J. Opt. Soc. Am.* **59**, 804 (1969).

6.23. W. D. Montgomery, *J. Opt. Soc. Am.* **57**, 772 (1967).

6.24. W. D. Montgomery, *J. Opt. Soc. Am.* **58**, 1112 (1968).

6.25. H. Osterberg, *J. Opt. Soc. Am.* **55**, 1467 (1965).

6.26. G. C. Sherman, *J. Opt. Soc. Am.* **57**, 546 (1967).

6.27. L. Mandel and E. Wolf, *Optical Coherence and Quantum Optics*, Cambridge University Press, Cambridge, to be published.

6.28. G. A. Maggi, *Ann. di Mat. IIa* **16**, 21 (1888).

6.29. A. Sommerfeld, *Math. Ann.* **47**, 317 (1896).

6.30. A. Rubinowicz, *Ann. Phys. (Leipzig)* **53**, 257 (1917).

6.31. A. Rubinowicz, in *Progress in Optics*, Vol. 4, E. Wolf, ed., North-Holland, Amsterdam, 1965.

6.32. K. Miyamoto and E. Wolf, (a) *J. Opt. Soc. Am.* **52**, 615 (1962); (b) *J. Opt. Soc. Am.* **52**, 626 (1962).

6.33. E. W. Marchand and E. Wolf, (a) *J. Opt. Soc. Am.* **52**, 761 (1962); (b) *J. Opt. Soc. Am.* **54**, 587 (1964); (c) *J. Opt. Soc. Am.* **59**, 79 (1969).

6.34. M. J. Ehrlich, S. Silver and G. Held, *J. Appl. Phys.* **26**, 336 (1955).

6.35. J. J. Stamnes, *Waves in Focal Regions*, Adam Hilger, Bristol, 1986.
6.36. P. Debye, *Ann. Phys. (Leipzig)* **30**, 755 (1909).
6.37. G. C. Sherman and W. C. Chew, *J. Opt. Soc. Am.* **72**, 1076 (1982).
6.38. Y. Li, *J. Opt. Soc. Am.* **72**, 770 (1982).
6.39. Y. Li, *J. Opt. Soc. Am.* **73**, 1101 (1983).
6.40. E. Wolf and Y. Li, *Opt. Comm.* **39**, 204 (1981).
6.41. Y. Li and E. Wolf, *Opt. Comm.* **39**, 211 (1981).
6.42. Y. Li and E. Wolf, *J. Opt. Soc. Am. A* **1**, 801 (1984).
6.43. E. Collett and E. Wolf, *Opt. Lett.* **5**, 264 (1980).
6.44. J. A. Stratton and L. J. Chu, *Phys. Rev.* **56**, 99 (1939).
6.45. F. Kottler, in *Progress in Optics*, Vol. 6, E. Wolf, ed., North-Holland, Amsterdam, 1971.
6.46. B. Karczewski and E. Wolf, *J. Opt. Soc. Am.* **56**, 1207 (1966).
6.47. G. Toraldo di Francia, *Electromagnetic Waves*, Interscience Publishers, New York, 1955.
6.48. P. M. Morse and P. J. Rubinstein, *Phys. Rev.* **54**, 985 (1938).
6.49. E. T. Copson, *Quart. J. Math.* **17**, 19 (1946).
6.50. H. Levine and J. Schwinger, *Phys. Rev.* **73**, 383 (1948).

7

Scattering from Rough Surfaces

The theory of diffraction finds a broad range of application in many problems of optics, electromagnetic theory, and acoustics in connection with the scattering of waves by conducting or dielectric bodies of arbitrary shape. For instance, the scattering of light and other electromagnetic waves by spheres constitutes a subject extensively studied (see, e.g. Van de Hulst [7.1], Born and Wolf [7.2], or Kong [7.3]).

In this chapter we shall address the problem of light scattering by rough surfaces. Although the emphasis will be mainly optical, the principles discussed here are also of straightforward application to radiowave theory and acoustics. The subject is of considerable interest because, ultimately in practice, all surfaces of bodies are rough to some degree which depends on the wavelength of the interacting field. We shall consider surfaces either with random corrugation or with periodic profile (reflection gratings).

Within the context of the vector theory, we shall discuss the Kirchhoff approximation, the method of small perturbations, and the Rayleigh method. Further, we shall address numerical methods recently put forward that permit the prediction and interpretation of new experimental results of current active research, like, for example, the phenomenon of *enhanced backscattering*. We shall consider, first, scattering from perfectly conductive surfaces, then, we shall address the case of real metals and dielectrics.

7.1. INTRODUCTION

Rough surface scattering is of fundamental interest in optics and other areas of physics like acoustics, radiowave propagation, or condensed matter. Reviews on scalar theories for sound waves can be found in Refs. 7.4–7.7. Vector theories and also particular emphasis to scalar cases are presented in the books by Beckmann and Spizzichino [7.8] and by Bass and Fuks [7.9] as well as in the report by McGinn and Sykes [7.10]. Applications to optical surfaces are reviewed in the recent book by Bennett and Mattsson [7.11].

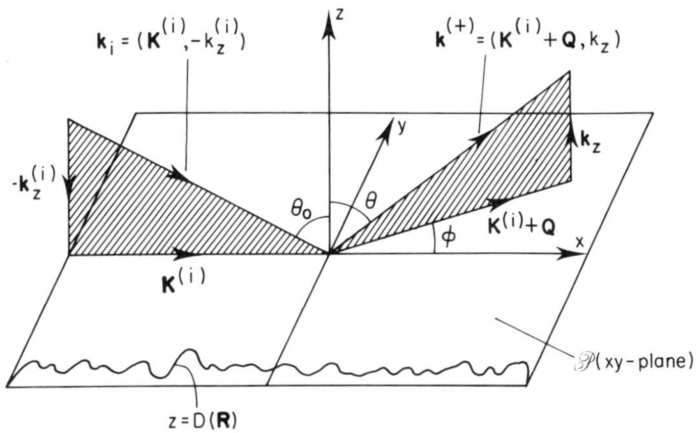

Figure 7.1. Scattering geometry.

Light scattering studies for solid surfaces and the role of surface wave excitations may be found in the review articles of the books edited by Agranovich and Mills [7.12] and by Wallis and Stegeman [7.13]. Concerning diffraction from periodic surfaces and its optical applications to gratings, extensive discussions can be found in Ref. 7.14. The earlier studies are dated back to Rayleigh [7.15] in connection with the scattering of sound from rough walls.

We shall address two main analytical approaches used in this problem: the Kirchhoff approximation and the method of small perturbations; the Rayleigh method, in relation to the latter, will also be discussed. More recently, the extinction theorem has been shown to provide an exact boundary condition subject to which the sources, induced in the surface by the incident wave, can be found by numerical methods. Differential methods, widely used in diffraction by gratings, are extensively reviewed, e.g. in Ref. 7.14.

The notation used will be in accordance with the geometry of scattering shown in Fig. 7.1. Namely, a monochromatic, linearly polarized, plane wave with electric vector:

$$\mathbf{E}^{(i)}(\mathbf{r}) = \mathbf{e} \exp[i(\mathbf{K}^{(i)} \cdot \mathbf{R} - k_z^{(i)} z)], \tag{7.1}$$

is incident on a surface that varies about the x, y-plane $\mathscr{P}$ with a corrugation which is characterized by the equation: $z = D(\mathbf{R})$. The position point is $\mathbf{r} = (\mathbf{R}, z)$, and $\mathbf{k}^{(i)} = (\mathbf{K}^{(i)}, -k_z^{(i)})$ denotes the wavevector. (We shall choose the system of coordinates such that $\mathbf{K}^{(i)}$ is along the x-axis.) The angle θ_0 and the plane made by $\mathbf{k}^{(i)}$ and the z-axis are the angle of incidence and the plane of incidence, respectively. A time dependence factor $\exp(-i\omega t)$ is omitted throughout. The constant vector $\mathbf{e} = (e_1, e_2, e_3)$ defines the linear polarization of the incident field and, of course, it is such that: $\mathbf{k}^{(i)} \cdot \mathbf{e} = 0$. In addition, $|\mathbf{k}^{(i)}|^2 = \omega^2/c^2$.

7.2. STATISTICAL CHARACTERIZATION OF RANDOM ROUGH SURFACES

Random rough surfaces are characterized either by their probability density or by their statistical moments. We shall consider surfaces with profile $z = D(\mathbf{R})$ being Gaussian distributed with zero mean, namely $\langle D \rangle = 0$, so that its statistics are sufficiently defined by the covariance $C(\mathbf{R}_1, \mathbf{R}_2)$:

$$C(\mathbf{R}_1, \mathbf{R}_2) = \langle D(\mathbf{R}_1) D(\mathbf{R}_2) \rangle. \tag{7.2}$$

In addition, we shall assume that the surface profile is statistically homogeneous and isotropic, then the covariance function is of the form:

$$C(\mathbf{R}_1, \mathbf{R}_2) = C(\tau), \qquad \tau = |\mathbf{R}_1 - \mathbf{R}_2|. \tag{7.3}$$

The *roughness* of the surface is defined by the variance $\sigma = C(0)$; on the other hand, the *scale* of the random corrugation is given by the correlation length T or width of the correlation function $B(\tau)$:

$$B(\tau) = \sigma^{-2} C(\tau). \tag{7.4}$$

Most works on rough surface scattering deal with Gaussian statistics; this is due both to its relative simplicity for analytical calculations and to its frequent occurrence in real practical cases. Nevertheless, other statistical models have also been studied (see, e.g. Refs. 7.16 and 7.17).

7.3. BOUNDARY CONDITION FOR SCATTERING FROM PERFECTLY CONDUCTIVE SURFACES

The application of Eqs. (1.35), (1.36), and (1.38) to the surface $z = D(\mathbf{R})$ separating vacuum from a perfect conductor gives:

(a) For points inside the conductor:

$$\mathbf{E}^{(i)}(\mathbf{r}) + \frac{ik}{c}\int_S \mathbf{J}(\mathbf{r}')\mathscr{G}(\mathbf{r}, \mathbf{r}')\, dS = 0, \qquad \text{for } z \leqslant D(\mathbf{R}). \tag{7.5}$$

(b) For points outside the conductor:

$$\mathbf{E}(\mathbf{r}) = \mathbf{E}^{(i)}(\mathbf{r}) + \frac{ik}{c}\int_S \mathbf{J}(\mathbf{r}')\mathscr{G}(\mathbf{r}, \mathbf{r}')\, dS, \qquad \text{for z} > D(\mathbf{R}). \tag{7.6}$$

The integrals of Eqs. (7.5) and (7.6) are extended to the illuminated portion of the surface. Equation (7.5) constitutes the boundary condition from which the electric current density **J**, induced by the incident field on the surface, can be calculated. The field scattered above the surface is then obtained from the second term of Eq. (7.6).

The solution of Eq. (7.5) for **J** is, however, not easy to calculate and, thus, either some analytical approximations or some numerical procedures have to be used.

It should be remarked that in the case of a plane wave illuminating a rough surface, infinity can be reached without leaving the vicinity of the surface; thus in the application of Eqs. (1.35), (1.36), and (1.38), the Sommerfeld radiation condition, Eq. (1.32), cannot be straightforwardly employed. We shall then say that the scattered field satisfies the radiation condition by meaning that it is outgoing above the surface; namely, its angular spectrum representation contains only plane wave components propagating into $z > 0$ (cf. Section 2.2). This difficulty in this kind of geometry has been discussed in detail, e.g. by Cadilhac in p. 53 of Ref. 7.14. On the other hand, by illuminating the surface with an optical beam, the sample becomes localized and then Eq. (1.32) can be used.

7.4. ANGULAR SPECTRUM REPRESENTATION

By using the result of Problem 1.1, we express Eq. (7.5), as well as the scattered field $\mathbf{E}^{(s)}(\mathbf{r})$ given by the second term of Eq. (7.6), in the form:

$$\mathbf{E}^{(i)}(\mathbf{r}) + \frac{i}{kc}\nabla \times \nabla \times \int_{\mathscr{P}} \mathbf{J}(\mathbf{R}')G(\mathbf{r}; \mathbf{R}', D(\mathbf{R}'))\frac{d^2R'}{n_{z'}} = 0, \qquad \text{for } z \leqslant D(\mathbf{R}), \tag{7.7}$$

and:

$$\mathbf{E}^{(s)}(\mathbf{r}) = \frac{i}{kc} \nabla \times \nabla \times \int_{\mathscr{P}} \mathbf{J}(\mathbf{R}')G(\mathbf{r}; \mathbf{R}', D(\mathbf{R}')) \frac{d^2R'}{n_{z'}}, \qquad \text{for } z > D(\mathbf{R}), \tag{7.8}$$

where: $\mathbf{R}' = (x', y')$.

In Eqs. (7.7) and (7.8) we take into account that the surface current density $\mathbf{J}$ and the local outward normal $\mathbf{n}$ are perpendicular to each other:

$$\mathbf{J}(\mathbf{R}')\cdot\mathbf{n}(\mathbf{R}') = 0. \tag{7.9}$$

Also, the following expression for $\mathbf{n}$, straightforwardly obtained, has been used:

$$\mathbf{n} = \left[1 + \left(\frac{\partial D}{\partial x}\right)^2 + \left(\frac{\partial D}{\partial y}\right)^2\right]^{-1/2} \left(-\frac{\partial D}{\partial x}, -\frac{\partial D}{\partial y}, 1\right), \tag{7.10}$$

so that the surface element dS has been expressed as:

$$dS = \left[1 + \left(\frac{\partial D}{\partial x}\right)^2 + \left(\frac{\partial D}{\partial y}\right)^2\right]^{1/2} dx\, dy = \frac{d^2R}{n_z}. \tag{7.11}$$

The integration of Eqs. (7.7) and (7.8) is now extended to the x, y-plane $\mathscr{P}$ about which the surface profile varies. In practice, the incident field is a beam and thus this integration is limited to a finite area S of the illuminated plane $\mathscr{P}$.

The angular spectrum representation of the scattered field above the surface consists of plane wave components with wavevector $\mathbf{k}^{(+)}$ that will be denoted as: $\mathbf{k}^{(+)} = (\mathbf{K}^{(i)} + \mathbf{Q}, k_z) = k(p, q, m)$; $k = \omega/c$; p, q, and m satisfying Eqs. (2.9). (Therefore, in this notation $\mathbf{Q}$ is the component, parallel to the plane $\mathscr{P}$, of the wavevector transfer $\mathbf{K} = \mathbf{k}^{(+)} - \mathbf{k}^{(i)}$.)

We now introduce Weyl's plane wave expansion (Eq. (2.51)) for $G(\mathbf{r}, \mathbf{r}')$ into Eqs. (7.7) and (7.8), using the following notation with reference to Fig. 7.1: $\mathbf{k}^{(-)} = (\mathbf{K}^{(i)} + \mathbf{Q}, -k_z)$ and $dp\, dq = d^2Q/k^2$.

Also, denoting by $D_{\min}$ and $D_{\max}$ the minimum and maximum values of the corrugation $D(\mathbf{R})$ in the surface illuminated by the incident wave, respectively, the absolute value $|z - D(\mathbf{R}')|$ that appears in Weyl's expansion when z' is substituted by its value: $z' = D(\mathbf{R}')$ on the surface, is:

$$|z - D(\mathbf{R}')| = z - D(\mathbf{R}'), \qquad \text{for } z > D_{\max}, \tag{7.12a}$$

$$|z - D(\mathbf{R}')| = D(\mathbf{R}') - z, \qquad \text{for } z < D_{\min}. \tag{7.12b}$$

Using Eq. (7.12b), Eq. (7.7) then becomes:

$$\mathbf{E}^{(i)}(\mathbf{r}) - \int_{-\infty}^{\infty} d^2Q \exp(i\mathbf{Q}\cdot\mathbf{R}) \exp(-ik_z z)$$
$$\times \frac{1}{k_z}\int_{\mathscr{P}} d^2R' \exp(-i\mathbf{Q}\cdot\mathbf{R}') \exp(ik_z D(\mathbf{R}'))(\mathbf{k}^{(-)} \times [\mathbf{k}^{(-)} \times \mathbf{F}(\mathbf{R}')]) = 0, \tag{7.13}$$

for $z \leqslant D_{\min}$.

In Eq. (7.13) the source function $\mathbf{F}(\mathbf{R}')$ is defined as

$$\mathbf{F}(\mathbf{R}') = -\frac{1}{2\pi kcn_z}\mathbf{J}(\mathbf{R}') \exp(-i\mathbf{K}^{(i)}\cdot\mathbf{R}'). \tag{7.14}$$

Similarly, for points $z > D_{\max}$, one can, on using Eq. (7.12a), write Eq. (7.8) as an expansion in plane waves propagating into $z > D(\mathbf{R})$:

$$\mathbf{E}^{(s)}(\mathbf{r}) = \int_{-\infty}^{\infty} \mathcal{E}(\mathbf{Q}) \exp(i\mathbf{k}^{(+)}\cdot\mathbf{r})\, d^2Q, \tag{7.15}$$

where the angular spectrum $\mathcal{E}(\mathbf{Q})$ of the scattered field is:

$$\mathcal{E}(\mathbf{Q}) = -\mathbf{k}^{(+)} \times [\mathbf{k}^{(+)} \times \mathbf{A}(\mathbf{Q})], \tag{7.16}$$

$\mathbf{A}(\mathbf{Q})$ being:

$$\mathbf{A}(\mathbf{Q}) = \frac{1}{k_z}\int_{\mathscr{P}} \exp(-i\mathbf{Q}\cdot\mathbf{R}') \exp(ik_z D(\mathbf{R}'))\mathbf{F}(\mathbf{R}')\, d^2R'. \tag{7.17}$$

On the other hand, assuming the electric vector of the incident field given by Eq. (7.1), the boundary condition, Eq. (7.13), becomes in $\mathbf{Q}$-space:

$$\delta(\mathbf{Q})\mathbf{e} - \frac{1}{k_z}\int_{\mathscr{P}} d^2R' \exp(-i\mathbf{Q}\cdot\mathbf{R}') \exp(ik_z D(\mathbf{R}'))(\mathbf{k}^{(-)} \times [\mathbf{k}^{(-)} \times \mathbf{F}(\mathbf{R}')]) = 0, \tag{7.18}$$

for $z \leqslant D_{\min}$.

7.5. THE KIRCHHOFF APPROXIMATION

In this section we shall discuss the Kirchhoff approximation for the scattering of a plane wavefield by a perfectly conductive rough surface. As shown in Section 6.16 (cf. Eq. (6.108a)), the scattered field may be expressed in terms of the tangential component of the magnetic vector as:

$$\mathbf{E}^{(s)}(\mathbf{r}) = -\frac{ik}{2\pi}\int_S [\mathbf{n} \times \mathbf{H}(\mathbf{r}')]\cdot\mathscr{G}(\mathbf{r}, \mathbf{r}')\, dS. \tag{7.19}$$

We recall that, as seen in Section 6.18, Eq. (7.19) is identical to the second term of Eq. (7.6) through the relationship: $\mathbf{J} = -(c/2\pi)\mathbf{n} \times \mathbf{H}$.

Using the result of Problem 1.1 (see also Problem 6.10), Eq. (7.19) may also be written as:

$$\mathbf{E}^{(s)}(\mathbf{r}) = \frac{i}{2\pi k}\nabla \times \nabla \times \int_S [\mathbf{n} \times \mathbf{H}(\mathbf{r}')]G(\mathbf{r}, \mathbf{r}')\, dS. \tag{7.20}$$

By means of Eq. (7.11), Eq. (7.20) may be expressed as:

$$\mathbf{E}^{(s)}(\mathbf{r}) = \frac{i}{2\pi k}\nabla \times \nabla \times \int_{\mathscr{P}} [\mathbf{n} \times \mathbf{H}(\mathbf{r}')]G(\mathbf{r}, \mathbf{r}')\frac{d^2R'}{n_{z'}}; \qquad \mathbf{R}' = (x', y'). \tag{7.21}$$

As in the discussion presented in Section 6.3, the *Kirchhoff approximation* [7.8], [7.18], consists of replacing the boundary value $\mathbf{n} \times \mathbf{H}(\mathbf{r})$ on the surface by that corresponding to the magnetic vector $\mathbf{n} \times \mathbf{H}^{(i)}(\mathbf{r})$ of the incident field. Under this approximation the scattered field electric vector, Eq. (7.21), is therefore:

$$\mathbf{E}_{KA}^{(s)}(\mathbf{r}) = \frac{i}{2\pi k}\nabla \times \nabla \times \int_{\mathscr{P}} [\mathbf{n} \times \mathbf{H}^{(i)}(\mathbf{r}')]G(\mathbf{r}, \mathbf{r}')\frac{d^2R'}{n_{z'}}, \tag{7.22}$$

the magnetic and electric vectors of the incident field being related by:

$$\mathbf{H}^{(i)}(\mathbf{r}) = \frac{1}{k}\mathbf{k}^{(i)} \times \mathbf{E}^{(i)}(\mathbf{r}). \tag{7.23}$$

The angular spectrum of the scattered field under this approximation can be easily obtained from Eq. (7.22) (see Problem 7.1).

In order to evaluate the far field, we approximate the Green function as (cf. Eq. (3.14)):

$$G(\mathbf{r}, \mathbf{r}') \sim \frac{\exp(ikr)}{r} \exp(-ik\mathbf{r}' \cdot \hat{\mathbf{r}}), \tag{7.24}$$

where $\hat{\mathbf{r}}$ is the unit vector in the direction of observation. Obviously, $\hat{\mathbf{r}}$ coincides with the scattering direction corresponding to the unit vector: $\mathbf{N} = \mathbf{k}^{(+)}/k = (1/k)(\mathbf{K}^{(i)} + \mathbf{Q}, k_z)$ of Fig. 7.1.

We now make use of the wavevector transfer K:

$$\mathsf{K} \equiv (\mathbf{Q}, v_z) = k\mathbf{N} - \mathbf{k}^{(i)}; \tag{7.25}$$

and substitute Eqs. (7.24), (7.23) and (7.1) into (7.22). We then obtain for the far field the expression:

$$\mathbf{E}_{KA}^{(s)}(r\mathbf{N}) \sim -\frac{\mathrm{i}}{2\pi k^2} \frac{\exp(ikr)}{r} \mathbf{k}^{(+)} \times \mathbf{k}^{(+)} \times \int_{\mathscr{P}} [\mathbf{n}(\mathbf{R}') \times (\mathbf{k}^{(i)} \times \mathbf{e})] \exp(-i\mathsf{K} \cdot \mathbf{r}') \frac{d^2R'}{n_{z'}}. \tag{7.26}$$

The Kirchhoff approximation amounts to assuming that the reflection at each point of the surface takes place by approximating the profile around this point by its tangent plane; therefore, the reflection is considered locally specular. As shown in Fig. 7.2, the local outward normal $\mathbf{n}(\mathbf{R})$ is under this assumption approximated by:

$$\mathbf{n}(\mathbf{R}) \sim \frac{1}{|\mathsf{K}|} \mathsf{K}. \tag{7.27}$$

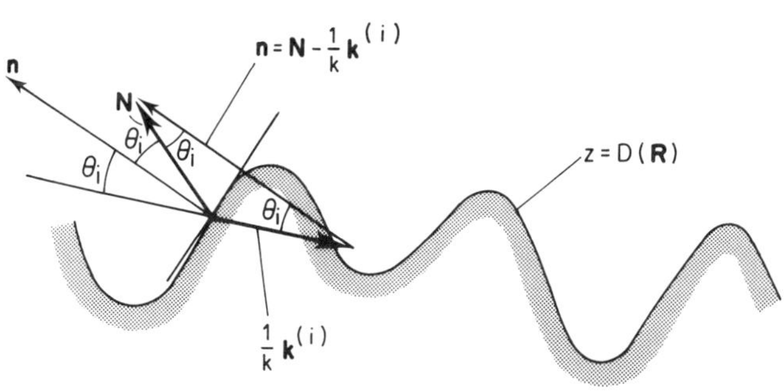

Figure 7.2. Illustration of the approximation, Eq. (7.27), for the assumption of local specular reflection.

In addition, Eq. (7.27) represents the stationary points in the asymptotic form of the integral Eq. (7.22).

On introducing Eq. (7.27) into (7.26) and using for $\mathbf{r}'$ its expression in terms of the corrugation $D(\mathbf{R}')$: $\mathbf{r}' = (\mathbf{R}', z' = D(\mathbf{R}'))$, we obtain:

$$\mathbf{E}_{KA}^{(s)}(r\mathbf{N}) \sim -\frac{i}{2\pi k^2}\frac{\exp(ikr)}{r}\frac{1}{v_z}[\mathbf{k}^{(+)} \times [\mathbf{k}^{(+)} \times [\mathbf{K} \times (\mathbf{k}^{(i)} \times \mathbf{e})]]]$$
$$\times \int_{\mathscr{P}} \exp(-i\mathbf{Q}\cdot\mathbf{R}')\exp(-iv_z D(\mathbf{R}'))\, d^2R'. \tag{7.28}$$

The integral in Eq. (7.28) is characteristic of the Kirchhoff approximation. Note that when the angular dispersion in the scattering is small, namely $v_z \sim k_z^{(i)}$, then this integral becomes a Fourier transform indicating that the surface behaves like a reflecting phase screen (cf. Problem 3.10); a formulation based on this scheme has been given by Welford [7.19].

Equation (7.28) shows that the Kirchhoff approximation predicts a depolarization of the scattered field which is independent of the surface profile, and is expressed by the angular factor $\mathscr{K}(\theta, \phi)$:

$$\mathscr{K}(\theta, \phi) = \frac{1}{k^3 e v_z}[\mathbf{k}^{(+)} \times [\mathbf{k}^{(+)} \times [\mathbf{K} \times (\mathbf{k}^{(i)} \times \mathbf{e})]]], \tag{7.29}$$

e being the modulus of $\mathbf{e}$.

7.5.1. Example: Scattering of a Linearly Polarized Plane Wave

Let us consider an s-polarized plane wave incident on the surface, namely, with its electric vector perpendicular to the plane of incidence. With reference to Fig. 7.1, this vector is then given by Eq. (7.1) with:

$$\mathbf{e} = (0, e, 0); \qquad \mathbf{k}^{(i)} = k(\sin\theta_0, 0, -\cos\theta_0). \tag{7.30}$$

On the other hand, the vectors $\mathbf{k}^{(+)}$ and $\mathbf{K}$ are:

$$\mathbf{k}^{(+)} = k(\sin\theta\cos\phi, \sin\theta\sin\phi, \cos\theta), \tag{7.31a}$$
$$\mathbf{K} = k(\sin\theta\cos\phi - \sin\theta_0, \sin\theta\sin\phi, \cos\theta + \cos\theta_0). \tag{7.31b}$$

We introduce now two unit vectors $\mathbf{s}$ and $\mathbf{p}$ that form with the vector $\mathbf{N} = (1/k)\mathbf{k}^{(+)}$ an orthonormal system, $\mathbf{s}$ being perpendicular to the plane of scattering (defined by the $\mathbf{k}^{(+)}$-vector and the z-axis), and $\mathbf{p}$ being contained in this plane. Hence, these two vectors define s- and p-polarization, respectively, for the scattered field observed in the **N**-direction. Their expressions are:

$$\mathbf{s} = \mathbf{N} \times \hat{\mathbf{z}}, \tag{7.32a}$$

$$\mathbf{p} = \mathbf{s} \times \mathbf{N}, \tag{7.32b}$$

$\hat{\mathbf{z}}$ being the unit vector along the z-direction.

From Eqs. (7.29)–(7.32) we obtain the angular factor $\mathscr{K}(\theta, \phi)$ for s-polarization of the incident and scattered fields:

$$\mathscr{K}_{ss}(\theta, \phi) = \frac{-\sin\theta_0 \sin\theta + \cos\phi(1 + \cos\theta_0 \cos\theta)}{\cos\theta_0 + \cos\theta}. \tag{7.33}$$

It is easy to check that, if $\mathscr{K}_{pp}(\theta, \phi)$ denotes the angular factor for a p-polarized scattered field when the wave incident on the surface is also p-polarized (namely, with the incident magnetic vector perpendicular to the plane of incidence, i.e., with: $\mathbf{e} = e(\cos\theta_0, 0, \sin\theta_0)$, then:

$$\mathscr{K}_{pp}(\theta, \phi) = \mathscr{K}_{ss}(\theta, \phi). \tag{7.34}$$

On the other hand, the reader can check that for the incident and scattered field cross-polarized to each other one has:

$$\mathscr{K}_{sp}(\theta, \phi) = \mathscr{K}_{ps}(\theta, \phi) = \frac{\sin\phi}{\cos\theta_0 + \cos\theta}. \tag{7.35}$$

Equations (7.33)–(7.35) show that the Kirchhoff approximation does not account for depolarization of the field scattered from a perfectly conductive rough surface and observed in the plane of incidence ($\phi = 0$). In this plane one has:

$$\mathscr{K}_{ss}(\theta) = \mathscr{K}_{pp}(\theta) = \frac{1 + (\cos\theta_0 + \cos\theta)}{\cos\theta_0 + \cos\theta}, \tag{7.36}$$

and:

$$\mathscr{K}_{sp}(\theta) = \mathscr{K}_{ps}(\theta) = 0. \tag{7.37}$$

In particular, this lack of depolarization is predicted both in the specular ($\theta = \theta_0$) and backscattering ($\theta = -\theta_0$) directions.

7.5.2. The Mean Scattered Intensity

The mean scattered intensity in the far zone, per unit of flow of energy of the incident field and per unit of illuminated area S of the $\mathscr{P}$-plane containing the random surface, is:

$$\langle I^{(s)}(\theta, \phi)\rangle = \frac{\langle|\mathbf{E}^{(s)}(r\mathbf{N})|^2\rangle r^2}{S \cos\theta_0 |\mathbf{E}^{(i)}|^2}. \tag{7.38}$$

On introducing Eqs. (7.1) and (7.28) into (7.38), and using the definition in Eq. (7.29) we obtain in the Kirchhoff approximation:

$$\langle I^{(s)}_{KA}(\theta, \phi)\rangle = \frac{k^2}{4\pi^2 S} F^2(\theta, \phi) \int_{\mathscr{P}} \int_{\mathscr{P}} d^2R\, d^2R' \exp[-i\mathbf{Q}\cdot(\mathbf{R} - \mathbf{R}')]$$
$$\times \langle \exp[-iv_z D(\mathbf{R})] \exp[iv_z D(\mathbf{R}')]\rangle. \tag{7.39}$$

The factor $F(\theta, \phi)$ in Eq. (7.39) is:

$$F(\theta, \phi) = \frac{\mathscr{K}(\theta, \phi)}{\cos\theta_0}. \tag{7.40}$$

Considering the random profile $D(\mathbf{R})$ to be a statistically homogeneous and isotropic Gaussian random function, we introduce the two-dimensional characteristic function of the joint probability density of the random variable at two points $\mathbf{R}_1$ and $\mathbf{R}_2$, given by the average $\langle\cdot\rangle$ in the integrand of Eq. (7.39); (cf., e.g. Papoulis [7.20]):

$$\langle \exp[-iv_z D(\mathbf{R})] \exp[iv_z D(\mathbf{R}')]\rangle = \exp[-\sigma^2 v_z^2[1 - B(\tau)]]; \tag{7.41}$$

where $\tau = |\mathbf{R} - \mathbf{R}'|$, and $B(\tau)$ is the correlation function of the surface (cf. Eq. (7.4)).

Assuming that the correlation length T is much smaller than S, and using polar coordinates τ and φ for the integration, it is straightforward, after some simple operations, to obtain from Eqs. (7.39) and (7.41) the following expression for the angular distribution of mean scattered intensity under the Kirchhoff approximation:

$$\langle I^{(s)}_{KA}(\theta, \phi)\rangle = \frac{k^2}{2\pi} F^2(\theta, \phi) \exp(-\sigma^2 v_z^2) \int_0^\infty d\tau\, \tau J_0(|\mathbf{Q}|\tau) \exp[\sigma^2 v_z^2 B(\tau)]. \tag{7.42}$$

It is convenient to express the scattered field as the sum of an average and a fluctuating part. For the electric vector this amounts to making the decomposition:

$$\mathbf{E}(\mathbf{r}) = \langle \mathbf{E}(\mathbf{r}) \rangle + \delta \mathbf{E}(\mathbf{r}), \tag{7.43}$$

$\langle \mathbf{E} \rangle$ and $\delta \mathbf{E}$ being the average and the fluctuating part of $\mathbf{E}$, respectively. Obviously, $\langle \delta \mathbf{E} \rangle = 0$.

On introducing the decomposition Eq. (7.43) into (7.38), the mean scattered intensity is then represented as the sum of two terms. The first one is proportional to $|\langle \mathbf{E} \rangle|^2$. This is the *coherent component*, which is proportional to the intensity of the diffraction pattern produced by the shape of the illuminated area S, and hence it is concentrated about the specular direction (if this area were assumed to be infinite this term would be proportional to $\delta(\mathbf{Q})$). The second term is proportional to $\langle |\delta \mathbf{E}|^2 \rangle$ (which is usually referred to as the *diffuse component* and represents a scattering halo extended over all angles of observation).

By expanding the factor $\exp[\sigma^2 v_z^2 B(\tau)]$ of the integrand in Eq. (7.42) into a power series of its exponent, one obtains a decomposition of $\langle I_{KA}^{(s)}(\theta, \phi) \rangle$ into the sum of a specular coherent component, given by the first term, and a diffuse component, represented by all the other terms (Beckmann and Spizzichino [7.8]) (see Problem 7.3). This expansion is illustrative as it shows that if the surface is slightly rough, namely $\sigma^2 v_z^2 \ll 1$, the series converges very fast and, thus, its second term is sufficient to describe the diffuse component. The angular distribution of mean scattered intensity then consists of a large peak, characterized by the coherent component, and a weaker diffuse halo over all angles. On the other hand, if the roughness is large, that is, if $\sigma^2 v_z^2 \gg 1$, the series converges too slowly to be useful; namely, the diffuse component is large, and there is no appreciable coherent specular component.

7.5.3. On the Validity of the Kirchhoff Approximation

The approximation involved in considering the scattering of the field as a specular reflection on the tangent plane at each point of the surface requires that the local curvature be small, this in turn means that the local slope not be large. A sufficient criterion broadly accepted is that the correlation length T be much larger than the wavelength λ of the scattered field. Since T is a measure of the size of the surface asperities, this criterion resembles others generally adopted in the theory of diffraction by slits or other arbitrary bodies. A more accurate criterion was proposed by Brekhovskikh [7.21] according to whom the local radius of curvature ρ_0 should be large

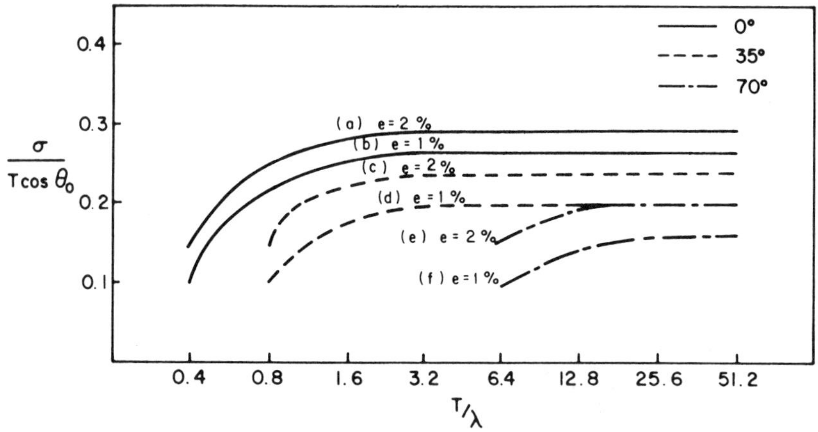

Figure 7.3. Plots of $\sigma/(T\cos\theta_0)$ versus T/λ at three different angles of incidence: $\theta_0 = 0°$, $35°$, and $70°$, marking the zones of validity of the Kirchhoff approximation. (After J. M. Soto-Crespo and M. Nieto-Vesperinas, *J. Opt. Soc. Am. A* **6**, 367 (1989).)

and such that: $\rho_0 \cos\theta_0 \gg \lambda/(4\pi)$; this is in agreement with the well known fact that the Kirchhoff approximation is progressively less accurate as the angle of incidence increases and at larger departures of the scattering angle from the specular direction.

The range of validity of the Kirchhoff approximation has been studied for light and other electromagnetic waves in Ref. 7.22, and in the scalar approach of acoustics in Ref. 7.23. The procedure consists of a direct comparison of this approximation with the results from Monte Carlo numerical calculations in one-dimensional rough surfaces (i.e., profiles that are constant in one direction). These numerical methods (to be addressed later in this chapter) can predict with great accuracy the scattered field from a wide variety of random surfaces, even with high slopes or small correlation distances.

The criterion of validity of the calculations used by the authors of Ref. 7.22 is the unitarity condition that the mean scattered intensity, normalized to the intensity of the incident field, has to satisfy. Figure 7.3 shows curves of $\sigma/(T\cos\theta_0)$ versus T/λ for three angles of incidence, $\theta_0 = 0°$, $35°$, and $70°$. They delimit the region (above these curves) in which the departure e from unitarity of the mean scattered intensity obtained with the Kirchhoff approximation differs from 1 by more than 1 or 2%.

For example, for $T = 4.8\lambda$ and $\sigma = \lambda$ at $\theta_0 = 0°$ the value of $\sigma/(\mathrm{T}\cos\theta_0)$ is 0.2. In Fig. 7.3 this value is appreciably down curve (b), below which the error in the unitarity e is less than 1%. Figure 7.4(a) shows the angular distribution of mean scattered intensity calculated from the Kirchhoff approximation and the exact one with no difference between *ss* and *pp* polarization. (Note that now all determinations are made in the plane of incidence, and thus no cross-polarization *sp* or *ps* occurs.) The coincidence with the Kirchhoff result is complete. It should be remarked that, in order to make the comparison easier, the Kirchhoff approximation has been done in two different ways: by using a smooth average equation (7.42) and by calculating the scattered field, Eq. (7.28), from each of 200 samples of the surface and then numerically averaging the corresponding 200 scattered intensities.

If one proceeds to larger angles of incidence for the same surface, good agreement is found between the Kirchhoff approximation and the exact Monte Carlo result up to $\theta_0 = 30°$; for this value of θ_0 one has $\sigma/(T \cos \theta_0) = 0.27$, which together with the value $T/\lambda = 4.8$, defines a point in the diagram of Fig. 7.3 that is in curve (c), namely, with an error of about 2%.

Figure 7.4(b) shows another comparison of the mean scattered intensity given by the Kirchhoff approximation with the exact numerical one at $\theta_0 = 30°$. The surface chosen has $\sigma = 0.2\lambda$ and $T = \lambda$; $\sigma/(T \cos \theta_0) = 0.21$, which defines a point for which the error e is 1.3%, i.e., between curves (*b*) and (*c*) of Fig. 7.3. The discrepancy now exists, although it is not large. On the other hand, at $\theta_0 = 0°$, $\sigma/(T \cos \theta_0) = 0.2$, the corresponding point in Fig. 7.3 is below curve (b) and, although not shown here, exhibits a close matching of the Kirchhoff result with the exact numerical one.

One can conclude, therefore, that the Kirchhoff approximation may be valid even for $T \leqslant \lambda$, provided that $\sigma/(T \cos \theta_0)$ is small enough. Of course, the larger the angle of incidence is, the smaller the coefficient $\sigma/(T \cos \theta_0)$ has to be at a given T/λ in order to maintain this validity. On the other hand, from a certain value of T/λ onwards, the quotient $\sigma/(T \cos \theta_0)$ below which this approximation is valid, remains constant.

A similar validity of the Kirchhoff approximation has been found with measured mean scattered intensities in experiments with controlled statistics (O'Donnell and Mendez [7.24]) (cf. Fig. 6 in that reference), for $T \sim 2\lambda$, $\sigma = 0.2\lambda$, and $\theta_0 = 20°$ ($\sigma/(T \cos \theta_0) = 0.11$).

7.6. THE METHOD OF SMALL PERTURBATIONS

In the case of perfectly conductive surfaces, the method of small perturbations [7.25]–[7.27] constitutes a procedure of solving the scattering equations (7.5) and (7.6), or their equivalent in **Q**-space Eqs. (7.18) and (7.17), which is useful for slightly rough surfaces, namely, when $k\sigma \ll 1$. It consists of writing $D(\mathbf{R})$ in the form $\alpha D(\mathbf{R})$ (α characterizing the strength of the corrugation) and

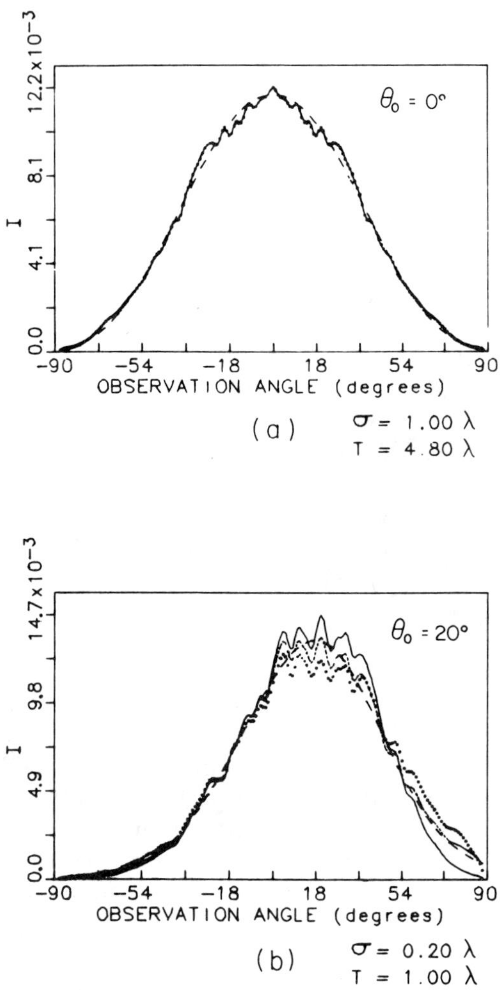

Figure 7.4. (a) Calculations of the angular distribution of mean scattered intensity from a one-dimensional random surface with $T = 4.8\lambda$, $\sigma = \lambda$, and $\theta_0 = 0°$. Solid line: Monte Carlo simulation, *s*-polarization. Dotted line: Monte Carlo simulation, *p*-polarization. Short-dashed line: Kirchhoff approximation averaging over 200 samples. Long-dashed line: Kirchhoff approximation from (7.42). (b) Same as (a), but with $T = \lambda$, $\sigma = 0.2\lambda$, and $\theta_0 = 20°$. (After J. M. Soto-Crespo and M. Nieto-Vesperinas, *J. Opt. Soc. Am. A* **6**, 367 (1989).)

then expanding the functions entering in Eqs. (7.15)–(7.18) in powers of α as follows:

$$\mathbf{F}(\mathbf{R}) = \sum_{n=0}^{\infty} \mathbf{F}^{(n)}(\mathbf{R}) \frac{\alpha^n}{n!}, \tag{7.44a}$$

$$\mathbf{A}(\mathbf{Q}) = \sum_{n=0}^{\infty} \mathbf{A}^{(n)}(\mathbf{Q}) \frac{\alpha^n}{n!}, \tag{7.44b}$$

$$\mathbf{n}(\mathbf{R}) = \sum_{n=0}^{\infty} \mathbf{n}^{(n)}(\mathbf{R}) \frac{\alpha^n}{n!}, \tag{7.44c}$$

$$\exp[ik_z \alpha D(\mathbf{R})] = \sum_{n=0}^{\infty} \frac{(ik_z \alpha D(\mathbf{R}))^n}{n!}. \tag{7.44d}$$

On introducing Eqs. (7.44a) and (7.44d) into (7.18), and since α is arbitrary, each term corresponding to each power of α should be zero; this yields:

$$\mathbf{k}^{(-)} \times [\mathbf{k}^{(-)} \times \tilde{\mathbf{F}}^{(0)}(\mathbf{Q})] = k_z^{(i)}\, \delta(\mathbf{Q})\mathbf{e}, \tag{7.45a}$$

$$\mathbf{k}^{(-)} \times [\mathbf{k}^{(-)} \times \tilde{\mathbf{F}}^{(1)}(\mathbf{Q})] = -\mathbf{k}^{(-)} \times [\mathbf{k}^{(-)} \times [ik_z D(\mathbf{R})\mathbf{F}^{(0)}(\mathbf{R})]_{\mathbf{Q}}], \tag{7.45b}$$

$$\mathbf{k}^{(-)} \times [\mathbf{k}^{(-)} \times \tilde{\mathbf{F}}^{(2)}(\mathbf{Q})] = -\mathbf{k}^{(-)} \times [\mathbf{k}^{(-)} \times [i2k_z D(\mathbf{R})\mathbf{F}^{(1)}(\mathbf{R}) + (ik_z D^2(\mathbf{R}))\mathbf{F}^{(0)}(\mathbf{R})]_{\mathbf{Q}}], \tag{7.45c}$$

and so on.

On introducing Eqs. (7.44a), (7.44b) and (7.44d) into (7.17), and using a similar procedure as before, we obtain:

$$\tilde{\mathbf{A}}^{(0)}(\mathbf{Q}) = \frac{1}{k_z} \tilde{\mathbf{F}}^{(0)}(\mathbf{Q}), \tag{7.46a}$$

$$\tilde{\mathbf{A}}^{(1)}(\mathbf{Q}) = -i\tilde{D}(\mathbf{Q}) * \tilde{\mathbf{F}}^{(0)}(\mathbf{Q}) + \frac{1}{k_z} \tilde{\mathbf{F}}^{(1)}(\mathbf{Q}), \tag{7.46b}$$

$$\tilde{\mathbf{A}}^{(2)}(\mathbf{Q}) = \frac{1}{k_z} \tilde{\mathbf{F}}^{(2)}(\mathbf{Q}) - 2i\tilde{D}(\mathbf{Q}) * \tilde{\mathbf{F}}^{(1)}(\mathbf{Q}) - k_z[D(\mathbf{R})]_{\mathbf{Q}}, \tag{7.46c}$$

and so on.

In Eqs. (7.45) and (7.46) we have introduced the two-dimensional Fourier transforms $\tilde{\mathbf{F}}^{(n)}(\mathbf{Q})$, $\tilde{\mathbf{A}}^{(n)}(\mathbf{Q})$, $\tilde{D}(\mathbf{Q})$ of $\mathbf{F}^{(n)}(\mathbf{R})$, $\mathbf{A}^{(n)}(\mathbf{R})$, and $D(\mathbf{R})$, respectively. The bracket $[\cdot]_{\mathbf{Q}}$ also denotes the two-dimensional Fourier transform of the function of $\mathbf{R}$ that is inside. The asterisk represents the two-dimensional convolution in $\mathbf{Q}$-space.

On the other hand, on introducing Eqs. (7.44a) and (7.44c) into (7.9) we obtain analogously:

$$\tilde{\mathbf{n}}^{(0)}(\mathbf{Q})\cdot\tilde{\mathbf{F}}^{(0)}(\mathbf{Q}) = 0, \tag{7.47a}$$

$$\tilde{\mathbf{n}}^{(0)}(\mathbf{Q})\cdot\tilde{\mathbf{F}}^{(1)}(\mathbf{Q}) = i\tilde{\mathbf{F}}^{(0)}(\mathbf{Q}) * \tilde{\mathbf{n}}^{(1)}(\mathbf{Q}), \tag{7.47b}$$

$$\tilde{\mathbf{n}}^{(0)}(\mathbf{Q})\cdot\tilde{\mathbf{F}}^{(2)}(\mathbf{Q}) = 2i\tilde{\mathbf{F}}^{(1)}(\mathbf{Q}) * \tilde{\mathbf{n}}^{(1)}(\mathbf{Q}), \tag{7.47c}$$

and so on.

In Eqs. (7.47) $\tilde{\mathbf{n}}^{(n)}(\mathbf{Q})$ is the two-dimensional Fourier transform of $\mathbf{n}^{(n)}(\mathbf{R})$. From Eq. (7.10) it is straightforward to see that:

$$\tilde{\mathbf{n}}^{(0)}(\mathbf{Q}) = \hat{\mathbf{z}}, \tag{7.48a}$$

$$\tilde{\mathbf{n}}^{(1)}(\mathbf{Q}) = i\tilde{D}(\mathbf{Q})\mathbf{Q}. \tag{7.48b}$$

Equations (7.47) are conditions that permit extraction of the source terms $\tilde{\mathbf{F}}^{(n)}(\mathbf{Q})$ from Eq. (7.45). For instance, from Eqs. (7.47a) and (7.45a) we get:

$$\tilde{\mathbf{F}}^{(0)}(\mathbf{Q}) = \mathbf{u}\,\delta(\mathbf{Q}), \tag{7.49}$$

the vector $\mathbf{u}$ being:

$$\mathbf{u} = -\left(\frac{k_z^{(i)}}{k^2}\,e_2, \frac{e_1}{k_z^{(i)}}, 0\right), \tag{7.50}$$

e_1 and e_2 being the Cartesian components of $\mathbf{e}$ along the x- and y-axes, respectively (cf. Eq. (7.1)). Analogously, Eqs. (7.47b) and (7.45b) yield:

$$\tilde{\mathbf{F}}^{(1)}(\mathbf{Q}) = -i\left[\mathbf{k}^{(+)}\frac{\tilde{D}(\mathbf{Q})}{k_z}\,\mathbf{Q}\cdot\mathbf{u} + k_z\tilde{D}(\mathbf{Q})\mathbf{u}\right]. \tag{7.51}$$

Higher order terms of the source function $\tilde{\mathbf{F}}(\mathbf{Q})$ can be found in a similar fashion. On substituting these successive terms into Eqs. (7.46) we obtain after some easy calculations:

$$\tilde{\mathbf{A}}^{(0)}(\mathbf{Q}) = \frac{1}{k_z^{(i)}}\,\delta(\mathbf{Q})\mathbf{u}, \tag{7.52a}$$

$$\tilde{\mathbf{A}}^{(1)}(\mathbf{Q}) = -2i\tilde{D}(\mathbf{Q})\mathbf{u} - i\mathbf{k}^{(+)}\frac{\mathbf{Q}\cdot\mathbf{u}}{k_z}\,\tilde{D}(\mathbf{Q}). \tag{7.52b}$$

Higher order terms are increasingly more complicated to obtain with similar operations.

7.6.1. Expansions for the Mean Scattered Intensity

By making use of Eqs. (7.15) and (7.16) in Eq. (7.38), and introducing in Eq. (7.15) the asymptotic expression of the angular spectrum representation as discussed in Section 2.12, we obtain in the far zone the following normalized angular distribution of mean scattered intensity:

$$\langle I^{(s)}(\mathbf{Q})\rangle = \frac{4\pi^2 k k_z^2}{S k_z^{(i)} e^2 \cos\theta_0} \langle |\mathcal{E}(\mathbf{Q})|^2\rangle. \tag{7.53}$$

For instance, for an s-polarized incident plane wave (cf. Eqs. (7.1) and (7.30)), the reader can check that on introducing Eqs. (7.16) and (7.52) into (7.53), we obtain the following expression up to the second power of $k\sigma$ for the copolarized mean scattered intensity (cf. the definition in Eq. (7.32a)):

$$\begin{aligned}\langle I_{ss}^{(s)}(\mathbf{Q})\rangle = {} & \frac{4\pi^2 k^6 \cos^2\theta}{\cos\theta_0} \\ & \times \Bigg[\Bigg[\left(\frac{1}{k^4} - \frac{\sigma^2\cos\theta_0}{k^3}\int_0^{2\pi} d\beta \int_0^{\infty} dQ' Q' \,\mathrm{Re}[k_z']\Phi(\mathbf{Q}')\right) \\ & \quad - \frac{\sigma^2\cos\theta_0}{k^3}\int_0^{2\pi} d\beta\cos^2\beta \int_0^{\infty} dQ' Q'^3\,\mathrm{Re}\left[\frac{1}{k_z'}\right]\Phi(\mathbf{Q}')\Bigg]\delta(\mathbf{Q}) \\ & \quad + \frac{4\pi^2\sigma^2}{k^2}\cos^2\theta_0\cos^2\phi\,\Phi(\mathbf{Q})\Bigg].\end{aligned} \tag{7.54}$$

In Eq. (7.54) $\Phi(\mathbf{Q})$ is the *spectral density* of the random profile $z = D(\mathbf{R})$ defined as:

$$\Phi(\mathbf{Q}) = \frac{1}{(2\pi)^2}\int_{-\infty}^{\infty} B(\boldsymbol{\tau}) \exp(-i\mathbf{Q}\cdot\boldsymbol{\tau})\, d^2\tau. \tag{7.55}$$

The integrals in Eq. (7.54) are performed in the polar coordinates $\mathbf{Q}' = Q'(\cos\beta, \sin\beta)$. Equation (7.54) contains a coherent specular term, proportional to $\delta(\mathbf{Q})$ (the illuminated surface S being assumed infinite), and a diffuse component proportional to $\Phi(\mathbf{Q})$.

The convergence of the perturbative series that would be obtained by adding to Eq. (7.54) the terms of higher order in $k\sigma$ has not yet been studied due to the difficulty in deriving these terms. When $k\sigma \ll 1$ the series is quickly convergent and thus Eq. (7.54) adequately describes copolarized scattering. (Note the analogy of the diffuse component of Eq. (7.54) with Eq. (3.103) of scattering by a weak random potential.)

From Eq. (7.54) the reader can check that the *ss*-mean scattered intensity in the plane of incidence is the same as the one obtained from the scalar theory [7.25], [7.26] (see Problem 7.4). It also exhibits a decrease of the diffuse component as the angle of incidence θ_0 increases. Eventually, at grazing incidence the reflection is only specular.

For instance, at normal incidence over a surface with a Gaussian correlation function: $B(\tau) = \exp(-\tau^2/T^2)$, there is no distinction between *s*- and *p*-polarization for the incident field; then, choosing $\mathbf{e} = (0, e, 0)$, Eq. (7.54) yields:

$$\begin{aligned}\langle I_s^{(s)}(\theta, \phi)\rangle = 4\pi^2 k^2 \cos^2\theta \Bigg[\Bigg[1 &- \frac{2}{3} T^2 k^4 \sigma^2 M\left(1, \frac{5}{2}, -\frac{k^2T^2}{4}\right) \\ &- \frac{2}{3} T^2 k^4 \sigma^2 M\left(2, \frac{5}{2}, -\frac{k^2T^2}{4}\right)\Bigg] \delta(\cos\theta)\,\delta(\phi) \\ &+ 4\pi^2 T^2 k^2 \sigma^2 \cos^2\phi \exp\left(-\frac{Q^2T^2}{4}\right)\Bigg]. \end{aligned} \tag{7.56}$$

$M(a, b, z)$ is the Kummer function [7.28].

In particular, in the plane perpendicular to the vector $\mathbf{e}$ (the *xz*-plane of incidence in our example, $\phi = 0$), Eq. (7.56) gives the following expression for the mean scattered intensity:

$$\begin{aligned}\langle I_s^{(s)}(\theta)\rangle = 4\pi^2 k^2 \cos^2\theta \Bigg[\Bigg[1 &- \frac{2}{3} T^2 k^4 \sigma^2 M\left(1, \frac{5}{2}, -\frac{k^2T^2}{4}\right)\Bigg] \delta(\cos\theta) \\ &+ 4\pi^2 T^2 k^2 \sigma^2 \cos^2\phi \exp\left(-\frac{Q^2T^2}{4}\right)\Bigg], \end{aligned} \tag{7.57}$$

which coincides with the result of the scalar theory [7.25], [7.26], as expected from the remark made above.

Equation (7.56) also indicates that a white noise random surface (namely, with $T = 0$) will reflect only specularly [7.26], [7.27]. This of course holds for any angle of incidence. We shall later see that Monte Carlo numerical calculations also suggest this result.

The diffuse term of Eq. (7.57) coincides with the corresponding one given by the Kirchhoff approximation (cf. Problem 7.1). Note, on the other hand, from the asymptotic behavior of $M(a, b, z)$ for large z, that the specular term in this approximation is equal to that of Eq. (7.57) for large kT, specifically for $k^2T^2/4 \gg 1$, while $T^2 \ll S$, since in this case the factor multiplying $\delta(\cos\theta)$ becomes $1 - 4k^2\sigma^2$ (this conclusion can be generalized for any angle of incidence [7.27]).

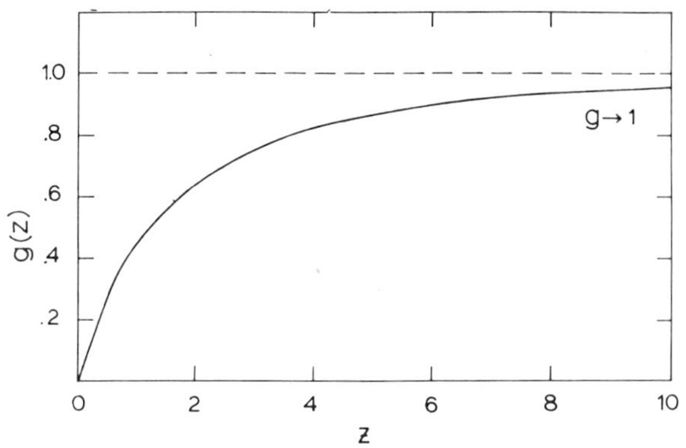

Figure 7.5. The function $\frac{2}{3}zM(1, \frac{5}{2}, -z)$ as a function of $z = k^2T^2/4$. When $T \gg \lambda$ the Kirchhoff approximation is accurate. (After M. Nieto-Vesperinas and N. Garcia, *Opt. Acta* **28**, 1651 (1981).)

The measurement of the specular component of the mean scattered intensity for low roughness is very important because this term can yield a direct evaluation of σ from knowledge of T.

Under normal incidence with $\mathbf{e} = (0, e, 0)$, and for measurements of the s-polarized scattered intensity in the plane $\phi = 0$, the complete expression of the specular component to all powers of $k\sigma$ is [7.26], [7.29], [7.30]:

$$\langle I_s^{(s)} \rangle_{\text{spec}} = \exp\left[-\frac{2}{3} T^2 k^4 \sigma^2 M\left(1, \frac{5}{2}, -\frac{k^2T^2}{4}\right)\right]. \tag{7.58}$$

Figure 7.5 displays the g-factor in the exponent of Eq. (7.58) versus $k^2T^2/4$. The region of validity of the Kirchhoff approximation is reached when this function is equal to one.

In Fig. 7.6 we plot the function (7.58) for several values of $k^2T^2/4$ versus $4k^2\sigma^2$. Experiments based on this formula have not yet been done; however, the diagrams of Fig. 7.6 offer a suggested means of evaluating σ from T, or vice versa.

7.6.2. On the Range of Validity of the Method of Small Perturbations

Globally, the accuracy of the expression for the mean scattered intensity by means of the first terms of the perturbative series requires that the diffuse component be small. This range of validity has been studied for light and other electromagnetic waves in Refs. 7.31 and 7.32, and in the context of acoustic waves in Ref. 7.33.

Assuming that the surface is one-dimensional, namely, constant in one direction, e.g. along the y-axis, its profile being: $z = D(x)$, then the results from the method of small perturbations can be directly compared with those obtained from Monte Carlo numerical calculations.

For an s-polarized plane wave, incident under an angle θ_0, the diffuse component of the angular distribution of mean scattered intensity in the far zone can be obtained following the procedure of Section 1.6. After some calculations this leads to:

$$\langle I_s^{(s)}(\theta)\rangle_d = \langle I_s^{(s)(2)}(\theta)\rangle_d + \langle I_s^{(s)(4)}(\theta)\rangle_d, \tag{7.59}$$

where:

$$\langle I_s^{(s)(2)}(\theta)\rangle_d = \frac{2k}{\sqrt{\pi}} k^2\sigma^2 \cos\theta_0 \cos^2\theta \exp\left[-\frac{k^2T^2}{4}(\sin\theta - \sin\theta_0)^2\right], \tag{7.60}$$

and:

$$\begin{aligned}
\langle \mathrm{I}_s^{(d)(4)}(\theta)\rangle_d &= \frac{2k^4\sigma^4}{\pi} \cos\theta_0 \cos^2\theta \\
&\times \left[\frac{k^2T^2}{2}\exp\left[-\frac{k^2T^2}{8}(\sin\theta - \sin\theta_0)^2\right]\left[\frac{\sqrt{2\pi}}{k^3T^3} + (\Lambda^2(\theta) - 1)\frac{\sqrt{2\pi}}{kT}\right.\right. \\
&+ 2\int_{-1-\Lambda(\theta)}^{1+\Lambda(\theta)} [1-(\xi + \Lambda(\theta))^2] \exp\left[-\frac{k^2T^2}{2}\xi^2\right] d\xi \\
&+ \left.\int_{-\infty}^{\infty} [(1-(\xi+\Lambda(\theta))^2)^*]^{1/2}[1-(\xi-\Lambda(\theta))^2]^{1/2} \exp\left[-\frac{k^2T^2}{2}\xi^2\right] d\xi\right] \\
&+ kT\exp\left[-\frac{k^2T^2}{4}(\sin\theta - \sin\theta_0)^2\right]\left[\sqrt{\pi}(\cos^2\theta + \cos^2\theta_0)\right. \\
&- \frac{4\sqrt{\pi}}{k^2T^2} - kT\cos\theta_0 \int_{-1}^{1} \sqrt{1-\xi^2}\exp\left[-\frac{k^2T^2}{4}(\sin\theta_0 - \xi)^2\right] d\xi \\
&- kT\cos\theta \int_{-1}^{1} \sqrt{1-\xi^2}\exp\left[-\frac{k^2T^2}{4}(\sin\theta - \xi)^2\right] d\xi \\
&- kT\,\mathrm{Re}\int_{-\infty}^{\infty} \sqrt{1-\xi^2}\sqrt{1-(\sin\theta - \sin\theta_0 - \xi)^2} \\
&\left.\left.\times \exp\left[-\frac{k^2T^2}{4}(\sin\theta - \xi)^2\right] d\xi\right]\right].
\end{aligned} \tag{7.61}$$

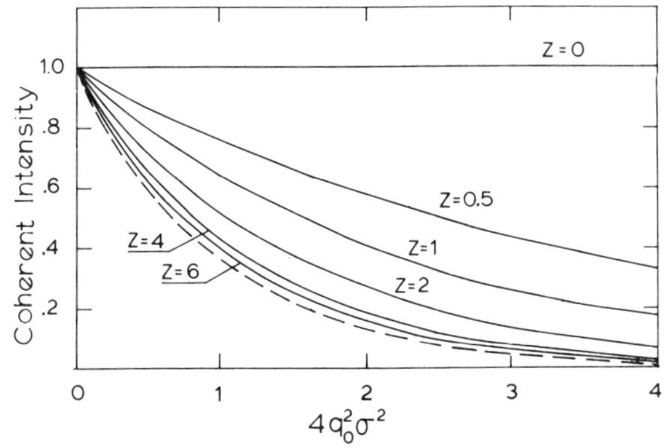

Figure 7.6. Dashed line: coherent component $\exp(-4k^2\sigma^2)$ of the mean scattered intensity in the Kirchhoff approximation as a function of $4k^2\sigma^2$. Continuous lines: coherent component from Eq. (7.58), which is exact both for $z = k^2T^2/4$ being zero (white noise surface), and for $z \to \infty$ (Kirchhoff approximation). (After M. Nieto-Vesperinas and N. Garcia, *Opt. Acta* **28**, 1651 (1981).)

In Eq. (7.61) $\Lambda(\theta)$ is:

$$\Lambda(\theta) = \tfrac{1}{2}(\sin\theta + \sin\theta_0), \tag{7.62}$$

and "Re" denotes "real part".

Analogously, for a p-polarized plane wave, incident at angle θ_0, the diffuse component of the angular distribution of mean scattered intensity is found to be:

$$\langle I_p^{(s)(2)}(\theta)\rangle_d = \frac{2k}{\sqrt{\pi}}\, k^2\sigma^2\, \frac{(1-\sin\theta_0)^2}{\cos\theta_0} \exp\left[-\left(\frac{k^2T^2}{4}(\sin\theta - \sin\theta_0)^2\right)\right]. \tag{7.63}$$

It has not been possible yet to calculate the term of fourth order in $k\sigma$. This is owing to nonintegrable singularities appearing at angles of 90° in the resulting integrals.

For a p-polarized incident plane wave, if the expansion of the scattered field in powers of $k\sigma$ is obtained by applying the scheme of Section 7.6, there appears a singular denominator at grazing emergence, namely, it contains a factor $\cos\theta$. On the other hand, Eq. (7.63) shows that the diffuse component of the mean scattered intensity also diverges at grazing incidence ($\theta_0 = 0$). This problem is not only inherent to the perturbative series, but is also connected with the existence of the resonance associated with the dispersion relation on the *light line* [7.34]–[7.36], namely, on the curve:

$$|\mathbf{K}^{(i)} + \mathbf{Q}|^2 = k^2, \tag{7.64}$$

which is the limit for a perfectly conducting surface ($\varepsilon \to \infty$) of the well-known flat surface *polariton* dispersion relation [7.34]–[7.37]:

$$|\mathbf{K}^{(i)} + \mathbf{Q}|^2 = k^2 \frac{\varepsilon}{\varepsilon + 1}, \tag{7.65}$$

ε being the dielectric permittivity of the material in $z < D(\mathbf{R})$. This suggests the difficulties of the model of a perfectly conducting surface at grazing angles.

Due to the formidably increasing difficulty in analytically calculating higher order terms for the intensity, the analysis on the validity of the perturbative series is restricted to the expansions in Eqs. (7.59) and (7.63).

As an illustration, Fig. 7.7 shows the angular distribution of the diffuse mean scattered intensity for an s-polarized incident plane wave scattered from surfaces with three different statistical parameters in the range $kT > 1$. As seen, the convergence of the expansion is faster the lower the total amount of this diffuse intensity is. This occurs of course the lower σ is, or the higher is the angle of incidence.

Figures 7.8(a) and 7.8(b) show the curves of σ/λ versus T/λ for four angles of incidence $\theta_0 = 0°, 40°, 70°$, and $80°$ for an s-polarized incident plane wave. They mark the region below which the following relation holds:

$$\frac{|E^{(4)} - E^{(2)}|}{E^{(4)}} \leqslant 0.05. \tag{7.66}$$

Where $E^{(2)}$ and $E^{(4)}$ are the total diffuse intensities to second and third order, respectively, namely:

$$E^{(2)} = \int_{-\pi/2}^{\pi/2} \langle I_s^{(s)(2)}(\theta)\rangle_d \, d\theta, \tag{7.67a}$$

$$E^{(4)} = E^{(2)} + \int_{-\pi/2}^{\pi/2} \langle I_s^{(s)(4)}(\theta)\rangle_d \, d\theta. \tag{7.67b}$$

Therefore, the region below these curves can be considered of validity in the perturbative expansion up to second order in $k\sigma$. Also, as T decreases, these demarcation lines increase monotonically.

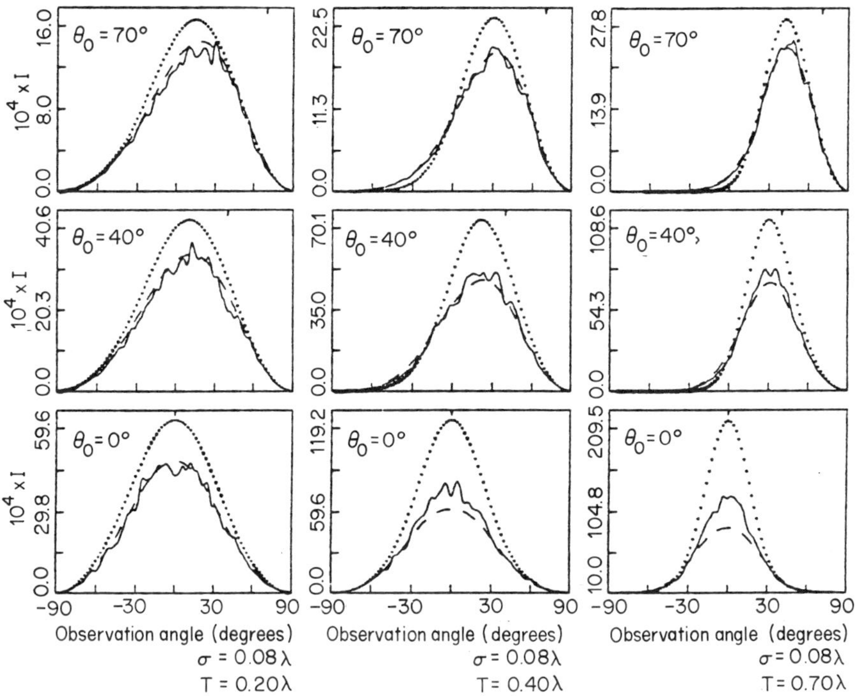

Figure 7.7. Calculations of the angular distribution of the diffuse component of the mean scattered intensity for *s*-polarization from a one-dimensional surface with statistical parameters as indicated under several angles of incidence. Solid line: Monte Carlo numerical solution. Dashed line: result from perturbation up to fourth order in $k\sigma$. Dotted line: result from perturbation up to second order in $k\sigma$. (After J. M. Soto-Crespo and M. Nieto-Vesperinas, *J. Opt. Soc. Am. A* 7, 1185 (1990).)

The above can be summarized by concluding that for $kT > 1$ the expansion up to fourth order in $k\sigma$ is a better approximation than the expansion up to second order. The latter yields good results providing that the total energy under all angles of scattering of the diffuse component (TISE) does not exceed 9% of the incident energy. This limit decreases as the value $kT = 1$ is approached [7.32].

However, for $kT < 1$ the perturbative series is no longer convergent. This manifests itself in a greater accuracy of the expansion up to second order than up to fourth order of $k\sigma$. As a matter of fact, comparisons with Monte Carlo calculations [7.32] show that as $kT \to 0$ the expansion up to fourth order in $k\sigma$ diverges, whereas the expansion up to second order yields better results

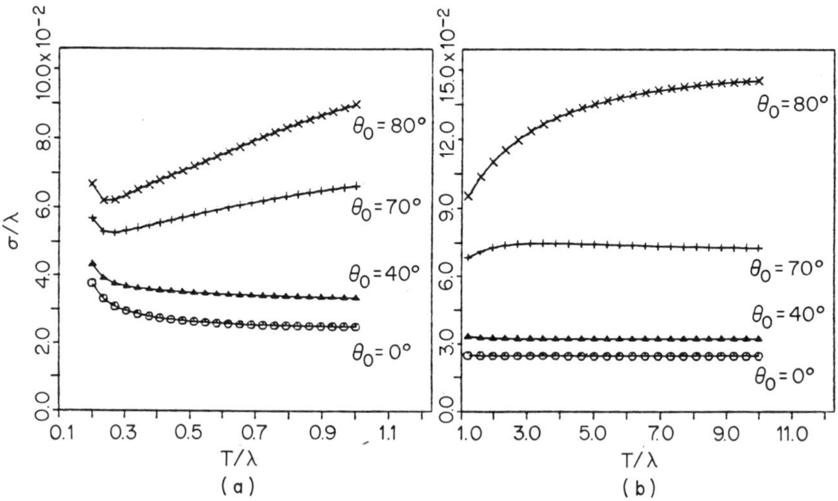

Figure 7.8. Plot of σ/λ versus T/λ below which the perturbative solution up to second order of $k\sigma$ is valid for s-polarization, according to Eq. (7.66). (After J. M. Soto-Crespo and M. Nieto-Vesperinas, *J. Opt. Soc. Am. A* **7**, 1185 (1990).)

(even at larger values of $k\sigma$) the closer T is to zero. This suggests that the perturbative series becomes asymptotic to second order of $k\sigma$ as $kT \to 0$. On the other hand, the increasing accuracy of the second order results (also linked to a lower TISE) as $kT \to 0$ indicates that a white noise surface ($T = 0$) reflects only specularly.

The considerations on the validity of the perturbative expression of the diffuse component of the mean scattered intensity for p-waves are similar to those for s-waves [7.32]. One has to bear in mind, however, that for T smaller than λ the angular distribution of mean diffuse scattered intensity is broader, and it also has larger area, for p-polarization than for s-polarization. This is illustrated in Figs. 7.9(a) and 7.9(b), showing from exact Monte Carlo calculations that for p-waves the diffuse halo contains more energy than for s-waves, and it is skewed towards the backscattering direction. Therefore, in this region of values of T the validity of Eq. (7.63) will be expected to be worse than the corresponding second order expansion for the diffuse mean scattered intensity for s-waves.

An expansion that achieves faster convergence is the so-called *phase perturbation method* [7.29], [7.30], [7.38], developed for the coherent component of scalar fields. However, it entails difficulties for the calculation of the diffuse component of the mean scattered intensity, and also for its extension to electromagnetic fields.

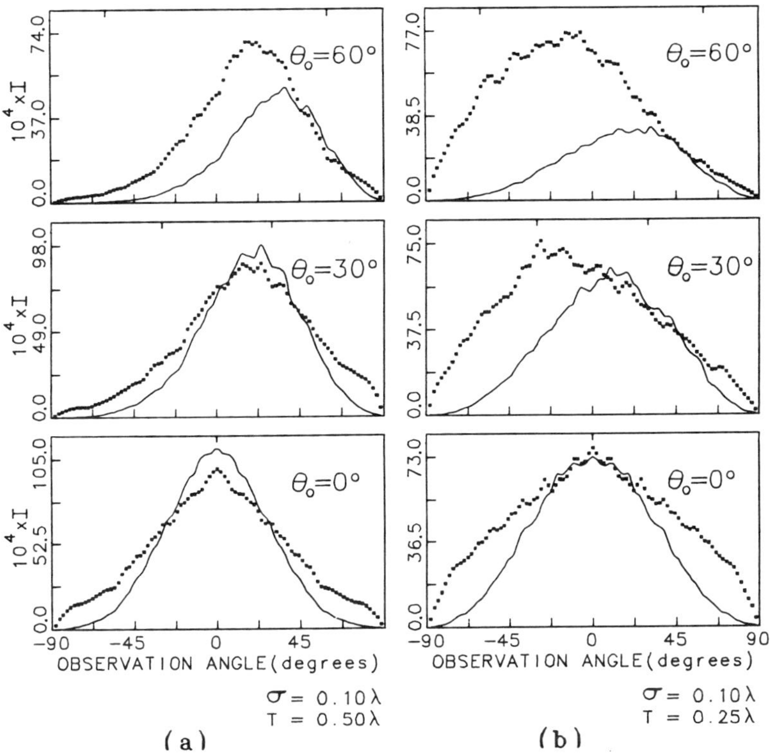

Figure 7.9. Monte Carlo calculations of the angular distribution of the diffuse component of the mean scattered intensity from a one-dimensional surface with $\sigma = 0.1\lambda$ and: (a) $T = 0.5\lambda$; (b) $T = 0.25\lambda$. Solid line: *s*-polarization. Dotted line: *p*-polarization. The upper tip marks the specular direction (right) and backscattering direction (left). (After M. Nieto-Vesperinas and J. M. Soto-Crespo, *Scattering in Volumes and Surfaces*, M. Nieto-Vesperinas and J. C. Dainty, eds., North-Holland, Amsterdam, 1990, p. 179.)

7.7. THE RAYLEIGH METHOD

We have seen in Section 7.4 that the use of the plane wave expansion for the Green function $G(\mathbf{r}, \mathbf{r}')$ permits one to write an angular spectrum representation for the scattered field at points above the surface, namely, for $z > D_{\max}$, with plane wave components propagating into the half-space $z > D(\mathbf{R})$. However, at points with $D_{\max} > z > D(\mathbf{R})$, the absolute value $|z - D(\mathbf{R})|$ coming from Eq. (2.51) cannot be expressed as Eq. (7.12a). This connects with the considerations of Section 2.2 (Fig. 2.2), and shows that this region is not source-free, thus involving a plane wave expansion for the scattered field given by the sum of fields of the form of Eqs. (2.12) and (2.13), namely, containing plane waves propagating both into $z > D(\mathbf{R})$ and into $z < D(\mathbf{R})$.

The *Rayleigh hypothesis* [7.15] assumes that the angular spectrum representation, Eq. (7.15), containing plane waves that propagate into $z > D(\mathbf{R})$ only, is also valid for describing the scattered field in the region $D(\mathbf{R}) < z < D_{\text{max}}$. On the basis of what has been seen before, this is clearly inaccurate in general, however, for low roughness it yields correct results and constitutes the starting point of an important class of perturbation methods for the scattered field. It was subsequently extended by Fano [7.39] to light diffraction from reflection gratings and by Rice [7.40] to the scattering of electromagnetic waves from perfectly conductive random surfaces.

From the continuity condition, Eq. (1.6a), and the fact that the field in the perfect conductor is zero, we derive that the tangential component of the electric vector is zero on the surface of the perfectly conducting surface. Decomposing the field into the sum of an incident and a scattered vector, $\mathbf{E}^{(i)}$ and $\mathbf{E}^{(s)}$, respectively; this leads to:

$$[\mathbf{n} \times \mathbf{E}^{(s)}]_S = -[\mathbf{n} \times \mathbf{E}^{(i)}]_S. \tag{7.68}$$

On using the angular spectrum representation for the scattered field (cf. Eqs. (7.15)–(7.17)), and on the basis of the Rayleigh hypothesis, one obtains from Eqs. (7.68) and (7.1):

$$\begin{aligned}\mathbf{n}(\mathbf{R}, D(\mathbf{R})) \times \mathbf{e} \exp[i(\mathbf{K}^{(i)}\cdot\mathbf{R} - k_z^{(i)}D(\mathbf{R}))] \\ = \mathbf{n}(\mathbf{R}, D(\mathbf{R})) \int_{-\infty}^{\infty} \exp(i\mathbf{Q}\cdot\mathbf{R}) \exp(ik_z D(\mathbf{R})) \\ \times [\mathbf{k}^{(+)} \times (\mathbf{k}^{(+)} \times \mathbf{A}(\mathbf{Q}))]\, d^2Q.\end{aligned} \tag{7.69}$$

Equation (7.69) constitutes a boundary condition for $\mathbf{A}(\mathbf{Q})$. A method based on this equation has the advantage over a procedure using the extinction theorem, in that it does not require one to find first the source field (namely, the electric current density $\mathbf{J}$ in the case of a perfect conductor). However, the analytical solution of the integral equation (7.69) for $\mathbf{A}(\mathbf{Q})$ requires a series expansion. A direct numerical solution, on the other hand, entails certain difficulties in connection with the number of homogeneous and evanescent components to be included in the integral, Eqs. (7.69) [7.32].

The series solution of Eq. (7.69) is based on the expansions, Eqs. (7.44). This is the procedure of Rice [7.40]. When one carries out the calculations, however, one discovers that up to fourth order of $k\sigma$ for s-waves, and up to second order of $k\sigma$ for p-waves, the diffuse component of the mean scattered intensity obtained from the Rayleigh method, Eq. (7.69), is identical to that obtained in Sections 7.6.1 and 7.6.2 from the extinction theorem [7.32], [7.38], [7.41]. Thus, one can expect that the accuracy of the result will be limited more by the truncation of the perturbative series than by the validity of Eq. (7.69). This has been proved, in fact, to be the case.

The Rayleigh method has been used by several authors in electromagnetic scattering [7.42] and in light interaction with metal and dielectric surfaces [7.43]–[7.45]. Its precise range of validity is not easy to establish. For diffraction gratings, this range has been studied by several authors, among them we should quote the works by Lipmann [7.46] and Urestky [7.47]. On the other hand, Petit and Cadilhac [7.48] and Millar [7.49] made an elegant study on the validity of the Rayleigh hypothesis based on the analytic properties of the scattered field.

For a sinusoidal grating of profile $z = \mathsf{h}\cos(2\pi x/a)$, Millar [7.49], using arguments based on the properties of analyticity of the scattered field, arrived at the following condition of validity:

$$\frac{\mathsf{h}}{a} \leqslant 0.0713. \tag{7.70}$$

Nevertheless, Wirgin [7.50], [7.51] found empirically the correctness of results for h and a surpassing the limit in Eq. (7.70). Similarly, in a one-dimensional rough surface scattering, the authors of Ref. 7.32, following a suggestion by Maystre and Rossi [7.52] on the equivalence of amplitude h and roughness σ and of period d and correlation length T, proposed a translation of the criterion (7.70) to random surfaces, which reads:

$$\frac{\sigma}{T} \leqslant 0.224. \tag{7.71}$$

It was found again empirically from comparison with Monte Carlo simulations [7.32], that the validity of Eq. (7.69) slightly surpassed the limit in Eq. (7.71).

7.8. ILLUSTRATION OF THE SCATTERING EQUATIONS FOR ONE-DIMENSIONAL SURFACES

We shall consider in this section one-dimensional surfaces, namely, those for which the profile $z = D(\mathbf{R})$ is constant in one variable (to be specific in what follows, we shall assume them to be constant along the y-axis, then the scattering problem is defined in the xz-plane of incidence). The interest of these surfaces stems from the possibility of directly obtaining an exact numerical solution to Eqs. (7.5) and (7.6).

The Green function $G(\mathbf{r}, \mathbf{r}')$ is in this case [7.53]:

$$G(\mathbf{r}, \mathbf{r}') = i\pi H_0^{(1)}(k|\mathbf{r} - \mathbf{r}'|), \tag{7.72}$$

$H_0^{(1)}$ being the Hankel function of the first kind and zero order.

For an s-polarized incident wave (cf. Eqs. (7.1) and (7.30)), using Eq. (7.72), the extinction theorem, Eq. (7.7), yields the boundary condition (cf. Problem 7.5):

$$e \exp[i(K^{(i)}x - k_z^{(i)}z)] = \frac{\pi k}{c} \int_{-\infty}^{\infty} J_y(x', D(x'))H_0^{(1)}(k|\mathbf{r} - \mathbf{r}'|) \frac{dx'}{n_{z'}}, \tag{7.73}$$

where $|\mathbf{r} - \mathbf{r}'| = \sqrt{(x - x')^2 + (z - D(x'))^2}$ and $n_{z'} = [1 + (dD/dx')^2]^{-1/2}$.

Similarly, Eq. (7.8) gives, for the electric vector of the field scattered above the surface:

$$E_y^{(s)}(x, z) = -\frac{\pi k}{c} \int_{-\infty}^{\infty} J_y(x', D(x'))H_0^{(1)}(k|\mathbf{r} - \mathbf{r}'|) \frac{dx'}{n_{z'}}. \tag{7.74}$$

Equations (7.73) and (7.74) are a consequence of the fact that in this case $J_x = J_z = 0$. They show that there is no cross-polarization sp or ps, as expected.

Equation (7.73) is solved for J_y by taking the limit as z tends to $D(x)$, i.e., the (x, z)-point is considered on the rough surface. In practice, the extent of the x'-integration is not infinite, but is restricted to a finite record L representing the portion of the surface illuminated by the incident wave.

For a p-polarized incident wave, it is more convenient to work with the magnetic field $\mathbf{H}$, which is now transversal. By following the same procedure as before, the reader can check that in this case $J_y = 0$ and that the extinction theorem yields the boundary condition (cf. Problem 7.6):

$$h \exp[i(K^{(i)}x - k_z^{(i)}z)] = -\frac{\pi i k}{c} \int_{-\infty}^{\infty} J_x(x', D(x'))H_1^{(1)}(k|\mathbf{r} - \mathbf{r}'|) \times \frac{(z - D(x')) - \dfrac{dD}{dx'}(x - x')}{|\mathbf{r} - \mathbf{r}'|} \frac{dx'}{n_{z'}}, \tag{7.75}$$

the vector $\mathbf{h}$ which defines the linear polarization of $\mathbf{H}^{(i)}$ being $(0, h, 0)$.

Similarly, the magnetic vector of the field scattered above the surface is:

$$H_y^{(s)}(x, z) = \frac{\pi i k}{c} \int_{-\infty}^{\infty} J_x(x', D(x'))H_1^{(1)}(k|\mathbf{r} - \mathbf{r}'|) \times \frac{(z - D(x')) - \dfrac{dD}{dx'}(x - x')}{|\mathbf{r} - \mathbf{r}'|} \frac{dx'}{n_{z'}}, \qquad (7.76)$$

where $H_1^{(1)}$ is the Hankel function of the first kind and first order. Once again, Eq. (7.75) is solved for J_z as z tends to $D(x)$.

The singularity of $H_1^{(1)}(k\sqrt{(x - x')^2 + (z - D(x'))^2})$ at $\mathbf{r} = \mathbf{r}'$ is not integrable (this is in contrast with that of $H_0^{(1)}(k\sqrt{(x - x')^2 + (z - D(x'))^2})$ appearing in Eq. (7.73) for s-polarization). This makes it necessary to evaluate the principal value of the x'-integral in Eq. (7.75). The procedure to accomplish this can be found in the work by Andreasen [7.54] and leads to the following expression for Eq. (7.75):

$$h \exp[i(K^{(i)}x - k_z^{(i)}D(x))] = \frac{2\pi}{c} J_x(x, D(x)), \qquad (7.77a)$$

for $\mathbf{r} = \mathbf{r}'$ in the integral of Eq. (7.75);

$$h \exp[i(K^{(i)}x - k_z^{(i)}D(x))] = -\frac{\pi i k}{c} \int_{-\infty}^{\infty} J_x(x', D(x'))H_1^{(1)}(k|\mathbf{r} - \mathbf{r}'|) \times \frac{(z - D(x')) - \dfrac{dD}{dx'}(x - x')}{|\mathbf{r} - \mathbf{r}'|} \frac{dx'}{n_{z'}}. \qquad (7.77b)$$

for $\mathbf{r} \neq \mathbf{r}'$ in the integral of Eq. (7.75).

7.9. NUMERICAL SOLUTION OF THE SCATTERING EQUATIONS

7.9.1. Numerical Generation of Random Surfaces

The numerical solution of the boundary condition, represented by the extinction theorem, by means of computer simulations of the process of scattering, constitutes a new tool capable of yielding data from surfaces with high slopes, which none of the approximation techniques, discussed previously, can yield. As discussed in Sections 7.5.3 and 7.6.2, this also permits a better understanding of the range of validity of the analytical approximations, since it systematically renders a broad collection of scattering data for different surface parameters. However, due to the limitations in memory and speed of computers, this procedure has been so far developed for one-dimensional profiles only.

It should be remarked that the computational solution of Eq. (7.7) is preferable to that of Eq. (7.18), since the inversion of the latter is numerically unstable.

Surfaces $z = D(x)$ with desired statistics are generated by a Monte Carlo method, then the profile is introduced into Eqs. (7.73)–(7.77).

For instance, surfaces with statistics as discussed in Section 7.2, with a Gaussian correlation function, are constructed in the following way:

1. A sequence $[r_n]$ of uncorrelated random numbers (typically 100 000 numbers), with normal statistics, zero mean value, and unit variance, is constructed from another sequence of pseudorandom numbers, uniformly distributed in the interval (0, 1), directly generated by a computer routine.
2. $[r_n]$ is scaled to the desired variance σ, then, the resulting sequence $[t_n]$ is correlated with the exponential function whose samples are:

$$f_n = \frac{2}{\sqrt{\pi T}} \exp\left[-2\left(\frac{n}{T}\right)^2\right], \tag{7.78}$$

   thus obtaining another sequence $[z_n]$ with correlation function $B(\tau) = \exp(-\tau^2/T^2)$ which has the desired correlation length T, each number z_n corresponds to a sampling point $z_z = D_n$, $(D_n = D(x_n))$ of the simulated surface profile.

Due to the limited capability of the computer for the numerical integration, a segment of length L is extracted out of the sequence $[z_n]$ (this represents a sample of the surface, and it contains a few hundred sampling points z_n). Therefore, the x-integration is in fact restricted to the length L, which is then considered as the effective record of illumination of the surface.

7.9.2. Numerical Calculation of the Scattering Integrals

We shall illustrate the numerical method for s-polarization. The case of p-polarization is worked out in a similar manner. Equation (7.73) with the left-hand side evaluated at $z = D(x)$ is solved in the following way:

The L-interval of x-integration is divided into N segments of length $\delta = L/N$. At the midpoint x_m of each segment one defines the corresponding sampling value J_m as:

$$J_m = \frac{\pi k}{c} J_y(x_m) \sqrt{1 + \left(\frac{\Delta D}{\Delta x}\right)_m^2}, \qquad m = 1, \ldots, N, \tag{7.79}$$

where $(\Delta D/\Delta x)_m$ is the slope of the surface in the corresponding segment around the sampling point x_m.

The discretization of the integral in Eq. (7.73) gives rise to the matrix elements:

$$A_{nm} = H_0^{(1)}(k\sqrt{(x_n - x_m)^2 + (D_n - D_m)^2})\delta, \qquad n \neq m, \quad n, m = 1, \ldots, N. \tag{7.80}$$

On the other hand, for $n = m$ the singularity of $H_0^{(1)}$ is integrable, this means that the diagonal terms of the $\mathbf{A}$-matrix are calculated as:

$$A_{nm} = \int_{(m-1)\delta}^{m\delta} H_0^{(1)}(k\sqrt{(x_n - x')^2 + (D_n - D(x'))^2})\, dx', \qquad n = m. \tag{7.81}$$

The left-hand side of Eq. (7.73) at each sampling argument x_n is:

$$E_n = e \exp[i(K^{(i)} x_n - k_z^{(i)} D_n)], \qquad n = 1, \ldots, N. \tag{7.82}$$

Some authors [7.55], [7.56] use a taper incident field, namely, they multiply the incident wavefield by a taper function, usually a Gaussian. This confers a beam structure to the incident field.

Therefore, from Eqs. (7.79)–(7.82), Eq. (7.73) leads to the algebraic system of equations:

$$E_n = A_{nm} J_m, \qquad n, m = 1, \ldots, N, \tag{7.83}$$

from which the sampling values J_m of the surface electric current density are to be determined (current computer routines control the stability of the solution of Eq. (7.83)). Once these values have been found, the angular distribution of mean scattered intensity is obtained from a similar discretization of Eq. (7.74) for the scattered field, and then from Eq. (7.38).

The averages are performed over several different samples of length L of the same sequence $[z_n]$ of given parameters T and σ. Typically, the number of samples is a few hundred, although an appreciable numerical convergence of the averages is usually obtained within one hundred samples, or even less.

The verification of the unitarity conditions by each of the scattered intensities obtained from each sample is a criterion that should be taken as a measure of the numerical consistency of the results. As a matter of fact, this method permits one to calculate in a straightforward way any statistical moment of the scattered field, other than its second order moment [7.57]. Also, surfaces with other statistics can be similarly studied [7.17].

Equation (7.80) shows that the **A**-matrix is symmetric. On the other hand, only the left-hand side terms E_n depend on the angle of incidence θ_0. Since it is the inversion of the **A**-matrix that makes the computations time consuming, there is not much difference in the calculation time between the obtention of **J** for one angle of incidence and for several of these angles. In addition, this fact permits one to duplicate the effective number of samples over which the averages are performed, since for each inverted matrix, calculations of the scattered intensity at angles of incidence θ_0 and $-\theta_0$ correspond to a surface sample and to its specular image, respectively.

It should be remarked, however, that these numerical procedures still have two main limitations; first, since the model is restricted to one-dimensional surfaces, for s or p-incident waves no depolarization of the scattered field can be accounted for; second, the number of sampling points that it is possible to deal with for each sample (owing to reasons of memory and computer time) is only of a few hundred; consequently, the illuminated length L of the surface is rather small; and thus the angular resolution of the scattered intensity is low. Nevertheless this method can still show important aspects of the angular distribution of mean scattered intensity and of other statistical moments of the scatter wavefunction.

To give an idea of the quantities involved in the numerical computations, for instance, for surface samples with 221 sampling points, if $T = \lambda$ then L is typically $L = 22\lambda$ ($\delta = 0.1\lambda$) (although for $\sigma/T \ll 1$ the value L can be made three times larger). For $T = 0.2\lambda$ and $T = 0.25\lambda$, L is about: $L = 12\lambda$ ($\delta = 0.05\lambda$). For $T > 3\lambda$, L can be made equal to 60λ ($\delta = 0.27\lambda$).

7.10. SCATTERING FROM DEEPLY ROUGH SURFACES. ENHANCED BACKSCATTERING

Scattering experiments from random rough surfaces with controlled statistics, made by Renau and coworkers [7.58], [7.59], have been able to show the angular dependence of the scattered intensity on the polarization and in part to test some of the existing theories such as the Kirchhoff approximation. Recently, experiments in highly sloped surfaces, done by Mendez and O'Donnell [7.24], [7.60], have shown new features concerning the dependence of the scattering process on the polarization, as well as the presence of a peak of the angular distribution of mean scattered intensity in the direction of backscattering; this is the *enhanced backscattering* phenomenon.

Figures 7.10(a)–(c) show the existence of this peak as well as a certain degree of depolarization for incident *s*-waves under three different angles of incidence.

The phenomenon of enhanced backscattering is characteristic of a strong interaction (multiple scattering) of the field with the random structure, and has a relationship with the effect of weak localization of photons in random media [7.61]–[7.68]. This is due to constructive interference between a wave outgoing from the scatterer as a result of a series of interactions with the surface and the wave emitted from the reverse process (in the sense of the reciprocity relations (5.60)) [7.63], [7.64]. The width of this peak is about λ/T (see the work by Maradudin *et al.* in p. 157 of Ref. 7.68).

These results are not obtained with any of the analytical approximations discussed in previous sections. However, the numerical solution of the scattering equations as described in Section 7.9 has permitted the derivation of those features that are predictable with a one-dimensional model and, as such, do not account for cross-polarization effects [7.22], [7.69].

For instance, the aforementioned Figs. 7.9(a) and (b), quoted in Section 7.6.2, show different mean scattered intensities under *s*- or *p*-polarization. On the other hand, Figs. 7.11(a)–(d) show the angular distribution of mean scattered intensity, calculated according to the numerical procedure of Section 7.9.2, from perfectly conductive surfaces, all with the same correlation length $T = 1.8\lambda$, but with four different values of the roughness σ. (Note that in these figures the positions of the specular and backscattering directions are interchanged with respect to those of Figs. 7.10.) As σ increases, keeping the same value for T, we see: first, in Fig. 7.11(a) a regime in which the Kirchhoff approximation is valid; then Fig. 7.11(b) shows a domain in which this approximation is no longer valid, the halo broadens and the surface behaves almost as a *Lambertian diffuser*, at least for angles θ of observation not too large. This range of angles θ at which the quasi-Lambertian regime occurs decreases with the angle of incidence. As σ increases further (Fig. 7.11(c)) a peak in the backscattering direction appears. This peak is higher the larger σ is (Fig. 7.11(d)). As the angle of incidence increases, the height of this peak decreases due to shadowing from the surface ridges.

In addition, the angular distribution of scattered intensity from one sample exhibits *speckle* fluctuations that swamp this peak (see Fig. 4 of Ref. 7.69 and Ref. 7.70). Consequently, the enhancement in the backscattering direction can only be observed after averaging this angular distribution over several samples.

As regards the quasi-Lambertian regime, this is found in surfaces with $\sigma/T \sim 0.35$ and T of the order of, or somewhat larger than, λ. On the other

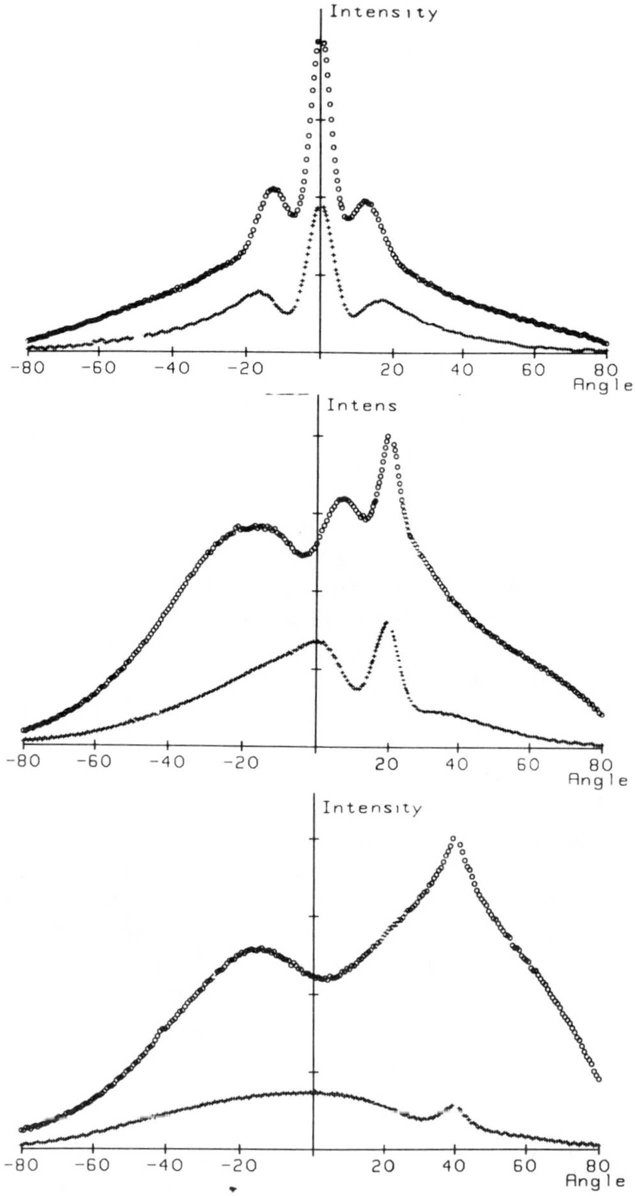

Figure 7.10. (a) Measurements of the angular distribution of mean scattered intensity from a random surface in a gold coated photoresist, with $T = 1.4$ μm and $\sigma = 1$ μm, illuminated with *s*-polarization under an angle of incidence $\theta_0 = 0°$, $\lambda = 0.633$ μm. Circles: *ss*-polarization. Pluses: *ps*-polarization. (b) Same as (a), but with $\theta_0 = 20°$. (c) Same as (a), but with $\theta_0 = 40°$. (After K. A. O'Donnell and E. R. Mendez, *J. Opt. Soc. Am. A* **4**, 1194 (1987).)

hand, keeping the values of σ and T, and decreasing the wavelength λ increases the enhanced backscattering peak (see, e.g. Refs. 7.22 and 7.71).

7.10.1. Comparison of Experimental Data and Numerical Results from One-Dimensional Surfaces

As regards the comparison of experiments and calculations, while the numerical computations apply to one-dimensional profiles, experiments such as those described in Section 7.10 are performed on two-dimensional samples.

The authors of Refs. 7.72–7.74 have made experimental measurements on specially fabricated metallic and dielectric one-dimensional surfaces of controlled statistics, following the procedure of Refs. 7.24 and 7.60. Figures 7.12 and 7.13 show the measured angular distribution of the diffuse component of the mean scattered intensity from a metal (gold) surface illuminated with a wavelength (infrared) at which this material behaves as a perfect conductor. Superimposed on these curves are the numerical results obtained on following the procedure of Section 7.9. The matching between both sets of data assesses the adequacy of the numerical model. It is to be remarked, however, that at shorter wavelengths (visible light), at which the same metal no longer behaves as a perfect conductor, the agreement between experiment and theory is poorer, as expected [7.72], [7.74]. The model of a perfect conductor should then be replaced by that corresponding to a real metal. When this is done the agreement with the experiment is better; nevertheless, some discrepancies still remain, especially at large angles of incidence; the cause is under current investigation (e.g., inaccuracy of the numerical result due to the finite sampling, subsurface scattering, or slight deviations in the estimation of the dielectric constant or in the calibration of the statistical parameters of the samples).

7.11. SCATTERING FROM METAL AND DIELECTRIC ROUGH SURFACES

We shall address the scattering of light and other electromagnetic waves from the surface of a penetrable, optically homogeneous (namely, with constant refractive index n), isotropic, nonmagnetic and spatially nondispersive medium. This material can be a (nonperfect conductive) metal or a dielectric. The solution to this problem constitutes an extension of the methods

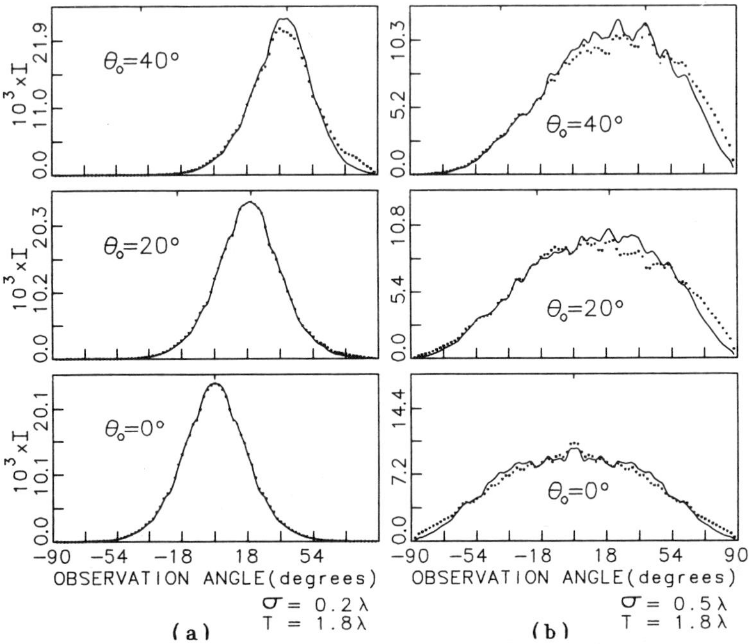

Figure 7.11. Calculations of the angular distribution of mean scattered intensity from a one-dimensional surface with $T = 1.8\lambda$. (a) $\sigma = 0.2\lambda$; (b) $\sigma = 0.5\lambda$.

discussed in previous sections, and can be addressed by using the Kirchhoff approximation [7.8], [7.18], the method of small perturbations based on the extinction theorem [7.37], [7.75], or the same kind of expansion based on the Rayleigh method [7.43], [7.45]. Also, numerical solutions have been obtained for this problem [7.56], [7.76], [7.77].

On application of the results of Section 1.6.2 and of Appendix 1.1 (cf. Eqs. (1.35) and (1.36), and Eqs. (28) and (29) of the Appendix), upon interaction with the surface of a field whose electric vector is given by Eq. (7.1), the boundary condition represented by the extinction theorem (the point **r** being inside the medium and **r**′ outside it) is:

$$\mathbf{E}^{(i)}(\mathbf{r}) + \frac{1}{4\pi k^2}\nabla \times \nabla \times \int_S \left[\mathbf{E}(\mathbf{r}')\frac{\partial G(\mathbf{r}, \mathbf{r}')}{\partial \mathbf{n}} - G(\mathbf{r}, \mathbf{r}')\frac{\partial \mathbf{E}(\mathbf{r}')}{\partial \mathbf{n}}\right] dS = 0, \quad (7.84)$$

for $z \leqslant D(\mathbf{R})$.

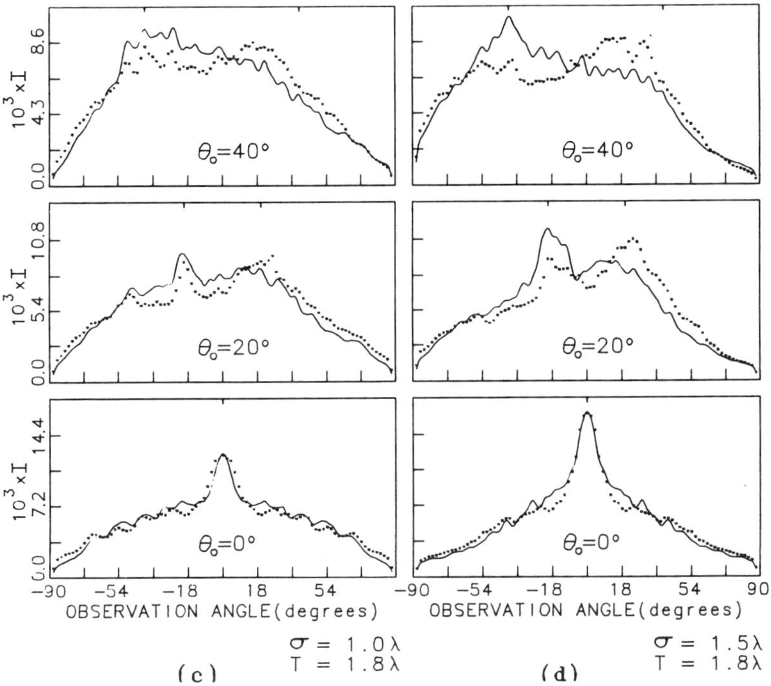

Figure 7.11 (*continued*). Calculations of the angular distribution of mean scattered intensity from a one-dimensional surface with $T = 1.8\lambda$. (c) $\sigma = \lambda$; (d) $\sigma = 1.5\lambda$. Solid line: *s*-polarization. Dotted line: *p*-polarization. (After M. Nieto-Vesperinas and J. M. Soto-Crespo, *Scattering in Volumes and Surfaces*, M. Nieto-Vesperinas and J. C. Dainty, eds., North-Holland, Amsterdam, 1990, p. 179.)

On the other hand, the field at points outside the medium consists of the sum of the incident field and the field scattered above the surface; namely, for $\mathbf{r}$ and $\mathbf{r}'$ outside the medium, this field is:

$$\mathbf{E}(\mathbf{r}) = \mathbf{E}^{(i)}(\mathbf{r}) + \frac{1}{4\pi k^2}\nabla \times \nabla \times \int_S \left[\mathbf{E}(\mathbf{r}')\frac{\partial G(\mathbf{r}, \mathbf{r}')}{\partial \mathbf{n}} - G(\mathbf{r}, \mathbf{r}')\frac{\partial \mathbf{E}(\mathbf{r}')}{\partial \mathbf{n}}\right] dS, \tag{7.85}$$

for $z > D(\mathbf{R})$.

The corresponding equations for the magnetic vector are obtained in a similar fashion.

In contrast to Eq. (7.5), Eq. (7.84) contains two functions to determine: the electric vector and its normal derivative on the surface. We shall now derive another pair of equations by using Green's theorem for the field inside the medium.

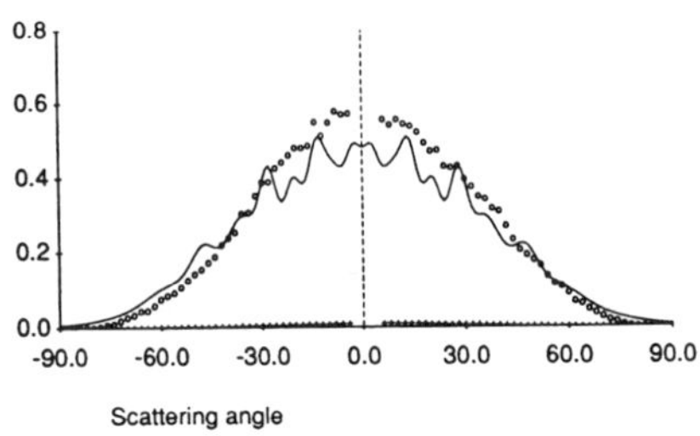

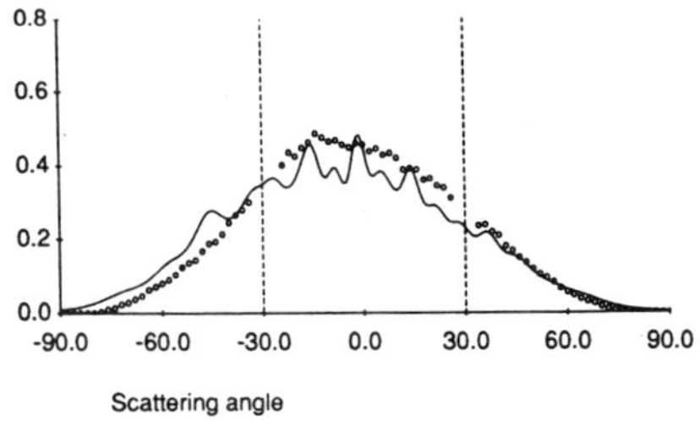

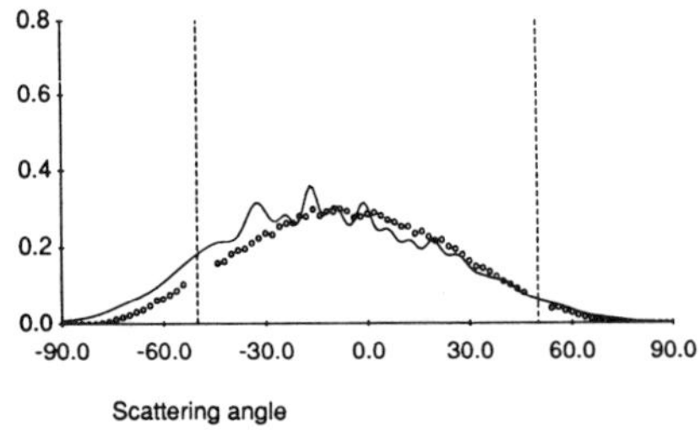

Figure 7.12. Angular distribution of the diffuse component of the mean scattered intensity from a one-dimensional rough surface in a gold coated photoresist with $\sigma = 1.2 \pm 0.1$ μm and $T = 2 \pm 0.2$ μm, $\lambda = 10.6$ μm, $\theta_0 = 0°$, 30°, and 50°. *s*-polarization; vertical lines indicate specular (left) and backscattering (right) directions. Circles: experimental measurements. Solid line: Numerical Monte Carlo calculations. (After J. C. Dainty, M. J. Kim, and A. J. Sant, *Scattering in Volumes and Surfaces*, M. Nieto-Vesperinas and J. C. Dainty, eds., North-Holland, Amsterdam, 1990, p. 151; M. J. Kim, J. C. Dainty, A. T. Friberg, and A. J. Sant, *J. Opt. Soc. Am. A* **7**, 569 (1990).)

The electric vector $\mathbf{E}^{(\text{in})}$ inside the material of permittivity $\varepsilon = n^2$ satisfies the wave equation:

$$\nabla \times \nabla \times \mathbf{E}^{(\text{in})}(\mathbf{r}) - k^2\varepsilon\mathbf{E}^{(\text{in})}(\mathbf{r}) = 0. \tag{7.86}$$

Let $\mathscr{G}^{(\text{in})}(\mathbf{r}, \mathbf{r}')$ be the Green function associated with Eq. (7.86), then $\mathscr{G}^{(\text{in})}(\mathbf{r}, \mathbf{r}')$ satisfies:

$$\nabla \times \nabla \times \mathscr{G}^{(\text{in})}(\mathbf{r}, \mathbf{r}') - k^2\varepsilon\mathscr{G}^{(\text{in})}(\mathbf{r}, \mathbf{r}') = -4\pi\delta(\mathbf{r} - \mathbf{r}'). \tag{7.87}$$

$\mathscr{G}^{(\text{in})}(\mathbf{r}, \mathbf{r}')$ is given by an equation identical to Eq. (1.27) (cf. also Eqs. (1.28)–(1.32)), apart from the wavenumber which now must be replaced by $k\sqrt{\varepsilon}$; and thus of the outgoing spherical wavefunction from which $\mathscr{G}^{(\text{in})}(\mathbf{r}, \mathbf{r}')$ is constructed, that now should be:

$$G^{(\text{in})}(\mathbf{r}, \mathbf{r}') = \frac{\exp(ik\sqrt{\varepsilon}\,|\mathbf{r} - \mathbf{r}'|)}{|\mathbf{r} - \mathbf{r}'|}. \tag{7.88}$$

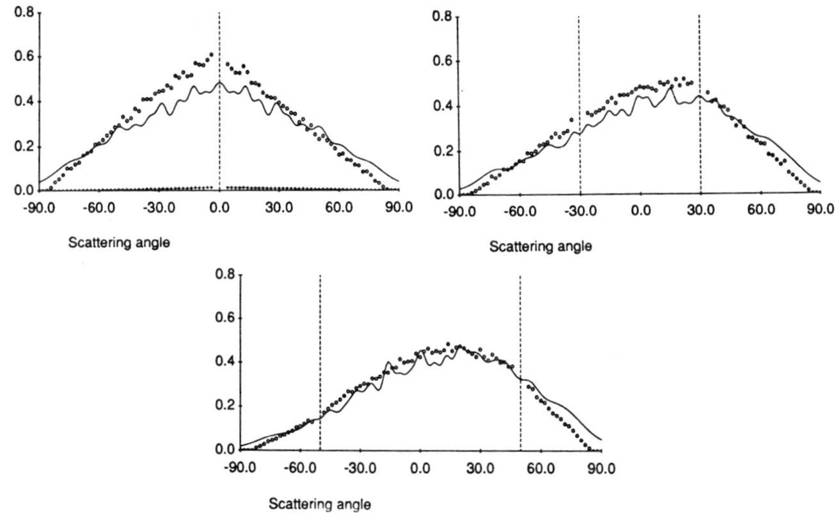

Figure 7.13. Same as Fig. 7.12, *p*-polarization. (After J. C. Dainty, M. J. Kim and A. J. Sant, *Scattering in Volumes and Surfaces*, M. Nieto-Vesperinas and J. C. Dainty, eds., North-Holland, Amsterdam, 1990, p. 151; M. J. Kim, J. C. Dainty, A. T. Friberg and A. J. Sant, *J. Opt. Soc. Am. A* **7**, 569 (1990).)

Therefore, on using Green's theorem, Eq. (1.33), for $\mathbf{E}^{(\mathrm{in})}(\mathbf{r})$ and $\mathscr{G}^{(\mathrm{in})}(\mathbf{r}, \mathbf{r}')$ one obtains, by a procedure similar to that leading to Eqs. (7.84) and (7.85), the following expressions:

For **r** outside the medium and **r**′ inside:

$$-\frac{1}{4\pi k^2 \varepsilon} \nabla \times \nabla \times \int_S \left[\mathbf{E}^{(\mathrm{in})}(\mathbf{r}') \frac{\partial G^{(\mathrm{in})}(\mathbf{r}, \mathbf{r}')}{\partial \mathbf{n}} - G^{(\mathrm{in})}(\mathbf{r}, \mathbf{r}') \frac{\partial \mathbf{E}^{(\mathrm{in})}(\mathbf{r}')}{\partial \mathbf{n}} \right] dS = 0,$$

$$z > D(\mathbf{R}). \quad (7.89)$$

On the other hand, when the medium is a dielectric, for **r** and **r**′ inside, we obtain for the field scattered by the rough surface inside the material (refracted field), the following expression:

$$\mathbf{E}^{(\mathrm{in})}(\mathbf{r}) = -\frac{1}{4\pi k^2 \varepsilon} \nabla \times \nabla \times \int_S \left[\mathbf{E}^{(\mathrm{in})}(\mathbf{r}') \frac{\partial G^{(\mathrm{in})}(\mathbf{r}, \mathbf{r}')}{\partial \mathbf{n}} - G^{(\mathrm{in})}(\mathbf{r}, \mathbf{r}') \frac{\partial \mathbf{E}^{(\mathrm{in})}(\mathbf{r}')}{\partial \mathbf{n}} \right] dS, \quad (7.90)$$

for $z \leqslant D(\mathbf{R})$.

The corresponding equations for the magnetic vectors are obtained in a similar way.

By using the continuity equations (1.6) ($\sigma = 0$ and $\mathbf{J} = 0$), one can obtain the field and its normal derivative from Eqs. (7.84), (7.85), (7.89), and (7.90) when the point $\mathbf{r}$ at which they are evaluated tends towards the surface. These expressions, however, are not independent of each other. For instance, one can solve for $\mathbf{E}$ and $\partial\mathbf{E}/\partial\mathbf{n}$ on the surface from Eqs. (7.85) and (7.89) as $z \to D(\mathbf{r})$ from outside the medium. Alternatively, one can solve for these boundary vaues from the pair of Eqs. (7.84) and (7.90) by making $z \to D(\mathbf{r})$ from inside the medium (see Problem 7.9).

7.11.1. Example: One-Dimensional Surfaces

We shall illustrate the use of the equations described above by applying them to the scattering of a plane wave from a one-dimensional surface with a profile $z = D(x)$.

For an s-polarized incident wave (Eqs. (7.1) and (7.30)) the problem is identical to one for scalar wavefields; the reader can check that in this case Eqs. (7.84) and (7.85) become, respectively:

$$e_2 \exp[i(K^{(i)}x - k_z^{(i)}z)] + \frac{i}{4}\int_{-\infty}^{\infty}\left[E(\mathbf{r}')\frac{\partial H_0^{(1)}(k|\mathbf{r}-\mathbf{r}'|)}{\partial \mathbf{n}} - H_0^{(1)}(k|\mathbf{r}-\mathbf{r}'|)\frac{\partial E(\mathbf{r}')}{\partial \mathbf{n}}\right]\frac{dx'}{n_{z'}} = 0, \quad z \leqslant D(x), \tag{7.91}$$

where

$$\mathbf{r}' = (x', D(x')) \qquad \text{and} \qquad n_{z'} = [1 + (\partial D/\partial x')^2]^{-1/2};$$

and:

$$E(\mathbf{r}) = e \exp[i(K^{(i)}x - k_z^{(i)}z)] + \frac{i}{4}\int_{-\infty}^{\infty}\left[E(\mathbf{r}')\frac{\partial H_0^{(1)}(k|\mathbf{r}-\mathbf{r}'|)}{\partial \mathbf{n}} - H_0^{(1)}(k|\mathbf{r}-\mathbf{r}'|)\frac{\partial E(\mathbf{r}')}{\partial \mathbf{n}}\right]\frac{dx'}{n_{z'}}, \quad z > D(x). \tag{7.92}$$

Similarly, Eqs. (7.89) and (7.90) yield, respectively:

$$-\frac{i}{4}\int_{-\infty}^{\infty}\left[E^{(\text{in})}(\mathbf{r}')\frac{\partial H_0^{(1)}(k\sqrt{\varepsilon}|\mathbf{r}-\mathbf{r}'|)}{\partial \mathbf{n}} - H_0^{(1)}(k\sqrt{\varepsilon}|\mathbf{r}-\mathbf{r}'|)\frac{\partial E^{(\text{in})}(\mathbf{r}')}{\partial \mathbf{n}}\right]\frac{dx'}{n_{z'}} = 0, \quad z > D(x), \tag{7.93}$$

and:

$$E^{(\text{in})}(\mathbf{r}) = -\frac{i}{4}\int_{-\infty}^{\infty}\left[E^{(\text{in})}(\mathbf{r}')\frac{\partial H_0^{(1)}(k\sqrt{\varepsilon}\,|\mathbf{r}-\mathbf{r}'|)}{\partial \mathbf{n}} - H_0^{(1)}(k\sqrt{\varepsilon}\,|\mathbf{r}-\mathbf{r}'|)\frac{\partial E^{(\text{in})}(\mathbf{r}')}{\partial \mathbf{n}}\right]\frac{dx'}{n_{z'}}, \quad z \leqslant D(x), \quad (7.94)$$

Once again, it should be remembered that, in practice, the x'-integrals are restricted to a finite record of length L.

On using Maxwell's equations (1.1), the continuity Eqs. (1.6a) and (1.6b) give, respectively:

$$[E(\mathbf{r})]_{z=D^{(+)}} = [E^{(\text{in})}(\mathbf{r})]_{z=D^{(-)}}, \quad (7.95)$$

and:

$$\left[\frac{\partial E(\mathbf{r})}{\partial \mathbf{n}}\right]_{z=D^{(+)}} = \left[\frac{\partial E^{(\text{in})}(\mathbf{r})}{\partial \mathbf{n}}\right]_{z=D^{(-)}}. \quad (7.96)$$

In Eqs. (7.95) and (7.96) $D^{(+)}$ and $D^{(-)}$ denote the value of z at the surface by approaching it from the regions $z > D(x)$ and $z < D(x)$, respectively.

On introducing Eqs. (7.95) and (7.96) into Eqs. (7.91) and (7.92) evaluated on the surface (namely, with $z = D(x)$), and denoting by $E(x)$ either side of Eq. (7.95), and by $n_z F(x)$ either side of Eq. (7.96), one obtains, respectively:

$$e\exp[i(K^{(i)}x - k_z^{(i)}D(x))] + \frac{i}{4}\int_{-\infty}^{\infty}\left[E(x')\left(\frac{\partial H_0^{(1)}(k|\mathbf{r}-\mathbf{r}'|)}{\partial z'} - D'(x')\frac{\partial H_0^{(1)}(k|\mathbf{r}-\mathbf{r}'|)}{\partial x'}\right) - H_0^{(1)}(k|\mathbf{r}-\mathbf{r}'|)F(x')\right]dx' = 0, \quad (7.97)$$

and:

$$E(\mathbf{r}) = e\exp[i(K^{(i)}x - k_z^{(i)}D(x))] + \frac{i}{4}\int_{-\infty}^{\infty}\left[E(x')\left(\frac{\partial H_0^{(1)}(k|\mathbf{r}-\mathbf{r}'|)}{\partial z'} - D'(x')\frac{\partial H_0^{(1)}(k|\mathbf{r}-\mathbf{r}'|)}{\partial x'}\right) - H_0^{(1)}(k|\mathbf{r}-\mathbf{r}'|)F(x')\right]dx'. \quad (7.98)$$

By a strictly identical procedure, Eqs. (7.93) and (7.94) give, respectively:

$$-\frac{i}{4}\int_{-\infty}^{\infty}\left[E(x')\left(\frac{\partial H_0^{(1)}(k\sqrt{\varepsilon}\,|\mathbf{r}-\mathbf{r}'|)}{\partial z'} - D'(x')\frac{\partial H_0^{(1)}(k\sqrt{\varepsilon}\,|\mathbf{r}-\mathbf{r}'|)}{\partial x'}\right) - H_0^{(1)}(k\sqrt{\varepsilon}\,|\mathbf{r}-\mathbf{r}'|)F(x')\right]dx' = 0, \quad (7.99)$$

and:

$$E(\mathbf{r}) = -\frac{i}{4}\int_{-\infty}^{\infty}\left[E(x')\left(\frac{\partial H_0^{(1)}(k\sqrt{\varepsilon}\,|\mathbf{r}-\mathbf{r}'|)}{\partial z'} - D'(x')\frac{\partial H_0^{(1)}(k\sqrt{\varepsilon}\,|\mathbf{r}-\mathbf{r}'|)}{\partial x'}\right)\right.$$
$$\left. - H_0^{(1)}(k\sqrt{\varepsilon}\,|\mathbf{r}-\mathbf{r}'|)F(x')\right]dx'. \quad (7.100)$$

In Eqs. (7.97)–(7.100) $\mathbf{r} = (x, D(x))$ and $\mathbf{r}' = (x', D(x'))$.

The pairs of Eqs. (7.98) and (7.99), or (7.97) and (7.100), permit the obtention of the source functions $E(x)$ and $F(x)$. Once these functions have been calculated, the scattered reflected field in the region $z > D(x)$ and the scattered refracted field in the region $z < D(x)$ are found from Eqs. (7.92) and (7.94), respectively. The procedure for determining the mean scattered intensities is done numerically, as in Section 7.9.2, from several surface samples. The x-integrals are extended over the finite record L of the sample. The far fields reflected and transmitted are obtained from the following expressions (cf. Problem 7.10).

Field scattered into $z > D(x)$:

$$E(\theta) = \frac{\exp[i(kr - \pi/4)]}{2\sqrt{2\pi kr}}\int_{-\infty}^{\infty}[k(\cos\theta - D'(x')\sin\theta)E(x') - iF(x')]$$
$$\times \exp(-i\mathbf{k}^{(+)}\cdot\mathbf{r}')\,dx', \quad (7.101)$$

with a similar equation for the field scattered (transmitted) into $z < D(x)$, where k is replaced by $k\sqrt{\varepsilon}$, the angle of scattering θ above the surface is substituted by the angle of scattering $-\theta$ below the surface, and $F(x')$ is replaced by $-F(x')$.

Once again, the unitarity condition may be adopted as a criterion of numerical consistency of the results:

$$R + T = 1. \quad (7.102)$$

In Eq. (7.102) R and T are the *reflectance* and *transmittance*, respectively; namely, the total reflected and transmitted intensities over all angles of scattering, normalized to the total incident intensity.

For p-polarized waves, it is convenient to work with the magnetic vector. If $h\exp[i(K^{(i)}x - k_z^{(i)}z)]$ denotes the magnetic vector of the field incident on the surface, then the corresponding equations are obtained in a similar fashion as before, to be:

For $\mathbf{r}'$ in vacuum:

$$H(\mathbf{r}) = h\exp[i(K^{(i)}x - k_z^{(i)}z)] + \frac{1}{4\pi}\int_{-\infty}^{\infty}\left[H(\mathbf{r}')\frac{\partial G(\mathbf{r},\mathbf{r}')}{\partial \mathbf{n}} - G(\mathbf{r},\mathbf{r}')\frac{\partial H(\mathbf{r}')}{\partial \mathbf{n}}\right]\frac{dx'}{n_{z'}}, \quad z > D(x). \quad (7.103a)$$

$$h\exp[i(K^{(i)}x - k_z^{(i)}z)] + \frac{1}{4\pi}\int_{-\infty}^{\infty}\left[H(\mathbf{r}')\frac{\partial G(\mathbf{r},\mathbf{r}')}{\partial \mathbf{n}} - G(\mathbf{r},\mathbf{r}')\frac{\partial H(\mathbf{r}')}{\partial \mathbf{n}}\right]\frac{dx'}{n_{z'}} = 0, \quad z \leqslant D(x); \quad (7.103b)$$

and for $\mathbf{r}'$ in the medium of permittivity ε:

$$-\frac{1}{4\pi}\int_{-\infty}^{\infty}\left[H^{(\mathrm{in})}(\mathbf{r}')\frac{\partial G^{(\mathrm{in})}(\mathbf{r},\mathbf{r}')}{\partial \mathbf{n}} - G^{(\mathrm{in})}(\mathbf{r},\mathbf{r}')\frac{\partial H^{(\mathrm{in})}(\mathbf{r}')}{\partial \mathbf{n}}\right]\frac{dx'}{n_{z'}} = 0, \quad z > D(x), \quad (7.104a)$$

$$H^{(\mathrm{in})}(\mathbf{r}) = -\frac{1}{4\pi}\int_{-\infty}^{\infty}\left[H^{(\mathrm{in})}(\mathbf{r}')\frac{\partial G^{(\mathrm{in})}(\mathbf{r},\mathbf{r}')}{\partial \mathbf{n}} - G^{(\mathrm{in})}(\mathbf{r},\mathbf{r}')\frac{\partial H^{(\mathrm{in})}(\mathbf{r}')}{\partial \mathbf{n}}\right]\frac{dx'}{n_{z'}}, \quad z \leqslant D(x). \quad (7.104b)$$

The magnetic field and its normal derivative satisfy the continuity equations (cf. Eqs. (1.6)):

$$[H(\mathbf{r})]_{z=D^{(+)}} = [H^{(\mathrm{in})}(\mathbf{r})]_{z=D^{(-)}}, \quad (7.105)$$

and:

$$\left[\frac{\partial H(\mathbf{r})}{\partial \mathbf{n}}\right]_{z=D^{(+)}} = \frac{1}{\varepsilon}\left[\frac{\partial H^{(\mathrm{in})}(\mathbf{r})}{\partial \mathbf{n}}\right]_{z=D^{(-)}}. \quad (7.106)$$

By then defining the source functions: $H(x)$ as either side of Eq. (7.105) and by $n_z(x)L(x)$ as either side of Eq. (7.106), and on introducing these equations with this notation into Eqs. (7.103b) and (7.103a) and evaluating them at the surface, we obtain, respectively:

$$h\exp[i(K^{(i)}x - k_z^{(i)}D(x))] + \frac{i}{4}\int_{-\infty}^{\infty}\left[H(x')\left(\frac{\partial H_0^{(1)}(k|\mathbf{r}-\mathbf{r}'|)}{\partial z'} - D'(x')\frac{\partial H_0^{(1)}(k|\mathbf{r}-\mathbf{r}'|)}{\partial x'}\right) - H_0^{(1)}(k|\mathbf{r}-\mathbf{r}'|)L(x')\right]dx' = 0, \quad (7.107)$$

and:

$$H(x) = h_2 \exp[i(K^{(i)}x - K_z^{(i)}D(x))] + \frac{i}{4}\int_{-\infty}^{\infty}\left[H(x)\left(\frac{\partial H_0^{(1)}(k|\mathbf{r}-\mathbf{r}'|)}{\partial z'} - D'(x')\frac{\partial H_0^{(1)}(k|\mathbf{r}-\mathbf{r}'|)}{\partial x'}\right) - H_0^{(1)}(k|\mathbf{r}-\mathbf{r}'|)L(x')\right]dx'. \quad (7.108)$$

By a strictly identical procedure, Eqs. (7.104a) and (7.104b) give, respectively:

$$-\frac{i}{4}\int_{-\infty}^{\infty}\left[H(x')\left(\frac{\partial H_0^{(1)}(k\sqrt{\varepsilon}|\mathbf{r}-\mathbf{r}'|)}{\partial z'} - D'(x')\frac{\partial H_0^{(1)}(k\sqrt{\varepsilon}|\mathbf{r}-\mathbf{r}'|)}{\partial x'}\right) - \varepsilon H_0^{(1)}(k\sqrt{\varepsilon}|\mathbf{r}-\mathbf{r}'|)L(x')\right]dx' = 0, \quad (7.109)$$

and:

$$H(\mathbf{r}) = -\frac{i}{4}\int_{-\infty}^{\infty}\left[H(x')\left(\frac{\partial H_0^{(1)}(k\sqrt{\varepsilon}|\mathbf{r}-\mathbf{r}'|)}{\partial z'} - D'(x')\frac{\partial H_0^{(1)}(k\sqrt{\varepsilon}|\mathbf{r}-\mathbf{r}'|)}{\partial x'}\right) - \varepsilon H_0^{(1)}(k\sqrt{\varepsilon}|\mathbf{r}-\mathbf{r}'|)L(x')\right]dx'. \quad (7.110)$$

In Eqs. (7.107)–(7.110) $\mathbf{r} = (x, D(x))$ and $\mathbf{r}' = (x', D(x'))$.

The source functions H and L may be found from Eqs. (7.108) and (7.109), or alternatively, from Eqs. (7.107) and (7.110). The scattered fields, and mean scattered intensities, both reflected above the surface and transmitted into the medium, are then calculated through Eq. (7.103a) and (7.104b), respectively.

Figure 7.14 shows numerical calculations by the authors of Ref. 7.56 of the mean scattered intensity for p-waves from a one-dimensional surface of silver with parameters $\sigma = 1.2\ \mu\text{m}$, $T = 2\ \mu\text{m}$, and $\lambda = 612.7$ nm; the dielectric permittivity is: $\varepsilon = -17.2 + i0.498$. A well-defined peak of enhanced backscattering is exhibited. On the other hand, Fig. 7.15 shows experimental measurements, carried out by the authors of Ref. 7.73, of the mean scattered intensity from a one-dimensional dielectric surface, exhibiting a remarkable difference between s- and p-polarization in the angular distribution.

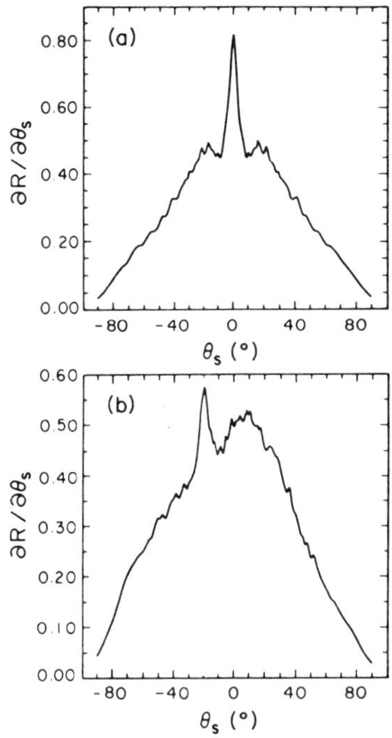

Figure 7.14. (a) Calculations of the angular distribution of mean scattered intensity from a one-dimensional silver surface with $T = 2\ \mu\text{m}$, $\sigma = 1.2\ \mu\text{m}$, $\lambda = 612.7$ nm, $\varepsilon = -17.2 + i0.498$. $L = 25.6\ \mu\text{m}$. Averages over 256 samples; *p*-polarization; $\theta_0 = 0°$. (b) Same as (a), but with $\theta_0 = 20°$. (After A. A. Maradudin, E. R. Mendez and T. Michel, *Opt. Lett.* **14**, 151 (1989).)

7.12. DIFFRACTION FROM PERIODIC SURFACES: REFLECTION GRATINGS

The scattering equations for one-dimensional surfaces have important applications in the case of light, and other electromagnetic waves, diffracted by periodic surfaces (gratings); namely, profiles $z = D(x)$ satisfying: $D(x) = D(x + a)$, a being the period. We shall illustrate the use of the equations addressed in Section 7.8 for perfectly reflecting periodic surfaces (*reflection gratings*). The extension to metal and dielectric surfaces is made in a similar way through the analysis of Section 7.11.1 [7.76].

From Eq. (7.18) it is seen that in the case of a grating the source function $\mathbf{F}(x)$ is also a periodic function with period a; therefore, one can easily see from Eq. (7.14) that the induced electric current density $\mathbf{J}(x)$ satisfies the pseudoperiodicity relation:

$$\mathbf{J}(x + a) = \mathbf{J}(x) \exp(iK^{(i)}x). \tag{7.111}$$

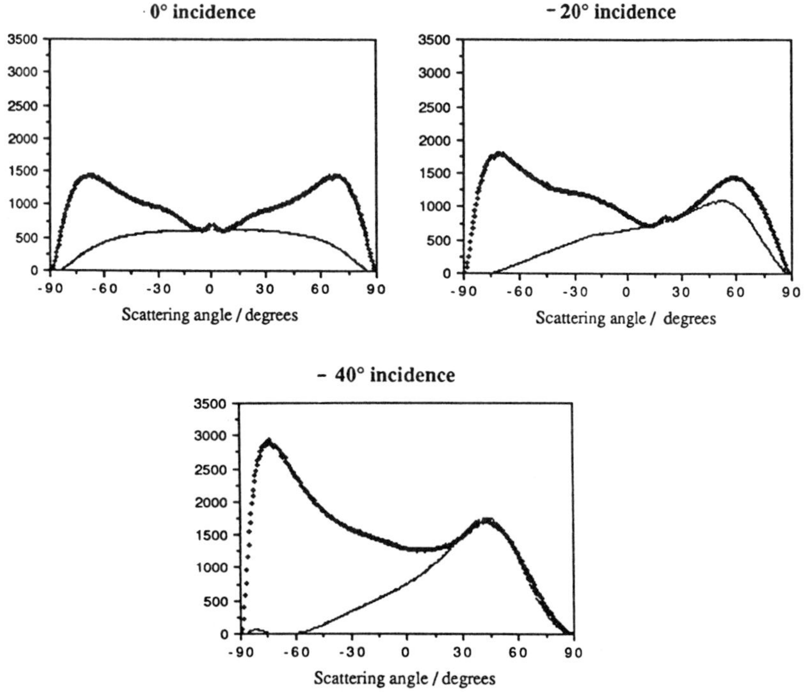

Figure 7.15. Measurements of the angular distribution of mean scattered intensity from a dielectric surface with $\varepsilon = 2.04$, $\sigma = 1.18 \pm 0.13$ µm, and $T = 2.97 \pm 0.05$ µm, $\lambda = 633$ nm for $\theta_0 = 0°$, 20°, and 40°. Bold line: *s*-polarization. Lighter line: *p*-polarization. (After A. J. Sant, J. C. Dainty and M. J. Kim, *Opt. Lett.* **14**, 1183 (1989).)

Equation (7.111) can be recognized as analogous to the Bloch–Floquet theorem of the theory of solids.

The Kirchhoff approximation [7.78], or perturbation techniques [7.79], can be applied to this problem as before.

Let us consider first the case of *s*-waves.

On introducing into Eq. (7.74) the two-dimensional plane wave expansion, analogous to Eq. (2.51) for the Green function (7.72), the x'-integral can be decomposed into an infinite sum of integrals, each of which is extended to one period; then, taking Eq. (7.111) into account, Eq. (7.74) is written as:

$$E_y^{(s)}(x, z) = -\frac{\pi k}{c}\int_0^a J_y(x')\frac{1}{\pi}\int_{-\infty}^{\infty} dK\,\frac{1}{k_z}\exp[i[K(x-x') + k_z(z - D(x'))]]$$

$$\times \sum_{l=-\infty}^{\infty}{}' \exp[i(K^{(i)} - K)la]\,\frac{dx'}{n_{z'}}, \quad (7.112)$$

where we have written $K = K^{(i)} + Q$.

Now, it is well known (see, e.g. Ref. 7.80), that the Fourier transform of a Dirac comb of period a is expressed as:

$$\sum_{l=-\infty}^{\infty} \exp[i(K^{(i)} - K)la] = \frac{2\pi}{a} \sum_{m=-\infty}^{\infty} \delta\left(K - \left(K^{(i)} + \frac{2\pi m}{a}\right)\right). \quad (7.113)$$

By means of Eq. (7.113), we can then write Eq. (7.112) as:

$$E_y^{(s)}(x, z) = \frac{2\pi k}{ca} \sum_{m=-\infty}^{\infty} \int_0^a \frac{J_y(x')}{q_m} \exp[iK_m(x - x')] \exp[iq_m|z - D(x')|] \frac{dx'}{n_{z'}}, \quad (7.114)$$

where:

$$K_m = K^{(i)} + \frac{2\pi m}{a}, \quad (7.115)$$

and q_m is:

$$q_m = \sqrt{k^2 - K_m^2}, \quad K_m \leqslant k, \text{ for homogeneous waves,} \quad (7.116a)$$

$$q_m = i\sqrt{K_m^2 - k^2}, \quad K_m > k, \text{ for evanescent waves.} \quad (7.116b)$$

By expressing $K^{(i)}$ and K_m in terms of the angle of incidence θ_0 and diffraction θ_m, respectively, Eq. (7.115) becomes:

$$\sin\theta_m = \sin\theta_0 + \frac{2\pi m}{ka}. \quad (7.117)$$

Equation (7.117), or its equivalent, Eq. (7.115), is the *grating equation.* It gives the position of the diffraction amplitudes A_m that represent the diffracted field above the surface:

$$E_y^{(s)}(x, z) = \sum_{m=-\infty}^{\infty} A_m \exp[i(K_m x + q_m z)], \qquad z > D(x). \quad (7.118)$$

The complex amplitudes A_m being (cf. Eq. (7.114)):

$$A_m = -\frac{2\pi k}{caq_m} \int_0^a J_y(x', D(x')) \exp[-i(K_m x' + q_m D(x'))] \frac{dx'}{n_{z'}}. \quad (7.119)$$

Also, the diffraction intensity corresponding to the mth diffraction order, normalized to the incident intensity $|e|^2$, is expressed as:

$$I_m^{(s)} = \frac{q_m}{k_z^{(i)}} \frac{|A_m|^2}{|e|^2}. \tag{7.120}$$

The electric current density J_y is calculated from the extinction theorem boundary condition, Eq. (7.73), which, on introducing the operations made above, becomes:

$$e \exp[i(K^{(i)}x - k_z^{(i)}z)] = \frac{\pi k}{c} \int_0^a J_y(x', D(x'))G^{(s)}(x, y; x', D(x')) \frac{dx'}{n_{z'}}. \tag{7.121}$$

In Eq. (7.121), $G^{(s)}$ is the Green function given by the following expansion:

$$G^{(s)}(x, y; x', D(x')) = \sum_{l=-\infty}^{\infty} H_0^{(1)}(k[[x - (x' + la)]^2 + [z - D(x')]^2]^{1/2}) \exp(iK_0 la). \tag{7.122}$$

Let us next consider the case of p-waves.

In a similar manner as before, Eq. (7.75) leads to the following boundary condition for the electric current density:

$$h \exp[i(K^{(i)}x - k_z^{(i)}z)] = -\frac{\pi ik}{c} \int_0^a J_x(x', D(x'))G^{(p)}(x, z; x', D(x')) \frac{dx'}{n_{z'}}. \tag{7.123}$$

In Eq. (7.123), $G^{(p)}$ is the Green function expressed by the following expansion:

$$G^{(p)}(x, y; x', D(x')) = \sum_{l=-\infty}^{\infty} H_1^{(1)}(k[[x - (x' + la)]^2 + [z - D(x')]^2]^{1/2}) \times \exp(iK_0 la) \frac{[z - D(x')] - \frac{\partial D}{\partial x'}[x - (x' + la)]}{[[x - (x' + la)]^2 + [z - D(x')]^2]^{1/2}}, \tag{7.124}$$

and the scattered field above the grating is expressed as (cf. Eq. (7.76)):

$$H_y^{(s)}(x, z) - \sum_{m=-\infty}^{\infty} B_m \exp[i(K_m x + q_m z)], \qquad z > D(x). \tag{7.125}$$

It is easy to see that the spectral amplitudes B_m are then given by:

$$B_m = -\frac{2\pi k}{caq_m} \int_0^a J_y(x', D(x')) \exp[-i(K_m x' + q_m D(x'))] \left(\frac{dD}{dx'} K_m - q_m \right) \frac{dx'}{n_{z'}}, \tag{7.126}$$

and the corresponding normalized diffracted intensities are given by:

$$I_m^{(p)} = \frac{q_m}{k_z^{(i)}} \frac{|B_m|^2}{|h|^2}. \tag{7.127}$$

Obviously, both $I_m^{(s)}$ and $I_m^{(p)}$ should satisfy the unitarity condition:

$$\sum_{m=-\infty}^{\infty} I_m^{(s)(p)} = 1. \tag{7.128}$$

7.13. VARIATION OF THE DIFFRACTED INTENSITIES

The equations presented in Section 7.12 constitute the procedure for finding in an exact way the angular distribution of diffracted intensities. Its solution, however, can be found only numerically. Obviously, the small perturbation method for slightly corrugated profiles, or the Kirchhoff approximation, can be employed in a strictly identical way, as indicated in previous sections, by taking now into account the periodicity of $D(x)$. As for differential procedures, these can be found in Ref. 7.14.

The case of deep grating, however, bears interest since it shows remarkable effects on the angular distribution of diffracted orders.

Figure 7.16 shows the diffracted intensities versus the amplitude $\mathbf{h}$, for s-polarization, corresponding to a sinusoidal grating with profile $D(x) = \mathbf{h}\cos(2\pi x/a)$, $a = 3.18\lambda$, for three different angles of incidence $\theta_0 = 9°$, $28.1°$, and $51.8°$. These angles have been chosen because they correspond to those cases in which one of the diffraction orders: m_0 is exactly antispecular, thus, being for this particular order: $\theta_{m_0} = -\theta_0$, and hence Eq. (7.117) becomes the Bragg condition:

$$ka \sin\theta_0 = \pi m_0. \tag{7.129}$$

In Fig. 7.16, m_0 is equal to -1, -3, and -5, respectively. It can be observed that there are values of $\mathbf{h}$ for a certain angle of incidence, for which either the specular order or the antispecular order is predominantly larger than the others. Also, the variation with $\mathbf{h}$ of the diffracted intensities is oscillatory. In fact, when the period of the grating is small enough, so that it supports only two propagating orders, this variation is almost sinusoidal [7.81]–[7.83]. This disappearance of the specular order and enhancement of the antispecular is sometimes referred to as a *Bragg angle anomaly* [7.81], [7.84], whose resemblance with the vanishing of the reflectivity under the Brewster angle has been sometimes remarked [7.81], [7.84]. In condensed matter theory, this oscillatory behavior of the diffracted intensities with the amplitude $\mathbf{h}$ is sometimes referred to as the *rainbow effect* [7.85], [7.86].

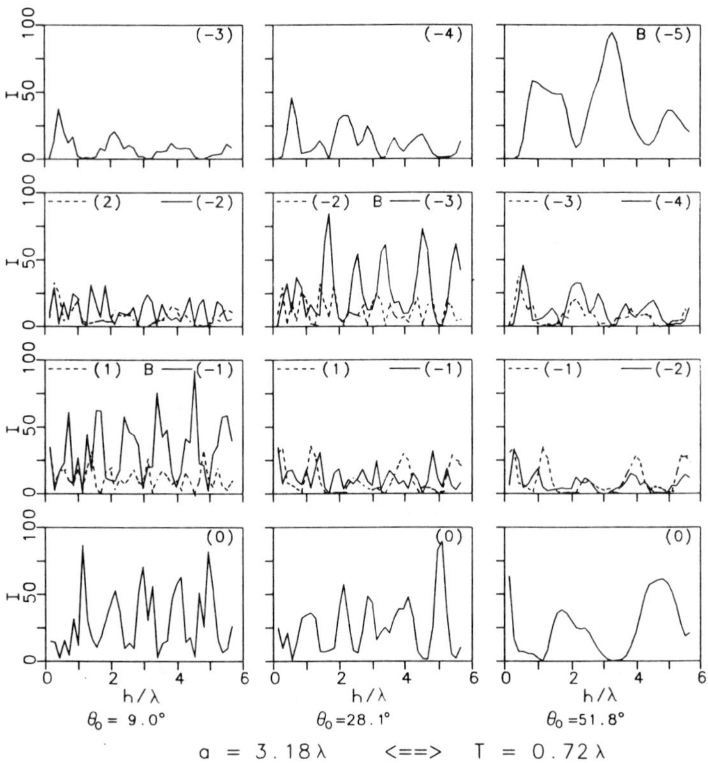

Figure 7.16. Diffracted intensities versus $\mathbf{h}/\lambda$ for a sinusoidal grating with $a = 3.18\lambda$ for three different angles of incidence. *s*-polarization. The character "**B**" indicates the antispecular order. (After J. M. Soto-Crespo and M. Nieto-Vesperinas, *Opt. Comm.* **69**, 185 (1989).)

The dependence of the diffracted intensities with the angle of incidence is shown in Fig. 7.17 for a sinusoidal grating with $a = 3.18\lambda$, $\mathbf{h} = 3.39\lambda$. For these parameters the antispecular intensity has a maximum in its diagram versus $\mathbf{h}$. Figure 7.17 shows that the intensities are maximum at those angles θ_0 at which they become antispecular [7.87], [7.88] (this configuration of the angle of incidence is sometimes called *Littrow mounting*). It should be remarked that this does not systematically occur in every deep grating; however, it happens very often. In other words, there is a high probability of choosing the values of the period and the (large) depth such that this effect occurs. In fact, the enhancement of one order over the others, and in particular, of the antispecular order, has been usually known as the *blaze effect*. It has a connection with the enhanced backscattering found in random surfaces [7.22], [7.88], [7.89] (basically, this can be understood considering a

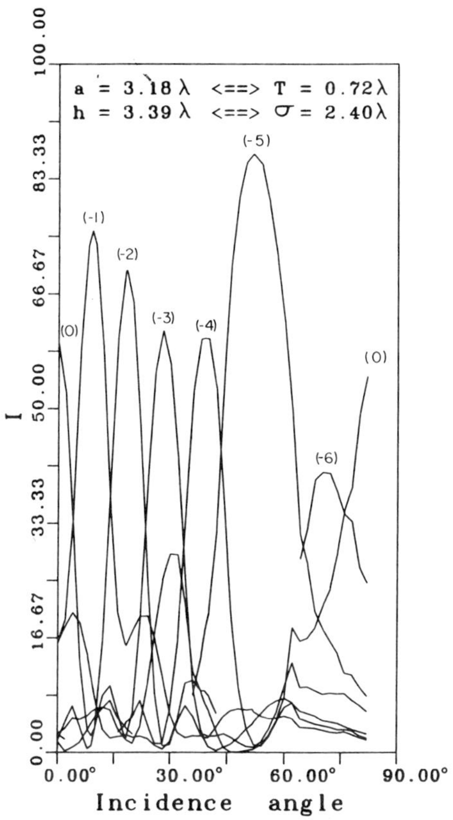

Figure 7.17. Diffracted intensities versus the angle of incidence for a sinusoidal grating with $a = 3.18\lambda$ and $\mathbf{h} = 3.39\lambda$. (After J. M. Soto-Crespo and M. Nieto-Vesperinas, *Opt. Comm.* **69**, 185 (1989).)

random surface as an infinite superposition of gratings). Also, the enhancement of the specular intensity for certain parameters of the grating, as seen in Fig. 7.16, is due to the evenness of the profile within a unit cell. As a matter of fact, and keeping the analogy with random surfaces, it has been shown [7.89] that one-dimensional random profiles with a center of symmetry also produce enhancement in the specular direction. However, this effect appears in any order of scattering and, hence, in the Kirchhoff approximation also. The width of this specular peak is λ/L, L being the illuminated extension of the surface.

The discontinuities in the diffracted intensities at those angles of incidence at which there is a passing off of an order, as shown in Fig. 7.17, are known as *Rayleigh anomalies.* Together with those at the Bragg angle, and with the *polariton anomalies* (Problem 7.12), they all constitute different kinds of those generically known as *Wood's anomalies* [7.39], [7.90]–[7.92].

The method of calculation described in this section also permits one to devise gratings that produce blaze in a high order [7.22]; this corresponds to a high angular dispersion, and hence, it is of interest in applications.

PROBLEMS

7.1. Find the angular spectrum representation for the field scattered from a perfectly conducting rough surface in the Kirchhoff approximation.

7.2. Prove Eqs. (7.33)–(7.37) for the angular factors corresponding to the Kirchhoff approximation.

7.3. Show that the angular distribution of mean scattered intensity from a rough surface with a Gaussian correlation function, with correlation distance T and variance σ, given by the Kirchhoff approximation is:

$$\langle I^{(s)}(Q)\rangle = \frac{k^2}{2\pi} F^2 \exp(-v_z^2\sigma^2) \int_0^a J_0(Q\tau)\tau \, d\tau$$

$$+ \frac{k^2}{4\pi} T^2 F^2 \exp(-v_z^2\sigma^2) \sum_{l=1}^{\infty} \frac{(\sigma v_z)^{2l}}{l!\,l} \exp\left[-\frac{Q^2T^2}{4l}\right],$$

where a is the radius of the illuminated area S, and $T \ll a$. F is the angular F-factor, Eq. (7.40), whose expression for the appropriate polarization should be included (cf. also Eqs. (7.33)–(7.35)). The first term represents the specular coherent part, whereas the second term corresponds to the diffuse component.

Prove that for large roughness: $v_z\sigma \gg 1$, the mean intensity per unit area S becomes:

$$\langle I^{(s)}(Q)\rangle = \frac{T^2F^2}{4\pi S\sigma^2 v_z^2} \exp\left[-\frac{Q^2T^2}{4\sigma^2 v_z^2}\right].$$

This expression shows that in this case there is only a diffuse distribution of scattered light.

7.4. Prove that Eq. (7.54), evaluated in the plane of incidence, becomes equal to that of the scalar theory. Prove also that for $k^2T^2/4 \gg 1$ while $T^2 \ll S$ the specular component of Eq. (7.57) becomes equal to that obtained by the series expansion of the Kirchhoff theory, given in Problem 7.3.

7.5. Prove Eqs. (7.73) and (7.74) for one-dimensional surfaces.

7.6. Prove Eqs. (7.75) and (7.76).

7.7. Find the specular and diffuse components of the mean scattered intensity up to $k^2\sigma^2$ in the backscattering direction following the procedure of Section 7.6.1. Prove that the cross-polarized intensities are zero; whereas the copolarized intensities for both s and p waves have the same expression for the coherent component, and differ by a factor $(1 + \sin^2 \theta_0)^2/(\cos^4 \theta_0)$ which appears in $\langle I_{ss}^{(s)}(Q)\rangle_d$.

Prove that in the limit $(k \cos \theta_0)^2 T^2 \gg 4$, while $T^2 \ll S$, this result does not predict depolarization and coincides with that of the Kirchhoff approximation.

7.8. Prove Eqs. (7.59)–(7.63) for one-dimensional surfaces.

7.9. Prove that the normal derivatives $\partial G/\partial \mathbf{n}$ and $\partial G^{(\text{in})}/\partial \mathbf{n}$ at $z = D(x)$ have a discontinuity $4\pi\delta(x - x')\delta(z - z')$ across the surface separating vacuum from a nonperfectly conductive medium. G and $G^{(\text{in})}$ are the exterior and interior (Eq. (7.88)) Green functions, respectively. Based on this result, prove the equivalence of the pair of equations (7.85) and (7.89) with the pair (7.84) and (7.90) [7.75].

7.10. Prove Eq. (7.101) for the field scattered into the far zone from a metal or dielectric surface.

7.11. *Surface polaritons* are p-polarized electromagnetic waves, bounded to the surface $z = D(\mathbf{R})$ separating a medium of permittivity ε from vacuum. As such, they are waves evanescent along the z-axis, both for $z > 0$ and $z < 0$. They can be excited by a homogeneous incident p-polarized incident plane wave only if there exists corrugation (why?). Show that the continuity of the tangential components of the electric and magnetic vector across a flat surface, for an incident p-polarized plane wave with magnetic vector amplitude: $\mathbf{h} = (0, h, 0)$, leads to the dispersion relation, Eq. (7.65), for the surface polaritons. Show also that the condition that they are evanescent waves requires that $\varepsilon < -1$.

Since surface polaritons are natural modes of the wave equation, and then, according to Section 1.6.4 they correspond to making the second term of Eq. (7.84) equal to zero. Prove, by assuming the surface to be the plane $z = 0$, and using the angular spectrum representation for the field, that this condition also leads to the dispersion relation, Eq. (7.65) [7.93].

7.12. By expanding the fields in Eqs. (7.84) and (7.85) according to the perturbational method of Section 7.6, prove that the dispersion relation

(7.65) appears as a resonant denominator of the zeroth and first order angular spectra of the fields under incident p-polarization (Agarwal [7.37]). These resonant excitations are responsible for reflectance anomalies observed at rough surfaces: *polariton anomalies* characterized by reflectance drops or field amplifications on the surface, leading through nonlinear interaction with the material to *enhanced Raman scattering* [7.12]–[7.14].

Prove also that the zeroth order angular spectra of the fields lead to the classical Fresnel coefficients for reflection and transmission of a plane electromagnetic wave at a plane interface (the obtention of these coefficients for a flat interface from the extinction theorem was originally done by Lalor and Wolf [7.94]).

7.13. The *Brewster modes* are natural modes of the wave equation associated with the Brewster angle: $\theta_0 = \pi/2 - \theta$. They correspond to a zero reflected field from a surface separating vacuum from a medium of permittivity $\varepsilon > 0$. They can be excited by a plane incident wave only if it has a p-polarized component. By equating to zero the second term of Eq. (7.85), prove, using the angular spectrum representation of the fields, and assuming the surface to be the plane $z = 0$, that these modes have a wavevector satisfying the dispersion relation, Eq. (7.65), with $\varepsilon > 0$ [7.93].

REFERENCES

7.1. H. C. Van de Hulst, *Light Scattering by Small Particles*, Dover, New York, 1981.

7.2. M. Born and E. Wolf, *Principles of Optics*, Pergamon Press, Oxford, 1980, 6th edition.

7.3. J. A. Kong, *Theory of Electromagnetic Waves*, J. Wiley, New York, 1975.

7.4. A. B. Shmelev, *Sov. Phys. Usp.*, **15**, 173 (1972).

7.5. J. A. De Santo and G. S. Brown, in *Progress in Optics*, Vol. 23, E. Wolf, ed., North-Holland, Amsterdam, 1965.

7.6. J. A. Ogilvy, *Rep. Prog. Phys.* **50**, 1553 (1987).

7.7. S. M. Rytov, Y. A. Kravtsov and V. I. Tatarskii, *Principles of Statistical Radiophysics*, Vol. 4, Springer-Verlag, Berlin, 1989.

7.8. P. Beckmann and A. Spizzichino, *The Scattering of Electromagnetic Waves from Rough Surfaces*, Pergamon Press, New York, 1963; P. Beckmann, in *Progress in Optics*, Vol. 6, E. Wolf, ed., North-Holland, Amsterdam, 1968.

7.9. F. B. Bass and I. M. Fuks, *Wave Scattering from Statistically Rough Surfaces*, Pergamon Press, Oxford, 1980.

7.10. A. McGinn and J. Sykes, *Microwave Scattering from Natural Targets*, U.K. Atomic Energy Authority, Report AERE-R. 9768, Harwell, 1980.

7.11. J. M. Bennett and L. Mattsson, *Introduction to Surface Roughness and Scattering*, Optical Society of America, Washington, DC, 1989.

7.12. V. M. Agranovich and D. L. Mills, eds., *Surface Polaritons*, North-Holland, Amsterdam, 1982.

7.13. R. F. Wallis and G. Stegeman, eds., *Electromagnetic Surface Excitations*, Springer-Verlag, Berlin, 1986.

7.14. R. Petit, ed., *Electromagnetic Theory of Gratings*, Topics in Current Physics, Vol. 22, Springer-Verlag, Berlin, 1980.

7.15. Lord Rayleigh, *Theory of Sound*, Vol. 2, Dover, New York, 1945.

7.16. P. Beckmann, *I.E.E.E. Trans. Antennas Propag.* **21**, 169 (1973).

7.17. T. Michel, A. A. Maradudin and E. R. Mendez, *J. Opt. Soc. Am. B* **6**, 2438 (1989).

7.18. J. C. Leader, *J. Appl. Phys.*, **42**, 4808 (1971).

7.19. W. T. Welford, *Opt. Quant. Electron.* **9**, 269 (1977).

7.20. A. Papoulis, *Probability, Random Variables and Stochastic Processes*, McGraw-Hill, New York, 1965.

7.21. L. M. Brekhovskikh, *Zh. Eksp. i Teor. Fiz.* **23**, 275 (1952).

7.22. J. M. Soto-Crespo and M. Nieto-Vesperinas, *J. Opt. Soc. Am. A* **6**, 367 1989.

7.23. E. I. Thorsos, *J. Acoust. Soc. Am.* **83**, 78 (1988).

7.24. K. A. O'Donnell and E. R. Mendez, *J. Opt. Soc. Am. A* **4**, 1194 (1987).

7.25. N. Garcia, V. Celli and M. Nieto-Vesperinas, *Opt. Comm.* **30**, 279 (1979).

7.26. M. Nieto-Vesperinas and N. Garcia, *Opt. Acta* **28**, 1651 (1981).

7.27. M. Nieto-Vesperinas, *J. Opt. Soc. Am.* **72**, 539 (1982).

7.28. L. J. Slater, in *Handbook of Mathematical Functions*, K. Abramowitz and I. Stegun, eds., Dover, New York, 1972.

7.29. J. Shen and A. A. Maradudin, *Phys. Rev. B* **22**, 4234 (1980).

7.30. D. Winebrenner and A. Ishimaru, *J. Opt. Soc. Am. A* **2**, 2285 (1985).

7.31. M. F. Chen and A. K. Fung, *Radio Sci.* **23**, 163 (1988).

7.32. J. M. Soto-Crespo, M. Nieto-Vesperinas and A. T. Friberg, *J. Opt. Soc. Am. A* **7**, 1185 (1990).

7.33. E. I. Thorsos and D. R. Jackson, *J. Acoust. Soc. Am.* **86**, 261 (1989).

7.34. D. N. Mirlin and A. A. Maradudin, in *Surface Polaritons*, V. M. Agranovich and D. L. Mills, eds., North-Holland, Amsterdam, 1982, Chapters 1 and 10.

7.35. F. Abeles and T. Lopez-Rios, in *Surface Polaritons*, V. M. Agranovich and D. L. Mills, eds., North-Holland, Amsterdam, 1982, Chapter 6; F. Abeles, in *Electromagnetic Surface Excitations*, R. F. Wallis and G. Stegeman, eds., Springer-Verlag, Berlin, 1986.

7.36. H. Raether, *Surface Plasmons*, Springer-Verlag, Berlin, 1988.

7.37. G. S. Agarwal, *Phys. Rev. B* **15**, 2371 (1977).

7.38. D. Winebrenner and A. Ishimaru, *Radio Sci.* **20**, 161 (1985).

7.39. U. Fano, *J. Opt. Soc. Am.* **31**, 213 (1941).
7.40. S. O. Rice, *Commun. Pure Appl. Math.* **4**, 351 (1951).
7.41. R. Schiffer, *Appl. Opt.* **26**, 704 (1987).
7.42. D. D. Barrick and W. H. Peake, *Radio Sci.* **3**, 865 (1968).
7.43. G. R. Valenzuela, *I.E.E.E. Trans. Antennas Propag.* **15**, 552 (1967).
7.44. J. M. Elson, *Phys. Rev. B* **12**, 2541 (1975).
7.45. V. Celli, A. Marvin and F. Toigo, *Phys. Rev. B* **11**, 1779 (1975).
7.46. B. A. Lippmann, *J. Opt. Soc. Am.* **43**, 408 (1953).
7.47. J. L. Urestky, *Ann. Phys. (New York)*, **33**, 400 (1965).
7.48. R. Petit and M. Cadilhac, *Compt. Rend. Acad. Sci. B* **262**, 468 (1966).
7.49. R. F. Millar, *Proc. Camb. Phil. Soc.* **65**, 773 (1969); **69**, 217 (1971); *Radio Sci.* **8**, 785 (1973).
7.50. A. Wirgin, *Opt. Acta* **27**, 1671 (1980).
7.51. A. Wirgin, *J. Opt. Soc. Am.* **73**, 1028 (1983).
7.52. D. Maystre and J. P. Rossi, *J. Opt. Soc. Am. A* **3**, 1276 (1986).
7.53. P. M. Morse and H. Feshbach, *Methods of Theoretical Physics*, McGraw-Hill, New York, 1953.
7.54. M. G. Andreasen, *I.E.E.E. Trans. Antennas Propag.* **31**, 885 (1983).
7.55. A. K. Fung and M. F. Chen, *J. Opt. Soc. Am. A* **2**, 2274 (1985).
7.56. A. A. Maradudin, E. R. Mendez and T. Michel, *Opt. Lett.* **14**, 151 (1989).
7.57. M. Nieto-Vesperinas and J. M. Soto-Crespo, *Opt. Comm.* **75**, 215 (1990).
7.58. J. Renau and J. A. Collinson, *Bell Syst. Tech. J.* **44**, 2203 (1965).
7.59. J. Renau, P. K. Cheo and H. G. Cooper, *J. Opt. Soc. Am.* **57**, 459 (1967).
7.60. E. R. Mendez and K. A. O'Donnell, *Opt. Comm.* **61**, 91 (1987).
7.61. S. J. Allen, *Phys. Today* **40**, S-50 (1986).
7.62. Y. Kuga and A. Ishimaru, *J. Opt. Soc. Am.* **1**, 831 (1984); L. Tsang and A. Ishimaru, *J. Opt. Soc. Am. A* **1**, 836 (1984).
7.63. M. P. Van Albada and A. Lagendijk, *Phys. Rev. Lett.* **56**, 1471 (1986).
7.64. P. E. Wolf and G. Maret, *Phys. Rev. Lett.* **55**, 2696 (1985).
7.65. M. Kaveh, M. Rosenbluh, M. Edrei and I. Freund, *Phys. Rev. Lett.* **57**, 2049 (1986).
7.66. E. Jakeman, *J. Opt. Soc. Am. A* **5**, 1638 (1988).
7.67. P. Sheng (ed.), *Scattering and Localization of Classical Waves in Random Media*, World Scientific, London, 1990.
7.68. M. Nieto-Vesperinas and J. C. Dainty (eds.), *Scattering in Volumes and Surfaces*, North-Holland, Amsterdam, 1990.
7.69. M. Nieto-Vesperinas and J. M. Soto-Crespo, *Opt. Lett.* **12**, 979 (1987).
7.70. S. Etemand, R. Thompson and M. J. Andrejco, *Phys. Rev. Lett.* **57**, 575 (1986).
7.71. M. Nieto-Vesperinas and J. M. Soto-Crespo, in *Scattering in Volumes and Surfaces*, M. Nieto-Vesperinas and J. C. Dainty, eds., North-Holland, Amsterdam, 1990, p. 175.

7.72. J. C. Dainty, M. J. Kim and A. J. Sant, in *Scattering in Volumes and Surfaces*, M. Nieto-Vesperinas and J. C. Dainty, eds., North-Holland, Amsterdam, 1990, p. 143.

7.73. A. J. Sant, J. C. Dainty and M. J. Kim, *Opt. Lett.* **14**, 1183 (1989).

7.74. M. J. Kim, J. C. Dainty, A. T. Friberg and A. J. Sant, *J. Opt. Soc. Am. A* **7**, 569 (1990).

7.75. F. Toigo, A. Marvin, V. Celli and N. R. Hill, *Phys. Rev. B* **15**, 5618 (1977).

7.76. N. Garcia and E. Stoll, *Phys. Rev. Lett.* **52**, 1798 (1984); *J. Opt. Soc. Am. A* **2**, 2240 (1985).

7.77. A. A. Maradudin, T. Michel, A. R. Mac Gurn and E. R. Mendez, *Ann. Phys. (New York)* **203**, 255 (1990).

7.78. N. R. Hill and V. Celli, *Surf. Sci.* **75**, 577 (1978).

7.79. C. Lopez, F. J. Yndurain and N. Garcia, *Phys. Rev. B* **18**, 970 (1978).

7.80. R. N. Bracewell, *The Fourier Transform and its Applications*, McGraw-Hill, New York, 1978.

7.81. D. Y. Tseng, A. Hessel and A. A. Oliner, in *Proc. U.R.S.I. Symp. on EM Waves* (*Alta Frequenza* **38**, 82 (1969), special issue).

7.82. D. Maystre, M. Cadilhac and J. Chandezon, *Opt. Acta* **28**, 457 (1981); D. Maystre, in *Progress in Optics*, Vol. 21, E. Wolf, ed., North-Holland, Amsterdam, 1984.

7.83. M. G. Moharam and T. K. Gaylord, *J. Opt. Soc. Am.* **72**, 1385 (1982).

7.84. A. Hessel and A. A. Oliner, *Opt. Comm.* **59**, 327 (1986).

7.85. V. Garibaldi, A. C. Levi, R. Spadacini and G. E. Tommei, *Surf. Sci.* **48**, 649 (1975).

7.86. N. Garcia and A. A. Maradudin, *Opt. Comm.* **45**, 301 (1983).

7.87. J. A. De Santo, *J. Math. Phys.* **12**, 1913 (1971); **13**, 336 (1972).

7.88. M. Nieto-Vesperinas and J. M. Soto-Crespo, *Phys. Rev. B* **38**, 7250 (1988).

7.89. M. Nieto-Vesperinas and J. M. Soto-Crespo, *Phys. Rev. B* **39**, 8193 (1989).

7.90. R. W. Wood, *Phylos. Mag.* **4**, 396 (1902).

7.91. A. Hessel and A. A. Oliner, *Appl. Opt.* **4**, 1275 (1965).

7.92. M. Nevière, in *Electromagnetic Theory of Gratings*, R. Petit, ed., Topics in Current Physics, Vol. 22, Springer-Verlag, Berlin, 1980, Chapter 5.

7.93. G. S. Agarwal, *Phys. Rev. B* **8**, 4768 (1973).

7.94. E. Lalor and E. Wolf, *J. Opt. Soc. Am.* **62**, 1165 (1972).

8

Propagation and Scattering of Phase-Conjugate Wavefields

Phase-conjugate optics is a fast developing subject that involves the generation, propagation and transmission of fields whose phase has been reversed. This effect is achieved in a certain material, usually by holographic means, or by nonlinear processes such as stimulated Brillouin scattering or three or four wave mixing.

The great interest of this phenomenon lies in the *healing effect* that phase-conjugation has on the wavefield upon propagation. This is the correction of distortions of a wavefront due to scattering, caused by, e.g. the inhomogeneities of the refractive index of the propagating medium, such as occurs in the atmosphere, the aberrations in a transmitting optical system, or the different phase velocities of modes in an optical fiber.

There exists a large number of references on techniques for generating phase-conjugate waves. In this chapter we shall address some of the properties of these wavefields, both in free propagation and suffering scattering through the medium that originally distorted the transmitted wavefront. This leads to some theorems and conclusions on how distortion correction is attained and what its limitations are.

We shall first present some theorems on reciprocity and symmetry pertaining to fields that are phase-conjugated in a certain plane. Formally, this phase-conjugation is dealt with, therefore, as a boundary condition at an idealized plane of conjugation. We shall then study the restoration of the original wavefront through scattering by means of the formalism of the S-matrix, introduced in Chapter 5. This will show that the so-called healing effect does not depend on the details of the particular interaction of wavefields with the distorting scatterer, but rather on universal characteristics of the scattering process, which are represented by the unitarity and reciprocity properties of the scattering matrix.

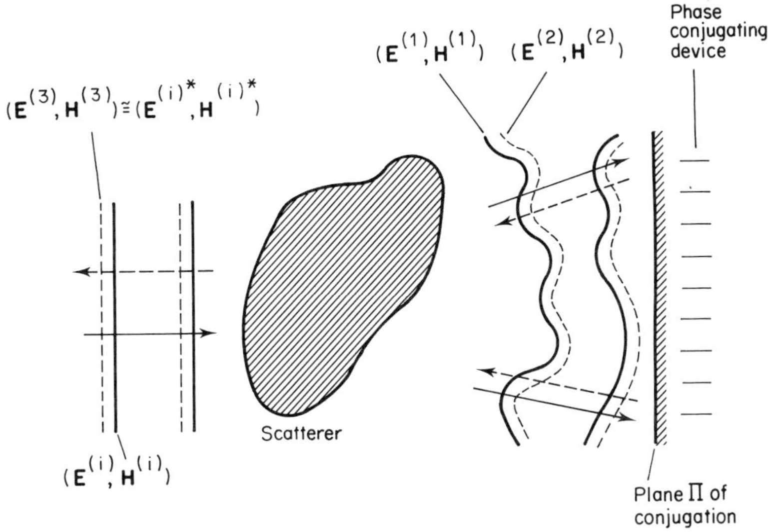

Figure 8.1. Phase conjugation of a field distorted by a scatterer.

8.1. INTRODUCTION

The potential usefulness of phase-conjugation mainly stems from the possibility of correcting wavefront distortions.

Let us consider an electromagnetic wave with electric and magnetic vectors $\mathbf{E}^{(i)}$ and $\mathbf{H}^{(i)}$, respectively. Suppose that this wave is distorted by a scatterer into a new wave $(\mathbf{E}^{(1)}, \mathbf{H}^{(1)})$ (see Fig. 8.1). If by some means, one can produce at a plane Π a phase-reversed replica $(\mathbf{E}^{(2)}, \mathbf{H}^{(2)})$ of $(\mathbf{E}^{(1)}, \mathbf{H}^{(1)})$, travelling backwards so that it passes again through the scatterer, then, under some conditions that will be specified later, a new wave $(\mathbf{E}^{(3)}, \mathbf{H}^{(3)})$ is created, which may be equal, or very similar, to the phase-conjugated $(\mathbf{E}^{(i)*}, \mathbf{H}^{(i)*})$ of the original wave.

The system occupying the region behind the plane Π at which conjugation and reflection occurs is usually referred to as a *phase-conjugate mirror*. Phase-conjugate devices can be very efficient, with reflectivities very close to unity, or even larger. Of course, reflectivities larger than unity are obtained by extracting energy from external fields that enter in the phase-conjugation process (for example, from the pumping fields in wave mixing).

As a brief outline of the works on generation of phase-conjugate waves, we shall mention that this effect was demonstrated by using *stimulated Brillouin Scattering* [8.1], [8.2], *three-wave mixing* in crystals [8.3], [8.4], *degenerate four-wave mixing* in *Kerr-like media* [8.5], [8.6], and *four-wave mixing* in *photorefractive materials* [8.7], [8.8].

The connection of the phase-conjugate wave with the pseudoscopic image in holography is discussed in Refs. 8.9–8.11, and the ability of real-time holography to compensate for phase distortions can be traced back to the works reported in Refs. 8.12–8.14. The characteristic of phase-conjugation in compensating distortions due to atmospheric turbulence has been discussed in, e.g. Ref. 8.15. The reader is referred to Refs. 8.16–8.18 for comprehensive reviews on the generation and applications of phase-conjugation.

Let $\mathbf{E}^{(i)}(\mathbf{r})$ be the electric vector of the wave incident on a phase-conjugate mirror (we shall assume from now on a harmonic time dependence $\exp(-i\omega t)$ which will be omitted). As regards polarization, the response of this mirror will depend on the experimental technique used in the generation of the conjugate field. A complete reversal of the polarization can be achieved by appropriately arranging the geometry of the setup, as well as the polarization of the several waves that interact in the device (Refs. 8.19–8.21). Hence, we shall assume that at the plane Π of conjugation, chosen to be $z = z_0$, the following boundary condition holds:

$$\mathbf{E}^{(1)}(\mathbf{r})|_{z=z_0} = \mu \mathbf{E}^{(i)*}(\mathbf{r})|_{z=z_0}, \tag{8.1}$$

where μ is a constant that represents the *reflectivity* of the phase-conjugate mirror, assuming that the response to a plane wave does not depend either on its state of polarization or on its angle of incidence.

Equation (8.1) implies that the polarization of the wave is completely reversed upon conjugation at the plane $z = z_0$. It should be stressed, however, that Eq. (8.1) is an idealization that so far has been only achieved in the paraxial regime. Nevertheless, we shall work with this assumption in the rest of this chapter for the sake of conceptual interpretation.

An equation equivalent to Eq. (8.1) can also be written for the magnetic vector. Therefore, we shall deal with each Cartesian component of either the electric or the magnetic vector. The electromagnetic generalization thus being either straightforward or very simple (see, e.g. Section 7 of Ref. 8.22).

It is easy to understand that for a homogeneous plane wave, propagating along the z-axis:

$$U^{(i)}(\mathbf{r}, t) = \mathrm{Re}[\psi(\mathbf{r}) \exp[i(kz - \omega t)]], \tag{8.2}$$

the wave conjugated at $z = z_0$, and reflected back, is:

$$U^{(1)}(\mathbf{r}, t) = \mathrm{Re}[\psi^*(\mathbf{r}) \exp[-i(kz + \omega t)]]. \tag{8.3}$$

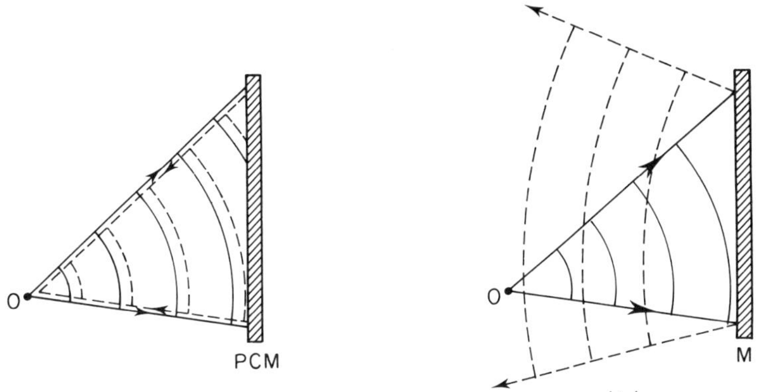

Figure 8.2. Illustration of reflection of a divergent spherical wave at: (a) a phase conjugate mirror PCM; (b) an ordinary mirror M.

Note that $U^{(1)}$ is obtained from $U^{(i)}$ by taking the complex conjugate of the spatial part only. In addition, it is evident that Eq. (8.3) is also obtained from Eq. (8.2) by *time-reversing* $U^{(i)}(\mathbf{r}, t)$ in Eq. (8.2), namely, by changing in $U^{(i)}(\mathbf{r}, t)$ the sign of the time variable from t into $-t$. This fact is usually accounted for in order to consider phase-conjugation as equivalent to time-reversal; it is then inferred that backward propagation of the phase-conjugate wave through its previous states will ultimately reconstruct its original structure. However, a word of caution is in order here: whereas this argument is certainly true for homogeneous waves, as the one given by Eq. (8.2), it does not apply to evanescent waves, and thus, in general, for a field that contains inhomogeneous components (in this connection, see a discussion in Refs. 8.23 and 8.24). The healing effect by phase-conjugation does not hold for evanescent waves, and therefore, distortion correction by phase-conjugation of fields that contain those components can be achieved only up to the limit of its inhomogeneous waves. We shall later discuss this point in greater detail.

For example, a monochromatic divergent spherical wave (see Fig. 8.2(a)), which undergoes phase-conjugation at a phase-conjugate mirror (PCM), propagates back towards the point source O as an approximated version of a convergent spherical wave (in contrast to the effect of an ordinary mirror M (Fig. 8.2(b)) that reflects it as a divergent spherical wave). The approximate spherical wave created by the phase-conjugate mirror *has not* the evanescent components that a spherical wave, convergent towards O, would have in accordance with the plane wave expansion, Eq. (2.52). (As seen in

Section 2.12, neglecting the evanescent components is not at all irrelevant for a spherical wave.) We shall discuss this point later (see also Problem 8.2).

Also, it will be seen that phase-conjugation of a wave at a certain plane does imply its conjugation throughout the free half-space in which it propagates (either by transmission or by reflection at that plane), if the wave is homogeneous [8.25], then holding, in particular, in the paraxial domain, as for example, for an optical beam. Nevertheless, again this is not verified by evanescent waves [8.26].

8.2. PHASE-CONJUGATION OF WAVEFIELDS PROPAGATING IN FREE SPACE

We shall consider two cases: phase-conjugation at a given plane, followed by transmission, and phase-conjugation at a plane, followed by reflection. The former case involves theorems concerning fields that propagate into the same half-space, whereas the latter pertains to fields that propagate into complementary half-spaces.

8.2.1. Phase-Conjugation of Fields that Propagate into the Same Half-Space

Let us consider two wavefields $U^{(1)}(\mathbf{r})$ and $U^{(2)}(\mathbf{r})$, obeying Sommerfeld's radiation condition, Eq. (2.6), and propagating into the source-free half-space $z > 0$. Under the conditions quoted in Sections 2.2 and 2.5, these fields may be represented by means of their angular spectra. We shall write them in the form:

$$U^{(j)}(x, y, z) = U_H^{(j)}(x, y, z) + U_I^{(j)}(x, y, z), \qquad (j = 1, 2), (z \geqslant 0). \quad (8.4)$$

In Eq. (8.4) the field $U^{(j)}$ has been split into two parts: $U_H^{(j)}$ corresponding to the contribution of the homogeneous components, and $U_I^{(j)}$ representing the contribution of the evanescent components, namely:

$$U_H^{(j)}(x, y, z) = \iint_{p^2+q^2 \leqslant 1} A^{(j)}(p, q) \exp[ik(px + qy + mz)]\, dp\, dq, \quad (8.5a)$$

$$U_I^{(j)}(x, y, z) = \iint_{p^2+q^2 > 1} A^{(j)}(p, q) \exp[ik(px + qy + mz)]\, dp\, dq, \quad (8.5b)$$

with p, q, and m related by Eqs. (2.9).

Let us assume that at every point of a plane $z = z_0 > 0$ the two fields are complex conjugates of each other, viz.:

$$U^{(2)}(x, y, z_0) = [U^{(1)}(x, y, z_0)]^*. \tag{8.6}$$

In terms of the angular spectrum representation, Eq. (8.6) means that:

$$\iint_{-\infty}^{\infty} A^{(2)}(p, q) \exp[ik(px + qy + mz_0)]\, dp\, dq$$
$$= \iint_{-\infty}^{\infty} [A^{(1)}(p, q)]^* \exp[-ik(px + qy + m^* z_0)]\, dp\, dq. \tag{8.7}$$

Observe that m^*, appearing on the right-hand side of Eq. (8.7), is real and equal to $\sqrt{1 - p^2 + q^2}$ for $p^2 + q^2 \leqslant 1$, and it is imaginary and equal to $-i\sqrt{p^2 + q^2 - 1}$ for $p^2 + q^2 > 1$.

If we change the variables of integration on the right-hand side of Eq. (8.7) from p, q to $-p, -q$, we find that:

$$\iint_{-\infty}^{\infty} A^{(2)}(p, q) \exp[ik(px + qy + mz_0)]\, dp\, dq$$
$$= \iint_{-\infty}^{\infty} [A^{(1)}(-p, -q)]^* \exp[ik(px + qy - m^* z_0)]\, dp\, dq. \tag{8.8}$$

By taking the two-dimensional Fourier transform in both sides of Eq. (8.8), we obtain the following relationship between the angular spectra of the two fields:

$$A^{(2)}(p, q) \exp(ikmz_0) = [A^{(1)}(-p, -q)]^* \exp(-ikm^* z_0). \tag{8.9}$$

Let us substitute for $A^{(2)}(p, q)$ from Eq. (8.9) into Eqs. (8.5a) and (8.5b) with $j = 2$, and write $z = z_0 + d$, $0 < d < \infty$. From Eq. (8.5a) we find, taking into account that $m^* = m$ for $p^2 + q^2 \leqslant 1$:

$$U_H^{(2)}(x, y, z_0 + d) = \iint_{p^2+q^2 \leqslant 1} [A^{(1)}(-p, -q)]^* \exp(-2ikmz_0)$$
$$\times \exp[ik(px + qy + m(z_0 + d))]\, dp\, dq,$$

$(m = \sqrt{1 - p^2 - q^2})$. Or, if we change the variables p, q into $-p, -q$:

$$U_H^{(2)}(x, y, z_0 + d) = \iint_{p^2+q^2 \leqslant 1} [A^{(1)}(p, q)]^* \times \exp[-ik(px + qy + m(z_0 - d))] \, dp \, dq, \quad (8.10)$$

$$(m = \sqrt{1 - p^2 + q^2}).$$

On comparing Eq. (8.10) with Eq. (8.5a) for $j = 1$, we see that the right-hand side of Eq. (8.10) is equal to $[U_H^{(1)}(x, y, z_0 - d)]^*$. Hence:

$$U_H^{(2)}(x, y, z_0 + d) = [U_H^{(1)}(x, y, z_0 - d)]^*. \quad (8.11)$$

On the other hand, on comparing Eqs. (8.9) and (8.5b), and recalling that $m^* = -i|m|$ for $p^2 + q^2 > 1$, we get:

$$U_I^{(2)}(x, y, z_0 + d) = \iint_{p^2+q^2>1} [A^{(1)}(-p, -q)]^* \times \exp[ik(px + qy)] \exp[-k|m|(z_0 + d)] \, dp \, dq.$$

Or, by changing the variables p, q into $-p, -q$:

$$U_I^{(2)}(x, y, z_0 + d) = \iint_{p^2+q^2>1} [A^{(1)}(p, q)]^* \times \exp[-ik(px + qy)] \exp[-k|m|(z_0 + d)] \, dp \, dq. \quad (8.12)$$

The comparison of Eqs. (8.12) and (8.5b) for $j = 1$ shows that the right-hand side of Eq. (8.12) is $[U_I^{(1)}(x, y, z_0 + d)]^*$. Hence, we conclude that:

$$U_I^{(2)}(x, y, z_0 + d) = [U_I^{(1)}(x, y, z_0 + d)]^*. \quad (8.13)$$

Equations (8.11) and (8.13) can be summarized in the following theorem (Wolf [8.25], Nieto-Vesperinas and Wolf [8.26]):

Theorem 1. *Let* $U^{(1)}(\mathbf{r}) \exp(-i\omega t)$ *and* $U^{(2)}(\mathbf{r}) \exp(-i\omega t)$ *be two wavefields that propagate into the half-space* $z > 0$, *and are bounded throughout that half-space. Suppose that at every point of a plane* $z = z_0$ *the two fields are the conjugate of each other, namely:*

$$U^{(2)}(x, y, z_0) = [U^{(1)}(x, y, z_0)]^*, \quad (8.14)$$

for all x, y; then, for any distance d, $0 < d < \infty$, and for all x, y:

$$U_H^{(2)}(x, y, z_0 + d) = [U_H^{(1)}(x, y, z_0 - d)]^*, \tag{8.15a}$$

$$U_I^{(2)}(x, y, z_0 + d) = [U_I^{(1)}(x, y, z_0 + d)]^*, \tag{8.15b}$$

where the subscripts H and I denote the homogeneous and the inhomogeneous parts, respectively, of the fields $U^{(j)}$, $(j = 1, 2)$, as defined in Eqs. (8.4) and (8.5), and $U_H^{(j)}(\mathbf{r})$, $(j = 1, 2)$ are to be interpreted for negative values of z, as the analytic continuations of the homogeneous parts of the corresponding fields into the half-space $z < 0$, in the sense of the theorem by Sherman, quoted in Section 2.7.2. ■

Equations (8.15) show that the homogeneous and inhomogeneous parts of the field behave quite differently from each other under phase conjugation. This is illustrated in Fig. 8.3.

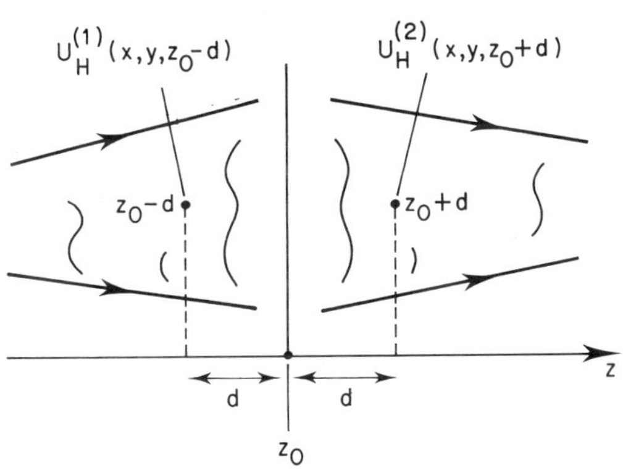

Figure 8.3. Illustration of Theorem 1.

Example 8.1 [8.26]. *Behavior of a homogeneous and an evanescent plane wave.*

Suppose that the field $U^{(1)}$ is a homogeneous plane wave that propagates in the direction characterized by the unit vector $(p, q, m > 0)$:

$$U_H^{(1)}(x, y, z) = A \exp[ik(px + qy + mz)], \qquad (z \geqslant 0), \tag{8.16}$$

where A is a complex constant.

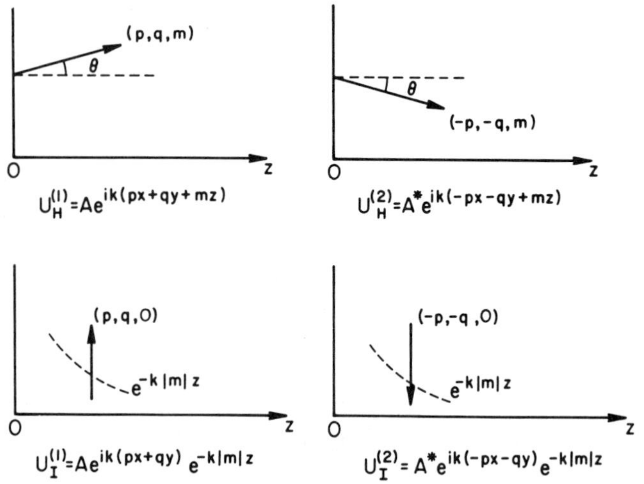

Figure 8.4. Illustration of phase-conjugation of a plane wave followed by transmission. (a) Homogeneous wave. (b) Evanescent wave. (Reproduced with permission from paper by M. Nieto-Vesperinas and E. Wolf, *J. Opt. Soc. Am. A* **2**, 1433 (1985).)

The field transmitted into $z = z + d$, obtained from $U_H^{(1)}$ after conjugation at the plane $z = z_0$, is according to Eqs. (8.16) and (8.15a):

$$U_H^{(2)}(x, y, z_0 + d) = A^* \exp[ik(-px - qy + m(d - z_0))], \qquad (z + d \geqslant 0). \tag{8.17}$$

This is a homogeneous wave whose wavevector is obtained from the wavevector of $U_H^{(1)}$ by reflection about the z-axis (see Fig. 8.4(a)).

Let us suppose now that the field $U^{(1)}$ is an evanescent plane wave:

$$U_I^{(1)}(x, y, z) = A \exp[ik(px + qy)] \exp(-k|m|z), \tag{8.18}$$

where $p^2 + q^2 > 1$ and $|m| = \sqrt{p^2 + q^2 - 1}$.

After conjugation at the plane $z = z_0$, the field transmitted into $z = z_0 + d$ is, following Eqs. (8.18) and (8.15b):

$$U_I^{(2)}(x, y, z_0 + d) = A^* \exp[-ik(px + qy)] \exp[-k|m|(z_0 + d)]. \tag{8.19}$$

Equation (8.19) represents an evanescent wave whose wavevector has the transversal component $k(-p, -q, 0)$, which is reversed with respect to the transversal component of the wavevector of $U_I^{(1)}$. The rate of decay of the amplitude of the two waves is, however, the same. This is illustrated in Fig. 8.4(b).

8.2.2. Phase-Conjugation of Fields that Propagate into Complementary Half-Spaces

Let us consider now two wavefields $U(\mathbf{r})$ and $V(\mathbf{r})$, propagating in a source-free half-space. The first wave propagates to the right, and the second wave to the left, of a certain plane $z = z_0$ of conjugation. Namely, in $z = z_0 > 0$ the two fields are complex conjugates of each other:

$$V(x, y, z_0) = [U(x, y, z_0)]^*, \tag{8.20}$$

for all x, y.

As before, the fields are written as:

$$U(x, y, z) = U_H(x, y, z) + U_I(x, y, z), \qquad (z \geqslant 0), \tag{8.21}$$

and:

$$V(x, y, z) = V_H(x, y, z) + V_I(x, y, z), \qquad (z \geqslant 0), \tag{8.22}$$

where once again the subindices H and I denote the contributions of the homogeneous and the inhomogeneous components, respectively. Thus, U_H and U_I are expressed in terms of the angular spectrum $A(p, q)$ as:

$$U_H(x, y, z) = \iint_{p^2+q^2\leqslant 1} A(p, q) \times \exp[ik(px + qy + mz)]\, dp\, dq, \qquad (z \geqslant 0); \tag{8.23a}$$

$$U_I(x, y, z) = \iint_{p^2+q^2> 1} A(p, q) \times \exp[ik(px + qy + mz)]\, dp\, dq, \qquad (z \geqslant 0). \tag{8.23b}$$

Similarly, V_H and V_I are expressed in terms of the angular spectrum $B(p, q)$ of V as:

$$V_H(x, y, z) = \iint_{p^2+q^2\leqslant 1} B(p, q) \times \exp[ik(px + qy - m(z - z_0))]\, dp\, dq, \qquad (z < z_0); \tag{8.24a}$$

$$V_I(x, y, z) = \iint_{p^2+q^2> 1} B(p, q) \times \exp[ik(px + qy - m(z - z_0))]\, dp\, dq, \qquad (z < z_0). \tag{8.24b}$$

Observe that in Eqs. (8.24) the angular spectrum representation of $V(\mathbf{r})$ has been expressed such that the integrals converge for $p^2 + q^2 > 1$ and $z < z_0$.

By using the boundary condition, Eq. (8.20), and Eqs. (8.23) and (8.24), it is straightforward to derive by a procedure similar to that employed before, the following theorem [8.25], [8.26]:

Theorem 2. *Let $U(\mathbf{r})\exp(-i\omega t)$ and $V(\mathbf{r})\exp(-i\omega t)$ be two wavefields, the first of which propagates into the half-space $z > 0$, and the second into the half-space $z < z_0$, each being bounded in its domain of definition. Suppose that at every point of a plane $z = z_0 \geqslant 0$ the two fields are conjugated with each other, namely:*

$$V(x, y, z_0) = [U(x, y, z_0)]^*, \qquad (z_0 \geqslant 0), \tag{8.25}$$

for all x, y; then, for any distance d, $0 < d < \infty$, and for all x, y:

$$V_H(x, y, z_0 - d) = [U_H(x, y, z_0 - d)]^*, \tag{8.26a}$$

$$V_I(x, y, z_0 - d) = [U_I(x, y, z_0 + d)]^*, \tag{8.26b}$$

where the subscripts H and I denote the homogeneous and the inhomogeneous parts, respectively, of the fields U and V, as defined in Eqs. (8.23) and (8.24), and U_H for negative z and V_H for $z > z_0$ are to be interpreted as the analytic continuations of the homogeneous parts of U and V into $z < 0$ and $z > z_0$, respectively. ■

Equations (8.26) show that the homogeneous and inhomogeneous parts behave quite differently under phase conjugation followed by reflection. This is illustrated in Fig. 8.5.

Example 8.2 [8.26]. Suppose that the field U is a homogeneous plane wave that propagates with wavevector $k(p, q, m > 0)$:

$$U_H(x, y, z) = A \exp[ik(px + qy + mz)], \qquad (z \geqslant 0), \tag{8.27}$$

where A is a complex constant.

The field conjugated at $z = z_0 > 0$ and reflected into $z = z_0 - d$ is, according to Eqs. (8.26a) and (8.27):

$$V_H(x, y, z_0 - d) = A^* \exp[ik(-px - qy - m(z_0 - d))], \qquad (z_0 \geqslant 0). \tag{8.28}$$

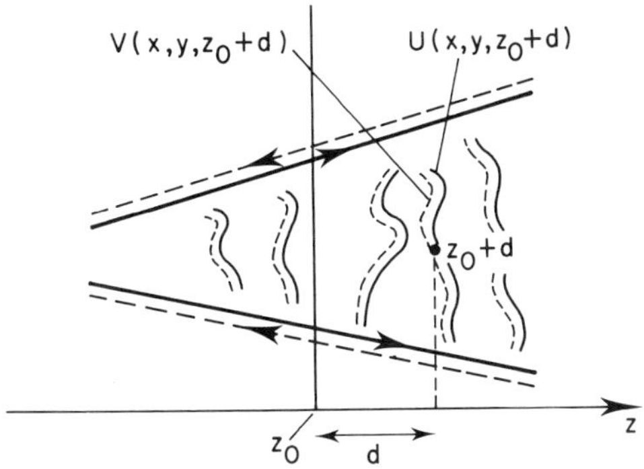

Figure 8.5. Illustration of Theorem 2.

This is a homogeneous wave whose wavevector is the reverse of the wavevector of U_H (see Fig. 8.6(a)).

Let us assume now that the evanescent plane wave:

$$U_I(x, y, z) = A \exp[ik(px + qy)] \exp(-k|m|z), \qquad (z \geqslant 0), \qquad (8.29)$$

where $p^2 + q^2 > 1$ and $|m| = \sqrt{p^2 + q^2 - 1}$.

After conjugation and reflection at the plane $z = z_0 \geqslant 0$, one obtains, according to Eqs. (8.26b) and (8.29):

$$V_I(x, y, z_0 - d) = A^* \exp[-ik(px + qy)] \exp[-k|m|(z_0 + d)], \qquad (z_0 \geqslant 0). \qquad (8.30)$$

Equation (8.29) represents an evanescent wave. The transverse component $-k(p, q, 0)$ of its complex wavevector is the reverse of the transverse component $k(p, q, 0)$ of the wavevector of U_I. The rate of decay of the amplitude of V_I in $z = z_0 - d$, however, is the same as that of U_I in $z = z_0 + d$ (see Fig. 8.6(b)).

Example 8.3. *Phase-conjugation of a spherical wave* (see also Problem 8.2).

Let us consider an outgoing spherical wave $G(\mathbf{r})$, originated at the point source O in $z = 0$ (Fig. 8.2). According to Eq. (2.51), its plane wave expansion is:

$$G(\mathbf{r}) \equiv \frac{\exp(ikr)}{r} = \frac{ik}{2\pi} \iint_{-\infty}^{\infty} \frac{1}{m} \exp[ik(px + qy + mz)]\, dp\, dq.$$

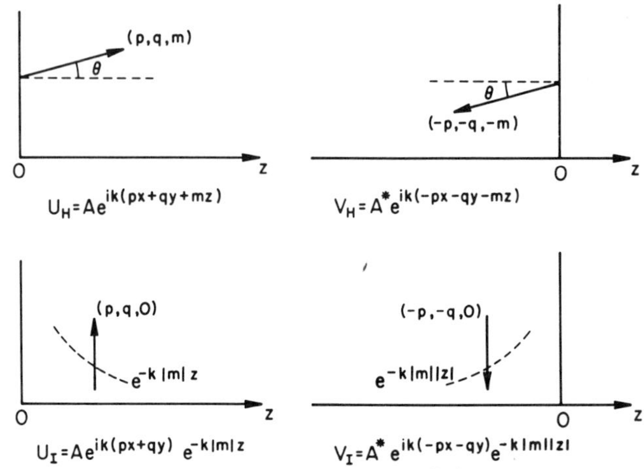

Figure 8.6. Illustration of phase-conjugation of a plane wave followed by reflection. (a) Homogeneous wave. (b) Evanescent wave. (Reproduced with permission from paper by M. Nieto-Vesperinas and E. Wolf, *J. Opt. Soc. Am. A* **2**, 1433 (1985).)

Suppose that this wave suffers phase-conjugation with reflection at a phase-conjugated mirror PCM in the plane $z = z_0$. The part G_H of the reflected wave, containing the homogeneous components, has at $z = z_0 - d$, $(0 < d < 0)$, the following expression in accordance with Theorem 2:

$$G_H(\mathbf{r})|_{z=z_0-d} = -\frac{ik}{2\pi}\iint_{p^2+q^2\leqslant 1}\frac{1}{m}$$
$$\times \exp[-ik(px + qy + m(z_0 - d))]\, dp\, dq. \qquad (8.31)$$

Equation (8.31) is identical to the homogeneous part, evaluated in $z = z_0 - d$, of an incoming spherical wave, converging towards O (cf. Eqs. (2.52) and (2.53a)). Therefore, under phase-conjugation with reflection, that part of a spherical wave, divergent from the point source O and corresponding to its homogeneous plane wave components, is transformed into the part of homogeneous components of a spherical wave converging towards O.

On the other hand, the part $G_I(\mathbf{r})$ containing the inhomogeneous components of the reflected wave, has at $z = z_0 - d$ the following expression according to Theorem 2:

$$G_I(\mathbf{r})|_{z=z_0-d} = -\frac{ik}{2\pi}\iint_{p^2+q^2>1}\frac{1}{|m|}\exp[-ik(px + qy)]$$
$$\times \exp[-k|m|(z_0 + d)]\, dp\, dq, \qquad (8.32)$$

with $|m| = \sqrt{p^2 + q^2 - 1}$. In contrast, the evanescent component part of a spherical wave, converging towards O, is (cf. Eqs. (2.52) and (2.53)):

$$G_I^{(c)}(\mathbf{r})|_{z=z_0-d} = -\frac{ik}{2\pi}\iint_{p^2+q^2>1} \frac{1}{|m|} \exp[-ik(px + qy)] \times \exp[-k|m|(z_0 - d)]\, dp\, dq, \tag{8.33}$$

which differs from Eq. (8.32) by a factor $\exp(-2k|m|d)$ in the integrand.

The above examples illustrate that phase-conjugation does not restore the original evanescent components of the wavefront. These inhomogeneous waves thus set a wavelength resolution limit (cf. Section 2.4) to the healing process achieved by this technique.

It should be remarked that the first part of Theorem 1, corresponding to homogeneous waves, was initially established by Lukosz [8.10], and by Wolf and Shewell [8.11] in connection with the nature of the pseudoscopic image in transmission holography and its relationship with inverse diffraction (we return to this point in Section 9.4). The proof presented here of this theorem for homogeneous compounds is due, however, to Wolf [8.25].

8.3. SCATTERING AND DISTORTION CORRECTION BY PHASE CONJUGATION

In this section we shall discuss the theory of distortion correction by phase-conjugation in the presence of a scatterer. We shall apply the scattering matrix formalism, discussed in Chapter 5. Therefore, it will be assumed that the scatterer is linear and nonabsorbing. On the other hand, we shall see that the interpretation of the healing effect does not require consideration of the details of the interaction between the wave and the scattering medium. On the contrary, it is a consequence of the properties of the generalized reflection and transmission coefficients. Therefore, the correction takes place irrespective of whether the scatterer is strong or weak, namely, of whether there is single or multiple scattering; it is also achieved independently of the existence of penetration or backscattering effects. This means that the wavefront correction will equally occur, for example, for a dielectric and a perfect conductor, and hence, it will be independent of the region of the electromagnetic spectrum under study.

We shall consider the scatterer between the planes $z = 0$ and $z = L$, according to the notation of Fig. 8.7, the phase-conjugate mirror PCM being at a plane z = constant in $\mathscr{R}^+$.

Using the notation of Chapter 5, the angular spectrum representation of

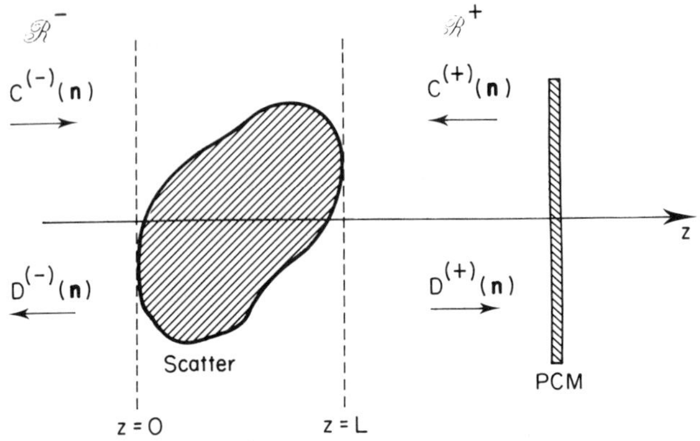

Figure 8.7. Geometry illustrating the distortion correction by phase-conjugation.

the total field, either in the half-spaces $\mathscr{R}^+$ or $\mathscr{R}^-$, is:

$$U(\mathbf{r}) = -\frac{ik}{2\pi}\int_{\sigma^{(+)}} C^{(-)}(\mathbf{n}) \exp(ik\mathbf{n}\cdot\mathbf{r})\, d\Omega + \frac{ik}{2\pi}\int_{\sigma^{(-)}} D^{(-)}(\mathbf{n}) \exp(ik\mathbf{n}\cdot\mathbf{r})\, d\Omega, \tag{8.34a}$$

in $\mathscr{R}^-$.

$$U(\mathbf{r}) = -\frac{ik}{2\pi}\int_{\sigma^{(-)}} C^{(+)}(\mathbf{n}) \exp(ik\mathbf{n}\cdot\mathbf{r})\, d\Omega + \frac{ik}{2\pi}\int_{\sigma^{(+)}} D^{(+)}(\mathbf{n}) \exp(ik\mathbf{n}\cdot\mathbf{r})\, d\Omega, \tag{8.34b}$$

in $\mathscr{R}^+$.

In Eqs. (8.34), **n** is a unit vector of components p, q and m, characterizing the direction of propagation of each plane wave component, and $\sigma^{(+)}$ and $\sigma^{(-)}$ are unit hemispheres defined in Eqs. (5.4).

The first term of Eq. (8.34a) represents the field incident on the scatterer, $U^{(i)}(\mathbf{r})$. This field propagates into $z \geqslant 0$, and thus, according to Section 2.7, it behaves asymptotically in $\mathscr{R}^-$ as a convergent spherical wave. On the other hand, the second term of Eq. (8.34a), which we shall denote by $U^{(r)}(\mathbf{r})$, represents the superposition of the field backscattered into $\mathscr{R}^-$ (and, therefore, propagating into $z < 0$) and of the field that, after reflection at the phase conjugate mirror PCM and after suffering multiple bounces between PCM and the scatterer, is finally scattered into the region $\mathscr{R}^-$ (evidently, it behaves asymptotically as a divergent spherical wave in $\mathscr{R}^-$).

When the healing effect takes place, $U^{(r)}$ should be the complex conjugate of the incident field $U^{(i)}$.

Concerning the fields in $\mathscr{R}^{(+)}$, the first term of Eq. (8.34b), which we shall denote by $U_2(\mathbf{r})$, represents the resulting field, which, after multiple bounces between the PCM and the scatterer, propagates into $z < 0$ and behaves asymptotically as a convergent spherical wave in $\mathscr{R}^+$. The second term of Eq. (8.34b), which we shall denote by $U_1(\mathbf{r})$, represents the field resulting from multiple bounces between the scatterer and the conjugator and it propagates into $z \geqslant 0$, behaving asymptotically as a divergent spherical wave in $\mathscr{R}^+$.

We assume that both the phase conjugate mirror and the detection plane in $\mathscr{R}^+$ are far from the scatterer, so that under the conditions discussed in Section 2.12, the fields do not contain evanescent components. Since the original structure of the evanescent waves is not recovered by phase-conjugation, this assumption is essential for obtaining perfect distortion correction (otherwise the correction would be achieved up to the wavelength resolution limit set by the inhomogeneous component cut-off, as discussed in Section 8.2). On the other hand, as quoted in Section 8.1, we shall assume that the response of the phase conjugate mirror does not depend either on the angle of incidence or on the polarization of the incident wave. Moreover, we consider here only those situations in which complete reversal of the state of polarization takes place upon conjugation, so that a scalar description can be used.

A scalar theory of the healing effect was initially put forward in the framework of the first Born approximation for weak scatterers by Agarwal et al. [8.27], [8.28]. The procedure used by those authors, by means of angular spectrum methods, is tutorial and clarifying, and thus we propose its study in Problems 8.5 to 8.9. A generalization for multiple scattering was performed by these authors, using the Born series [8.22], [8.29].

A method based on the S-matrix formalism of waveguide theory, was originally carried out by Mittra and Habashy [8.30] for interpreting the distortion correction. We shall next modify their model in order to adapt it to the notation of Chapter 5.

8.3.1. Self-Consistent Formulation

With reference to Fig. 8.7, the boundary condition at the plane of conjugation $z =$ constant implies that:

$$U_2(\mathbf{r})|_{z=\text{constant}} = \mu U_1^*(\mathbf{r})|_{z=\text{constant}}, \tag{8.35}$$

where U_2 and U_1 are the fields in $\mathscr{R}^+$, after multiple bounces between the scatterer and the plane PCM of conjugation, propagating into $z < 0$ and $z \geqslant 0$, respectively. As mentioned above, U_2 and U_1 are represented by the first and second terms of Eq. (8.34b), and μ is a constant denoting the reflectivity of the phase conjugate mirror (see Eq. (8.1)).

On introducing Eq. (8.34b) into (8.35), one obtains immediately (see also Theorem 2), that:

$$C^{(+)}(\mathbf{n}) = \mu[D^{(+)}(-\mathbf{n})]^*. \tag{8.36}$$

We now use Eq. (5.41a) and the definitions in Eqs. (5.58) for the generalized reflection and transmission coefficients. By taking the complex conjugate in Eq. (5.41a), and introducing Eq. (8.36) into its left-hand side, we get:

$$\begin{aligned} C^{(+)}(\mathbf{n}) = & -\mu \int_{\sigma^{(+)}} t^*(-\mathbf{n}, \mathbf{n}')C^{(-)*}(\mathbf{n}')\, d\Omega' \\ & -\mu \int_{\sigma^{(-)}} \rho^*(-\mathbf{n}, \mathbf{n}')C^{(+)*}(\mathbf{n}')\, d\Omega'. \end{aligned} \tag{8.37}$$

By taking the complex conjugate in Eq. (8.36), and introducing it into the integrand of the second term of Eq. (8.37), one obtains:

$$\begin{aligned} C^{(+)}(\mathbf{n}) = & -\mu \int_{\sigma^{(+)}} t^*(-\mathbf{n}, \mathbf{n}')C^{(-)*}(\mathbf{n}')\, d\Omega' + |\mu|^2 \int_{\sigma^{(-)}} d\Omega' \rho^*(-\mathbf{n}, \mathbf{n}') \\ & \times \left[\int_{\sigma^{(+)}} d\Omega'' t(-\mathbf{n}', \mathbf{n}'')C^{(-)*}(\mathbf{n}'') + \int_{\sigma^{(-)}} d\Omega'' \rho(-\mathbf{n}', \mathbf{n}'')C^{(+)}(\mathbf{n}'')\right], \end{aligned} \tag{8.38}$$

where, once again, Eq. (5.41a) has been used.

Rearranging terms, Eq. (8.38) may be expressed as:

$$\begin{aligned} C^{(+)}(\mathbf{n}) - |\mu|^2 & \int_{\sigma^{(-)}} d\Omega'' C^{(+)}(\mathbf{n}'') \int_{\sigma^{(-)}} d\Omega' \rho^*(-\mathbf{n}, \mathbf{n}')\rho(-\mathbf{n}', \mathbf{n}'') \\ & = -\mu \int_{\sigma^{(+)}} d\Omega' t^*(-\mathbf{n}, \mathbf{n}')C^{(-)*}(\mathbf{n}') \\ & \quad + |\mu|^2 \int_{\sigma^{(+)}} d\Omega'' C^{(-)}(\mathbf{n}'') \int_{\sigma^{(-)}} d\Omega' \rho^*(-\mathbf{n}, \mathbf{n}')t(-\mathbf{n}', \mathbf{n}''). \end{aligned} \tag{8.39}$$

By introducing the reciprocity relations, Eqs. (5.60a) and (5.60c), into the fourth and second terms of Eq. (8.39), respectively, we get:

$$C^{(+)}(\mathbf{n}) - |\mu|^2 \int_{\sigma^{(-)}} d\Omega'' C^{(+)}(\mathbf{n}'') \int_{\sigma^{(-)}} d\Omega' \rho^*(-\mathbf{n}, \mathbf{n}') \rho(-\mathbf{n}'', \mathbf{n}')$$

$$= -\mu \int_{\sigma^{(+)}} d\Omega' t^*(-\mathbf{n}, \mathbf{n}') C^{(-)*}(\mathbf{n}')$$

$$+ |\mu|^2 \int_{\sigma^{(+)}} d\Omega'' C^{(-)}(\mathbf{n}'') \int_{\sigma^{(-)}} d\Omega' \rho^*(-\mathbf{n}, \mathbf{n}') \tau(-\mathbf{n}'', \mathbf{n}'). \quad (8.40)$$

By using now the generalized Stokes reciprocity relations, Eqs. (5.92a) and (5.90c), in the second and fourth terms of Eq. (8.40), respectively, we can express this equation as:

$$C^{(+)}(\mathbf{n}) - |\mu|^2 \int_{\sigma^{(-)}} d\Omega'' C^{(+)}(\mathbf{n}'') \left[\Delta(-\mathbf{n} + \mathbf{n}'') - \int_{\sigma^{(+)}} d\Omega' t^*(-\mathbf{n}, \mathbf{n}') t(-\mathbf{n}'', \mathbf{n}') \right]$$

$$= -\mu \int_{\sigma^{(+)}} d\Omega' t^*(-\mathbf{n}, \mathbf{n}') C^{(-)*}(\mathbf{n}')$$

$$- |\mu|^2 \int_{\sigma^{(+)}} d\Omega'' C^{(-)}(\mathbf{n}'') \int_{\sigma^{(+)}} d\Omega' t^*(-\mathbf{n}, \mathbf{n}') r(-\mathbf{n}'', \mathbf{n}'). \quad (8.41)$$

Making use of the properties of integration of the spherical delta function $\Delta(-\mathbf{n} + \mathbf{n}'')$ and the reciprocity relation (Eq. (5.60a)) in the third term of Eq. (8.41), we can write, after recalling Eqs. (5.41b) and (5.58), and rearranging terms:

$$|\mu|^2 \int_{\sigma^{(+)}} d\Omega' t^*(-\mathbf{n}, \mathbf{n}') D^{(-)}(-\mathbf{n}') - \mu \int_{\sigma^{(+)}} d\Omega' t^*(-\mathbf{n}, \mathbf{n}') C^{(-)*}(-\mathbf{n}')$$

$$= (1 - |\mu|^2) C^{(+)}(\mathbf{n}). \quad (8.42)$$

Equation (8.42) has no obvious solution unless $|\mu| = 1$. In this case, writing $\mu = \exp(i\phi)$, it becomes:

$$\int_{\sigma^{(+)}} d\Omega' t^*(-\mathbf{n}, \mathbf{n}') [D^{(-)}(-\mathbf{n}') - \exp(i\phi) C^{(-)*}(\mathbf{n}')] = 0. \quad (8.43)$$

Since $t^*(-\mathbf{n}, \mathbf{n}')$ is not zero, Eq. (8.43) yields:

$$D^{(-)}(\mathbf{n}) = \exp(i\phi) D^{(-)*}(-\mathbf{n}). \quad (8.44)$$

Considering the definitions of $U^{(r)}$ and $U^{(i)}$ in terms of the representations in Eqs. (8.34), we obtain from Eq. (8.44):

$$U^{(r)}(\mathbf{r}) = \exp(i\phi)U^{(i)*}(\mathbf{r}); \tag{8.45}$$

for any point $\mathbf{r}$ of $\mathscr{R}^-$ (far enough from the scatterer so that there is no evanescent component contribution).

Equations (8.44) and (8.45) show that when the phase conjugate mirror does not introduce either gains or losses (namely, when $|\mu| = 1$), a complete correction of the wavefront distortions, due to scattering, is achieved.

8.3.2. Multiple Bounce Approach

This is an alternative method to the self-consistent formulation. It consists of adding the multiple reflections that occur both in the scatterer and the phase conjugator.

When this procedure is carried out, the first term gives the angular spectrum $D^{(-)}(\mathbf{n})$ of the field backscattered into $\mathscr{R}^+$, which is (cf. Eqs. (5.41b) and (5.58)):

$$\begin{aligned} D^{(-)(1)}(\mathbf{n}) = &-\int_{\sigma^{(+)}} r(\mathbf{n}, \mathbf{n}')C^{(-)}(\mathbf{n}')\, d\Omega' \\ &-\int_{\sigma^{(-)}} \tau(\mathbf{n}, \mathbf{n}')C^{(+)(1)}(\mathbf{n}')\, d\Omega', \end{aligned} \tag{8.46}$$

where $C^{(+)(1)}(\mathbf{n}')$ is the angular spectrum of the field incident on the scatterer after the first reflection at the conjugator PCM. Taking into account the boundary condition, Eq. (8.36), which holds for every reflection at the conjugator, we can write:

$$\begin{aligned} D^{(-)(1)}(\mathbf{n}) = &-\int_{\sigma^{(+)}} r(\mathbf{n}, \mathbf{n}')C^{(-)}(\mathbf{n}')\, d\Omega' \\ &-\mu\int_{\sigma^{(-)}} \tau(\mathbf{n}, \mathbf{n}')D^{(+)(1)*}(-\mathbf{n}')\, d\Omega', \end{aligned} \tag{8.47}$$

where $D^{(+)(1)}(-\mathbf{n}')$ is the first order of the angular spectrum of the field scattered into $\mathscr{R}^+$. This is given by the first term of Eq. (5.41a). Therefore,

Eq. (8.47) may be expressed as:

$$D^{(-)(1)}(\mathbf{n}) = -\int_{\sigma^{(+)}} r(\mathbf{n}, \mathbf{n}')C^{(-)}(\mathbf{n}')\, d\Omega' - \mu \int_{\sigma^{(-)}} d\Omega' \tau(\mathbf{n}, \mathbf{n}') \int_{\sigma^{(+)}} t(-\mathbf{n}', \mathbf{n}'')C^{(-)}(\mathbf{n}''). \qquad (8.48)$$

The next order $D^{(-)(2)}(\mathbf{n})$ is obtained from Eq. (5.41b), where now the spectral amplitude $C^{(+)(2)}(\mathbf{n})$ entering in the second term, includes a double bounce between the scatterer and the conjugator PCM. In this way, one obtains a Neumann series. By summing all orders of $D^{(-)}(\mathbf{n})$, when $|\mu| = 1$ one again gets Eq. (8.44) (see Problem 8.10). The complete distortion correction is thus obtained only when the multiple bounces are considered, and their corresponding terms are added.

Example 8.4. *Fabry–Perot interferometer with a phase-conjugate mirror.*

We consider the situation in which the scatterer has been replaced by an ordinary (lossless) stratified dielectric slab M (see Fig. 8.8). This system is equivalent to a Fabry–Perot interferometer in which one of the mirrors has been replaced by a conjugator PCM (see Refs. 8.31–8.34). Following these works, we assume that the incident field onto M is a plane wave:

$$U_i(\mathbf{r}) = A_i \exp(ik\mathbf{n}_i \cdot \mathbf{r}), \qquad (8.49)$$

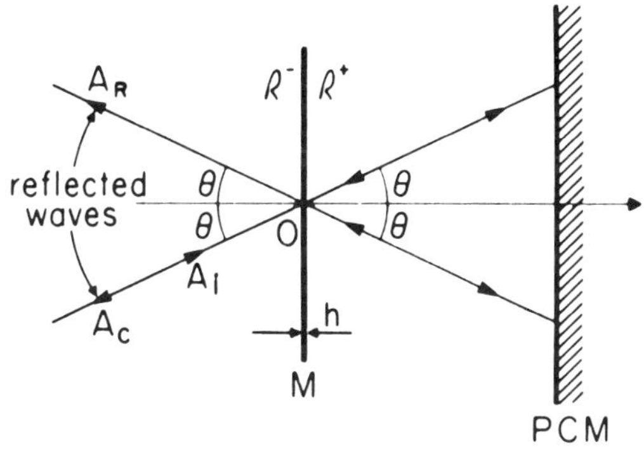

Figure 8.8. Illustration of fields in a Fabry–Perot interferometer with a phase-conjugate mirror. (Reproduced with permission from paper by A. T. Friberg and P. D. Drummond, *J. Opt. Soc. Am.* **73**, 1216 (1983).)

$\mathbf{n}_i$ being the unit vector along the direction of incidence, and making an angle θ with the normal to the slab M. We next follow the notation of Section 5.12.

The interaction of first order of the incident wave with this system produces a wave reflected from M with amplitude $\bar{r}A_i$, and a transmitted wave with amplitude $\bar{t}A_i$. This second wave is reflected at the PCM producing a conjugate wave $\mu\bar{t}^*A_i^*$, which is afterwards transmitted by M into $\mathscr{R}^+$ as a wave with amplitude: $\mu\bar{t}^*\bar{\tau}A_i^*$. We shall next calculate the contributions to both the wave of amplitude A_R, specularly reflected at M, and the conjugate one of amplitude A_C, emerging from M with a wavevector in the opposite direction to the incident wave (see Fig. 8.8).

In first order we have:

$$A_R^{(1)} = \bar{r}A_i, \qquad \text{for the reflected wave;} \tag{8.50a}$$

$$A_C^{(1)} = \mu\bar{t}^*\bar{\tau}A_i^*, \qquad \text{for the conjugate wave.} \tag{8.50b}$$

In a similar way, after the second bounce, we have for the second-order amplitudes:

$$A_R^{(2)} = |\mu|^2\bar{\rho}^*\bar{\tau}\bar{t}A_i, \qquad \text{for the reflected wave;} \tag{8.51a}$$

$$A_C^{(2)} = \mu\bar{t}^*\bar{\tau}|\mu|^2|\bar{\rho}|^2A_i^*, \qquad \text{for the conjugate wave.} \tag{8.51b}$$

The multiple bounces are represented by iterating this procedure. The resulting amplitudes are then given by the following geometric series:

$$A_R = [\bar{r} + |\mu|^2\bar{\rho}^*\bar{\tau}\bar{t}(1 + |\mu|^2|\bar{\rho}|^2 + \cdots)]A_i,$$

$$\text{for the resulting reflected wave;} \tag{8.52a}$$

$$A_C = [\mu\bar{t}^*\bar{\tau}(1 + |\mu|^2|\bar{\rho}|^2 + \cdots)]A_i^*,$$

$$\text{for the resulting conjugate wave.} \tag{8.52b}$$

The series of Eqs. (8.52) can be added provided that $|\mu|^2|\bar{\rho}|^2 < 1$, then giving:

$$A_R = \left(\bar{r} + \frac{|\mu|^2\bar{\rho}^*\bar{\tau}t}{1 - |\mu|^2|\bar{\rho}|^2}\right)A_i, \qquad \text{for the resulting reflected wave;} \tag{8.53a}$$

$$A_C = \left(\frac{\mu\bar{t}^*\bar{\tau}}{1 - |\mu|^2|\bar{\rho}|^2}\right)A_i^*, \qquad \text{for the resulting conjugate wave.} \tag{8.53b}$$

By using the Stokes relations, Eqs. (5.102)–(5.104), Eqs. (8.53) are finally expressed as:

$$A_R = \frac{\bar{r}(1 - |\mu|^2)}{1 - |\mu|^2|\bar{r}|^2} A_i, \qquad \text{for the resulting reflected wave;} \qquad (8.54a)$$

$$A_C = \frac{\mu(1 - |\bar{r}|^2)}{1 - |\mu|^2|\bar{r}|^2} A_i^*, \qquad \text{for the resulting conjugate wave.} \qquad (8.54b)$$

Equations (8.54) show that when $|\mu| = 1$, A_R becomes identically zero and A_C reduces to:

$$A_C = \exp(i\phi)A_i^*. \qquad (8.55)$$

This demonstrates that the only wave which remains in $\mathscr{R}^-$ is A_C, given by Eq. (8.55). The specularly reflected wave A_R (which plays the role of the "backscattered" wave in this degenerate situation) is cancelled out when perfect conjugation (i.e., without gains or losses) occurs at the conjugator.

This example shows dramatically the healing property of a perfect conjugate mirror. Also, as seen above, this is a consequence of the multiple bounces between the distorting body and the conjugator.

8.4. OTHER STUDIES

Phase-conjugation of evanescent waves has been studied by Ujihara [8.35], [8.36] in connection with the wave mixing of surface plasmons. Fields inside a Fabry–Perot interferometer with a phase-conjugate mirror have been addressed in Refs. 8.34 and 8.37; in this device, the interference fringes depend on the phase of the incident field. This property of interferometers with a phase-conjugator has also been discussed by Feinberg [8.38].

A study of the classical experiment of Wiener [8.39], [8.40], concerning the interference pattern produced on reflection of a wave on a metallic mirror, has been carried out in Refs. 8.41–8.42, replacing this mirror by a phase-conjugator.

Optical resonators with phase-conjugate mirrors have a variety of properties and applications (see, for instance Ref. 8.43).

Several symmetry properties related to phase-conjugation have been investigated in the framework of the theory of data-processing systems [8.44].

The effect of wavefront profile propagation and aberrations of the pumping fields in phase-conjugation by wave mixing in Kerr-like media has been investigated in Refs. 8.46–8.49.

The relationship between enhanced backscattering and partial phase conjugation is discussed in Refs. 8.50–8.52.

PROBLEMS

8.1. Using the boundary condition, Eq. (8.20), and the field representation, Eqs. (8.21)–(8.24), prove Eqs. (8.26a) and (8.26b) of Theorem 2 (Wolf [8.25]; Nieto-Vesperinas and Wolf [8.26]).

8.2. Using the results of Theorem 2, prove that phase-conjugation with reflection does not restore the inhomogeneous wave contribution in the plane wave expansion of a spherical wave.

8.3. Let $U^{(1)}(x, y, z)$ and $U^{(2)}(x, y, z)$ be two wavefields satisfying at every point of a plane $z = z_0$ the boundary condition:

$$U^{(2)}(x, y, z_0) = [U^{(1)}(-x, -y, z_0)]^*.$$

Prove that for $0 < d < \infty$, one has for the homogeneous part of these fields:

$$U_H^{(2)}(x, y, z_0 + d) = [U_H^{(1)}(-x, -y, z_0 - d)]^*.$$

(Wolf, [8.25].)

Discuss the connection of this result with the symmetry relations of Eq. (6.96) for focused wavefields. Find the corresponding relationship for the contribution of the evanescent components.

8.4. Show that if the two fields of Problem 8.3 are rotationally symmetric about the z-axis at a certain plane $z = z_0$, and satisfy:

$$U^{(2)}(R, z_0) = [U^{(1)}(R, z_0)]^*,$$

then both fields remain rotationally symmetric at any distance $0 < d < \infty$, and satisfy:

$$U_H^{(2)}(R, z_0 + d) = [U_H^{(1)}(R, z_0 - d)]^*; \qquad (R = \sqrt{x^2 + y^2}).$$

(Wolf, [8.25].) Discuss the behavior of the parts corresponding to the inhomogeneous components.

8.5. Consider the scattering of a scalar wavefield by a weak potential $F(\mathbf{r})$, contained in a volume V and limited by the planes $z = 0$ and $z = L > 0$ (cf. Section 3.10). Let us express the total field $U(\mathbf{r})$ as:

$$U(\mathbf{r}) = U^{(i)}(\mathbf{r}) + U^{(s)}(\mathbf{r}),$$

where $U^{(i)}$ denotes the field incident on the scatterer, and $U^{(s)}$ represents the field scattered outside V (Eq. (3.114)).

Suppose that at a plane $z = z_1 > L$ the field is phase-conjugated and reflected according to the boundary condition:

$$V^{(i)}(\mathbf{R}_1, z_1) = \mu[U(\mathbf{R}_1, z_1)]^*,$$

$$\mathbf{r}_1 = (\mathbf{R}_1, z_1), \qquad \mathbf{R}_1 = (x_1, y_1),$$

μ being a constant that denotes the reflectivity at $z = z_1$.

Show that, neglecting evanescent waves, the reflected phase-conjugate wavefield in the space between the planes $z = z_1$ and $z = L$ may be written as (Agarwal and Wolf [8.27]):

$$V^{(i)}(\mathbf{r}) = V^{(0)}(\mathbf{r}) + V^{(1)}(\mathbf{r}),$$

where:

$$V^{(0)}(\mathbf{r}) = \mu[U^{(i)}(\mathbf{r})]^*,$$

and:

$$V^{(1)}(\mathbf{r}) = \mu \iint_{K^2<k^2} [A^{(+)}(-\mathbf{K})]^* \exp[i(\mathbf{K}\cdot\mathbf{R} - k_z z)]\, d^2K;$$

where: $k = 2\pi/\lambda$, $\mathbf{k} = (\mathbf{K}, k_z)$, $\mathbf{r} = (\mathbf{R}, z)$. The angular spectrum is given in the first Born approximation (cf. Section 3.10.1) as:

$$[A^{(+)}(-\mathbf{K})]^* = \frac{i}{8\pi^2 k_z} \int_V F^*(\mathbf{r}')[U^{(i)}(\mathbf{r}')]^* \exp[-i(\mathbf{K}\cdot\mathbf{R}' - k_z z')]\, d^3r'.$$

8.6. Suppose that the field $V^{(i)}(\mathbf{r})$ of Problem 8.5 interacts again with the same scatterer, giving rise to:

$$V(\mathbf{r}) = V^{(i)}(\mathbf{r}) + V^{(s)}(\mathbf{r}),$$

$V^{(s)}$ representing the scattered field.

Show that in the half-space $z < 0$, and neglecting evanescent components, the wavefield $V^{(s)}$ is expressed within the first Born approximation as (Agarwal and Wolf [8.27]):

$$V^{(s)}(\mathbf{r}) = \mu \iint_{K^2<k^2} B^{(-)}(\mathbf{K}) \exp[i(\mathbf{K}\cdot\mathbf{R} - k_z z)]\, d^2K;$$

where:

$$B^{(-)}(\mathbf{K}) = -\frac{i}{8\pi^2 k_z}\int_V F(\mathbf{r}')[U^{(i)}(\mathbf{r}')]^* \exp[-i(\mathbf{K}\cdot\mathbf{R}' - k_z z')]\, d^3r'.$$

8.7. Suppose that the scattering potential $F(\mathbf{r})$ is real, i.e., nonabsorbing. Show that in this case the angular spectra of Problems 8.5 and 8.6 are related by:

$$[A^{(+)}(-\mathbf{K})]^* = -B^{(-)}(\mathbf{K}),$$

and therefore:

$$V^{(s)}(\mathbf{r}) = -V^{(1)}(\mathbf{r}), \qquad z < 0.$$

Show also that the resulting field $V(\mathbf{r})$ in $z < 0$ is (Agarwal and Wolf [8.27]):

$$V(\mathbf{r}) = \mu[U^{(i)}(\mathbf{r})]^*, \qquad z < 0,$$

namely, the scattering effect is cancelled out.

8.8. Suppose that under the conditions addressed in Problems 8.5–8.7, there exists backscattering into both half-spaces $z < 0$ and $z \geqslant L$. Show, then, that the scattered field in $z < 0$ is (Agarwal *et al.* [8.28]):

$$V(\mathbf{r}) = \mu[U^{(i)}(\mathbf{r})]^* + W^{(BS)}(\mathbf{r}), \qquad z < 0,$$

where $W^{(BS)}$ is the field backscattered into $z < 0$, given by:

$$W^{(BS)}(\mathbf{r}) = (1 - |\mu|^2)\iint_{K^2<k^2} A^{(-)}(\mathbf{K}) \exp[i(\mathbf{K}\cdot\mathbf{R} - k_z z)]\, d^2K.$$

Also, $A^{(-)}(\mathbf{K})$ is given within the first Born approximation as:

$$A^{(-)}(\mathbf{K}) = -\frac{i}{8\pi^2 k_z}\int_V F(\mathbf{r}')[U^{(i)}(\mathbf{r}')]^* \exp[-i(\mathbf{K}\cdot\mathbf{R}' - k_z z')]\, d^3r'.$$

Therefore, when $|\mu| = 1$, i.e., when there are neither losses nor gains, the wavefront distortions introduced by the scatterer are eliminated.

8.9. Make a discussion of the process studied in Problems 8.5–8.7 when the scattered fields are considered to contain evanescent components.

8.10. Find higher order terms from Eqs. (8.46)–(8.48) and prove Eq. (8.44) for $|\mu| = 1$ by summing the resulting series. Each term of this series corresponds to a bounce between the scatterer and the phase-conjugator. Use the reciprocity relations of Sections 5.8 and 5.11 for the generalized reflection and transmission coefficients.

REFERENCES

8.1. B. Ya. Zel'dovich, V. I. Popovichev, V. V. Ragul'skii and F. S. Faizullov, *Sov. Phys. J.E.T.P.* **15**, 109 (1972).

8.2. O. Yu. Nosach, V. I. Popovichev, V. V. Regul'skii and F. S. Faizullov, *Sov. Phys. J.E.T.P.* **16**, 435 (1972).

8.3. A. Yariv, *Appl. Phys. Lett.* **28**, 88 (1976).

8.4. A. Yariv, *J. Opt. Soc. Am.* **66**, 301 (1976).

8.5. R. W. Hellwarth, *J. Opt. Soc. Am.* **67**, 1 (1977).

8.6. A. Yariv and D. M. Pepper, *Opt. Lett.* **1**, 16 (1977).

8.7. J. P. Huignard, J. P. Herriau, P. Auborg and E. Spitz, *Opt. Lett.* **4**, 21 (1979).

8.8. J. Feinberg, D. Heiman, A. R. Tanguay and R. W. Hellwarth, *J. Appl. Phys.* **51**, 1297 (1980).

8.9. H. Kogelnik and K. S. Pennington, *J. Opt. Soc. Am.* **58**, 273 (1968).

8.10. W. Lukosz, *J. Opt. Soc. Am.* **58**, 1084 (1968).

8.11. E. Wolf and J. R. Shewell, *J. Math. Phys.* **11**, 2254 (1970).

8.12. H. J. Gerritsen, *Appl. Phys. Lett.* **10**, 237 (1967).

8.13. J. P. Woerdman, *Opt. Comm.* **2**, 212 (1971).

8.14. B. I. Stepanov, E. V. Ivakin and A. S. Rubanov, *Sov. Phys. Dokl.* **16**, 46 (1971).

8.15. V. Wang and C. R. Giuliano, *Opt. Lett.* **2**, 5 (1978).

8.16. D. M. Pepper, *Opt. Eng.* **21**, 156 (1982).

8.17. D. M. Pepper, in *Laser Handbook*, Vol. 4, M. L. Stitch and M. Bass, eds., North-Holland, Amsterdam, 1985.

8.18. R. A. Fisher, ed., *Optical Phase Conjugation*, Academic Press, New York, 1982.

8.19. B. Ya. Zel'dovich and V. V. Shkunov, *Sov. J. Quant. Electron.* **9**, 379 (1979).

8.20. B. Ya. Zel'dovich and T. V. Yakovleva, *Sov. J. Quant. Electron.* **10**, 501 (1980).

8.21. J. F. Lam, D. G. Steel, R. A. McFarlane and R. C. Lind, *Appl. Phys. Lett.* **38**, 977 (1981).

8.22. G. S. Agarwal, A. T. Friberg and E. Wolf, *J. Opt. Soc. Am.* **73**, 529 (1983).

8.23. E. Wolf and W. H. Carter, *Opt. Comm.* **40**, 397 (1982).

8.24. A. Yariv, *Opt. Comm.* **40**, 401 (1982).

8.25. E. Wolf, *J. Opt. Soc. Am.* **70**, 1311 (1980).

8.26. M. Nieto-Vesperinas and E. Wolf, *J. Opt. Soc. Am. A* **2**, 1429 (1985).

8.27. G. S. Agarwal and E. Wolf, *J. Opt. Soc. Am.* **72**, 321 (1982).
8.28. G. S. Agarwal, A. T. Friberg and E. Wolf, *J. Opt. Soc. Am.* **72**, 861 (1982).
8.29. G. S. Agarwal, A. T. Friberg and E. Wolf, *Opt. Comm.* **43**, 446 (1982).
8.30. R. Mittra and T. M. Habashy, *J. Opt. Soc. Am. A* **1**, 1103 (1984).
8.31. P. D. Drummond and A. T. Friberg, *J. Appl. Phys.* **54**, 5618 (1983).
8.32. A. T. Friberg and P. D. Drummond, *J. Opt. Soc. Am.* **73**, 1216 (1983).
8.33. M. Nazarathy, *Opt. Comm.* **45**, 117 (1983).
8.34. M. Nieto-Vesperinas, *J. Opt. Soc. Am. A* **2**, 427 (1985).
8.35. K. Ujihara, *Opt. Comm.* **42**, 1 (1982).
8.36. K. Ujihara, *Opt. Comm.* **43**, 225 (1982).
8.37. M. Nieto-Vesperinas, *Opt. Lett.* **9**, 555 (1984).
8.38. J. Feinberg, *Opt. Lett.* **8**, 569 (1983).
8.39. M. Born and E. Wolf, *Principles of Optics*, Pergamon Press, Oxford, 1980, 6th edition.
8.40. A. Sommerfeld, *Optics*, Academic Press, New York, 1964.
8.41. R. W. Boyd, T. M. Habashy, A. Jacobs, L. Mandel, M. Nieto-Vesperinas and E. Wolf, *Opt. Lett.* **12**, 42 (1987).
8.42. E. Wolf, L. Mandel, R. W. Boyd, T. M. Habashy and M. Nieto-Vesperinas, *J. Opt. Soc. Am. B* **34**, 1266 (1987); A. A. Jacobs, W. R. Tompkin, R. W. Boyd and E. Wolf, *J. Opt. Soc. Am. B* **4**, 1266 (1987).
8.43. A. E. Siegman, P. A. Belanger and A. Hardy, in *Optical Phase Conjugation*, R. A. Fisher, ed., Academic Press, New York, 1982, Chapter 13.
8.44. M. Nazarathy and J. Shamir, *J. Opt. Soc. Am.* **73**, 910 (1983).
8.45. E. Bochove, *J. Opt. Soc. Am.* **73**, 1330 (1983).
8.46. B. Crosignani and A. Yariv, *J. Opt. Soc. Am. A* **1**, 1034 (1984).
8.47. M. D. Skeldon, P. Narum and R. W. Boyd, *Opt. Lett.* **12**, 343 (1987).
8.48. J. W. Haus, C. M. Bowden and C. C. Sung, *Phys. Rev. A* **35**, 3398 (1987).
8.49. G. Lera and M. Nieto-Vesperinas, *Phys. Rev. A* **41**, 6400 (1990).
8.50. Y. A. Kravtsov and A. I. Saichev, *Sov. Phys. J.E.T.P.* **56**, 291 (1983); *Sov. Phys. Usp.* **25**, 494 (1983).
8.51. Y. A. Kravtsov and A. I. Saichev, *J. Opt. Soc. Am. A* **2**, 2100 (1985).
8.52. E. Jakeman, *J. Opt. Soc. Am. A* **5**, 1638 (1988).

9

Inverse Diffraction

In Chapter 6 we have addressed several theories on the *direct diffraction problem*. In this chapter we shall deal with the question of determining the field distribution on a boundary surface from knowledge of the field on another surface to which the field has propagated, when there is free space between both surfaces. This is the *inverse diffraction problem*.

We shall assume that the surfaces are planes; thus, for a field that propagates into the half-space $z \geqslant 0$, we shall study the problem of determining this field at a plane $z = z_0 \geqslant 0$ from the field at another plane $z = z_1 > z_0$. This field may have been directly obtained in this plane by some detection means, or it may have been determined from its angular spectrum, by finding the spectral amplitude of the field in the far zone with respect to a certain region of sources of this field. (Of course, this second procedure will yield the angular spectrum of the field up to the evanescent wave cut-off only. In any case, however, the determination of the field beyond the wavelength resolution limit is a very difficult subject, as seen in Section 2.4, and it will be further discussed.) The field determination on arbitrary surface geometries would constitute a generalization of the methods presented in this chapter.

Since we deal with free-space propagation, we shall not have to consider depolarization effects. Therefore, as far as electromagnetic fields are concerned, we shall deal with only one transversal component of either the electric or the magnetic vector, bearing in mind that the corresponding angular spectra are related through Eqs. (2.44) and (2.45).

It will be shown that the solution to the inverse diffraction problem involves a singular kernel (Shewell and Wolf [9.1]) (a fact also noted by other authors [9.2]–[9.4]). This singularity is in fact inherent to all inverse problems. As shown in Ref. 9.1, for fields that are bandlimited to the low spatial frequency domain ($p^2 + q^2 < 1$, p and q being the directional cosines satisfying Eqs. (2.9)), the singular kernel may be replaced by another kernel that is well behaved. This also leads to one of the reciprocity theorems studied in Chapter 8.

We shall also present an outline of two other important topics associated with the inverse diffraction problem, and which are common to all *inverse problems.* These refer to the question of reconstructing bandlimited signals (which, in turn, is linked to the *resolution* with which the original data can be retrieved) and to the *phase retrieval problem.*

9.1. THE INVERSE WAVEFIELD PROPAGATOR

We shall consider a monochromatic wavefield, whose spatial part is represented by the scalar wavefunction $U(\mathbf{r})$. We assume that this field propagates in the free half-space $z \geqslant 0$ (Fig. 9.1). Hence, all sources of this field are in the region $z < 0$. $U(\mathbf{r})$ thus obeys the homogeneous Helmholtz equation (2.1) in $z \geqslant 0$. It will be also assumed to hold Sommerfeld's radiation condition for outgoing fields, Eq. (2.6).

Let us represent by $U(x_0, y_0, z_0)$ the wavefield at the plane $z = z_0 > 0$, then, according to Eqs. (2.18) and (2.19), the wavefield at any other plane $z = z_1 > 0$ is:

$$U(x_1, y_1, z_1) = \iint_{-\infty}^{\infty} U(x_0, y_0, z_0)\mathscr{K}(x_1 - x_0, y_1 - y_0, z_1 - z_0)\, dx_0\, dy_0, \quad (9.1)$$

where $\mathscr{K}$ is the *direct* wavefield *propagator*:

$$\mathscr{K}(x_1 - x_0, y_1 - y_0, z_1 - z_0) = \left(\frac{k}{2\pi}\right)^2 \times \iint_{-\infty}^{\infty} \exp[ik[p(x_1 - x_0) + q(y_1 - y_0) + m(z_1 - z_0)]]\, dp\, dq; \quad (9.2)$$

p, q, and m satisfy Eqs. (2.9).

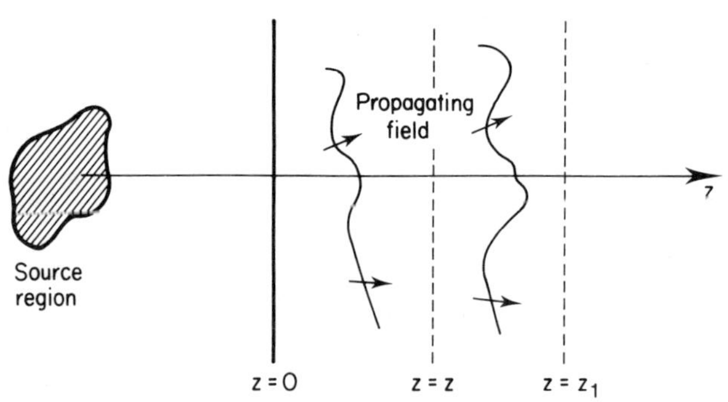

Figure 9.1. Geometry of propagation.

In an identical manner, it is immediately apparent that the wavefield $U(x_0, y_0, z_0)$ in $z = z_0$ is given from the values $U(x_1, y_1, z_1)$ in $z = z_1$ by means of the expression:

$$U(x_0, y_0, z_0) = \iint_{-\infty}^{\infty} U(x_1, y_1, z_1)\mathscr{K}^{-1}(x_0 - x_1, y_0 - y_1, z_0 - z_1)\, dx_1\, dy_1, \tag{9.3}$$

where the *inverse propagator* $\mathscr{K}^{-1}$ in free space is represented by:

$$\mathscr{K}^{-1}(x_0 - x_1, y_0 - y_1, z_0 - z_1) = \left(\frac{k}{2\pi}\right)^2 \times \iint_{-\infty}^{\infty} \exp[ik[p(x_0 - x_1) + q(y_0 - y_1) + m(z_0 - z_1)]]\, dp\, dq. \tag{9.4}$$

As shown by Eq. (6.28), the direct propagator $\mathscr{K}$ is connected to the outgoing spherical wave $G(r)$ by:

$$\mathscr{K}(x_0 - x_1, y_0 - y_1, z_0 - z_1) = \frac{1}{2\pi}\frac{\partial}{\partial z_0}\left(\frac{\exp(ikr)}{r}\right), \tag{9.5}$$

where:

$$r = \sqrt{(x_1 - x_0)^2 + (y_1 - y_0)^2 + (z_1 - z_0)^2}. \tag{9.6}$$

We now decompose the inverse propagator $\mathscr{K}^{-1}$, given by Eq. (9.4), as the sum of two parts containing the contribution of the homogeneous and of the evanescent components, respectively [9.1]:

$$\mathscr{K}^{-1}(x_0 - x_1, y_0 - y_1, z_0 - z_1) = \mathscr{K}_H^{-1}(x_0 - x_1, y_0 - y_1, z_0 - z_1) + \mathscr{K}_I^{-1}(x_0 - x_1, y_0 - y_1, z_0 - z_1), \tag{9.7}$$

where:

$$\mathscr{K}_H^{-1}(x_0 - x_1, y_0 - y_1, z_0 - z_1) = \left(\frac{k}{2\pi}\right)^2 \iint_{p^2+q^2\leqslant 1} \times \exp[ik[p(x_0 - x_1) + q(y_0 - y_1) + m(z_0 - z_1)]]\, dp\, dq, \tag{9.8a}$$

$$\mathscr{K}_I^{-1}(x_0 - x_1, y_0 - y_1, z_0 - z_1) = \left(\frac{k}{2\pi}\right)^2 \iint_{p^2+q^2> 1} \times \exp[ik[p(x_0 - x_1) + q(y_0 - y_1) + m(z_0 - z_1)]]\, dp\, dq. \tag{9.8b}$$

Equation (9.8a) shows that the integral representing $\mathscr{K}_H^{-1}$ converges since its integrand is a continuous unimodular function of p and q, and, moreover, its domain of integration is finite. However, inside the integral representation (Eq. (9.8b)) of $\mathscr{K}_I^{-1}$ there are factors of the form $\exp[-k\sqrt{p^2 + q^2 - 1}(z_0 - z_1)]$. Since $z_1 > z_0$, these factors diverge over the domain $p^2 + q^2 > 1$. Therefore, the integral of Eq. (9.8b) diverges. This implies that the inverse kernel $\mathscr{K}^{-1}$ is singular.

9.2. ALTERNATIVE FORM OF THE INVERSE PROPAGATOR

The inverse propagator (Eq. (9.7)) may be written [9.1], [9.4]:

$$\mathscr{K}^{-1}(x_0 - x_1, y_0 - y_1, z_0 - z_1) = \mathscr{K}_1^{-1}(x_0 - x_1, y_0 - y_1, z_0 - z_1) + \mathscr{K}_2^{-1}(x_0 - x_1, y_0 - y_1, z_0 - z_1), \quad (9.9)$$

where, in analogy with Eq. (9.5), $\mathscr{K}_1^{-1}$ is expressed as:

$$\mathscr{K}_1^{-1}(x_0 - x_1, y_0 - y_1, z_0 - z_1) = -\frac{1}{2\pi}\frac{\partial}{\partial z_0}\left(\frac{\exp(-ikr)}{r}\right). \quad (9.10)$$

On the other hand, $\mathscr{K}_1^{-1}$ is given by:

$$\mathscr{K}_2^{-1}(x_0 - x_1, y_0 - y_1, z_0 - z_1) = \left(\frac{k}{2\pi}\right)^2 \iint_{p^2+q^2>1} \exp[ik[p(x_0 - x_1) + q(y_0 - y_1)]] \times (\exp[ikm(z_0 - z_1)] - \exp[ikm'(z_0 - z_1)])\, dp\, dq. \quad (9.11)$$

In Eq. (9.11), m' is:

$$m' = -i\sqrt{p^2 + q^2 - 1}, \qquad p^2 + q^2 > 1. \quad (9.12)$$

While Eq. (9.10) represents a well-behaved function, the integral of Eq. (9.11) diverges; manifesting the singular contribution of $\mathscr{K}_2^{-1}$.

9.3. INVERSION FORMULA FOR FIELDS WITHOUT EVANESCENT COMPONENTS

If the field $U(\mathbf{r})$ is bandlimited to homogeneous waves; i.e., its angular spectrum vanishes for $p^2 + q^2 > 1$ in the plane $z = z_1$, then the term $\mathscr{K}_2^{-1}$, Eq. (9.11), does not contribute to the inverse propagator. In this case, and considering Eqs. (9.10) and (9.11), one obtains from Eq. (9.3) the following inversion formula [9.1]:

$$U(x_0, y_0, z_0) = -\frac{1}{2\pi}\iint_{-\infty}^{\infty} U(x_1, y_1, z_1)\frac{\partial}{\partial z_1}\left[\frac{\exp(-ikr)}{r}\right] dx_1\, dy_1. \quad (9.13)$$

Equation (9.13), the compact form of the inverse diffraction integral, is valid when the field $U(\mathbf{r})$ does not contain evanescent components in the plane $z = z_1$. It has a remarkable symmetry with the direct diffraction equation, as expressed by the first Rayleigh–Sommerfeld integral, Eq. (6.18). According to Eq. (9.13), the field U is retrieved in the plane $z = z_0$ by phase-conjugating the field propagated to the plane $z = z_1 > z_0$, imposing the boundary condition of phase conjugation of the form of Eq. (8.20), with V equal to U, at every point of the plane $z = z_1$, and propagating back into $z < 0$ (observe that the derivative $\partial/\partial z_1$ changes its sign when the direction of propagation of the field changes). The source-free field of the inversion equation (9.13) is incoming and converges towards the region occupied by the source that originated it in $z < 0$. There is, however, the possibility of interpreting it also as a converging field in $z > 0$, as shown next.

9.4. CONNECTION WITH THE RECIPROCITY THEOREM OF PHASE-CONJUGATED WAVEFIELDS

Equation (9.13) leads to the result expressed by the first part of Theorem 1 of Section 8.2.1. To prove this, let us consider the direct diffraction integral, Eq. (6.18), which we shall write as:

$$U(x_1, y_1, z_1) = \frac{1}{2\pi}\iint_{-\infty}^{\infty} U(x_0, y_0, z_0)\frac{\partial}{\partial z_0}\left[\frac{\exp(ikr)}{r}\right] dx_0\, dy_0, \quad (9.14)$$

and let us exclude in Eq. (9.14) the evanescent components. If we now substitute in the integrand of Eq. (9.14) $U(x_0, y_0, z_0)$ by $U^*(x_1, y_1, z_1)$ and we take the complex conjugate of both sides of Eq. (9.14), we obtain:

$$U^*(x_1, y_1, z_1) = \frac{1}{2\pi}\iint_{-\infty}^{\infty} U(x_1, y_1, z_1)\frac{\partial}{\partial z_0}\left[\frac{\exp(-ikr)}{r}\right] dx_0\, dy_0. \quad (9.15)$$

On comparing Eq. (9.15) with the inverse diffraction integral of Eq. (9.13), we see that the field distribution in the plane $z = z_0$, which constitutes the solution to the inverse diffraction problem for the boundary value:

$$U = U(x_1, y_1, z_1), \qquad \text{in } z = z_1 > z_0 \geqslant 0,$$

is equal to the complex conjugate of the field distribution in the plane $z = z_1$ for the boundary value:

$$U = U^*(x_1, y_1, z_1), \qquad \text{in } z = z_0 \geqslant 0.$$

This is expressed in the reciprocity theorem (Shewell and Wolf [9.1]; see also Mittra and Ramson or Lukosz [9.3]):

Theorem. *Let $U^{(1)}(x, y, z)$ and $U^{(2)}(x, y, z)$ represent two fields, obeying Sommerfeld's radiation condition in the free half-space $z > 0$ and being bandlimited to homogeneous components, $p^2 + q^2 \leqslant 1$ (i.e., source-free), in a plane $z = z_0 \geqslant 0$. If the two fields are conjugates of each other in $z = z_1 > 0$, namely:*

$$U^{(2)}(x, y, z_1) = U^{(1)*}(x, y, z_1), \tag{9.16}$$

then:

$$U^{(2)}(x, y, z_1 + d) = U^{(1)*}(x, y, z_1 - d), \tag{9.17}$$

for any distance $0 < d \leqslant z_1$ and for all x and y. ■

As mentioned above, this theorem is identical to the first half of Theorem 1 of Section 8.2.1 (cf. Eq. (8.15a)).

9.5. CONNECTION WITH THE PSEUDOSCOPIC IMAGE IN HOLOGRAPHY

The theorem of Section 9.4 bears a close connection with the formation of the pseudoscopic image in transmission holography. This fact was remarked by Wolf and Shewell [9.5], and their discussion is presented next as follows:

In transmission holography, a wave $U(\mathbf{r}_0)$, ($\mathbf{r}_0 = (x_0, y_0, z_0)$), considered at a plane $z = z_0$, emerges from an illuminated object. This wave is incident on a photographic plate situated in the plane $z = z_1 > z_0$, so that the wavefield in a generic point $\mathbf{r}_1 = (x_1, y_1, z_1)$ in the plane of the plate is given by Eq. (9.1).

The wave $U(\mathbf{r}_1)$ interferes with a reference wave $U^{(r)}(\mathbf{r}_1)$ in the photographic plate, so that the total field $U^{(\text{int})}(\mathbf{r}_1)$ in the plane $z = z_1$ is given by:

$$U^{(\text{int})}(\mathbf{r}_1) = U(\mathbf{r}_1) + U^{(r)}(\mathbf{r}_1). \tag{9.18}$$

We shall assume that the developed plate has an amplitude transmission function $T(\mathbf{r}_1)$ which is proportional to the exposure intensity $I^{(\text{int})}$ of the hologram: $I^{(\text{int})}(\mathbf{r}_1) = |U^{(\text{int})}(\mathbf{r}_1)|^2$ (see, e.g. Section 8.3 of Ref. 9.6):

$$T(\mathbf{r}_1) = a + bI^{(\text{int})}(\mathbf{r}_1). \tag{9.19}$$

If the processed hologram is placed in the plane $z = z_1$, and is illuminated by a monochromatic wave of the same frequency, and whose spatial part is $U^{(\text{rec})}(\mathbf{r})$, the resulting field, emerging from the hologram, is determined by its boundary value in $z = z_1$:

$$\begin{aligned} T(\mathbf{r}_1)U^{(\text{rec})}(\mathbf{r}_1) &= [a + b|U^{(r)}(\mathbf{r}_1)|^2 + b|U(\mathbf{r}_1)|^2]U^{(\text{rec})}(\mathbf{r}_1) \\ &\quad + bU^{(\text{rec})}(\mathbf{r}_1)U^{(r)*}(\mathbf{r}_1)U(\mathbf{r}_1) + bU^{(\text{rec})}(\mathbf{r}_1)U^{(r)}(\mathbf{r}_1)U^{*}(\mathbf{r}_1). \end{aligned} \tag{9.20}$$

Assuming that the reference wave $U^{(r)}(\mathbf{r})$ and the reconstructing wave $U^{(\text{rec})}(\mathbf{r})$ are both plane waves propagating with the same wavevector (this hypothesis is not essential, but it simplifies the argument that follows; a more general analysis can be found in Ref. 9.5), we may write:

$$U^{(r)}(\mathbf{r}) = A^{(r)} \exp[ik(p_r x + q_r y + m_r(z - z_1))], \tag{9.21a}$$

$$U^{(\text{rec})}(\mathbf{r}) = A^{(\text{rec})} \exp[ik(p_{\text{rec}} x + q_{\text{rec}} y + m_{\text{rec}}(z - z_1))], \tag{9.21b}$$

with $A^{(r)}$ and $A^{(\text{rec})}$ being constants, and the directional cosines satisfying: $p_{\text{rec}} = p_r$ and $q_{\text{rec}} = q_r$.

From Eqs. (9.21) one can see that the second term of Eq. (9.20) satisfies:

$$bU^{(\text{rec})}(\mathbf{r}_1)U^{(r)*}(\mathbf{r}_1)U(\mathbf{r}_1) = bA^{(\text{rec})}(\mathbf{r}_1)A^{(r)*}(\mathbf{r}_1)U(\mathbf{r}_1). \tag{9.22}$$

However, according to Eq. (9.1), this term, which we shall denote by $U^{(1)}(\mathbf{r}_1)$, is proportional to the outgoing field $U(\mathbf{r})$ emerging from the hologram, evaluated in the plane $z = z_1$, and appears created in this plane as if light emerged from the original object in the absence of any reference wave (of course, neglecting the diffraction effects introduced by the finite extent of the holographic plate).

Therefore, the term given by Eq. (9.22) represents an outgoing field that

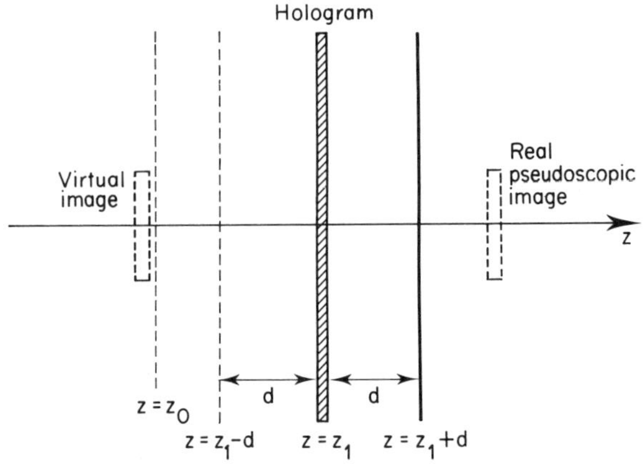

Figure 9.2. The virtual image and the pseudoscopic image in holography.

propagates into the half-space $z > z_1$ forming a *virtual image* of the object. This image is located at the same position as the original object with respect to the plane $z = z_1$ (namely, behind the plane $z = z_0$; see Fig. 9.2).

Let us now examine the third term of Eq. (9.20) when the reconstructing wave $U^{(\text{rec})}$ propagates in a direction *conjugate* with respect to the direction of the reference wave $U^{(r)}$, namely, when: $p_{\text{rec}} = -p_r$ and $q_{\text{rec}} = -q_r$. In this case, the third term of Eq. (9.20), which we shall denote by $U^{(2)}(\mathbf{r}_1)$, satisfies:

$$bU^{(\text{rec})}(\mathbf{r}_1)U^{(r)}(\mathbf{r}_1)U^*(\mathbf{r}_1) = bA^{(\text{rec})}(\mathbf{r}_1)A^{(r)}(\mathbf{r}_1)U^*(\mathbf{r}_1). \qquad (9.23)$$

On comparing Eq. (9.23) with Eq. (9.22) we obtain at once:

$$A^{(\text{rec})*}U^{(2)}(\mathbf{r}_1) = A^{(\text{rec})}U^{(1)*}(\mathbf{r}_1). \qquad (9.24)$$

Equation (9.24) is exactly the boundary condition of conjugation, Eq. (9.16). Therefore, according to Eq. (9.17) of the Theorem of Section 9.4, we get:

$$A^{(\text{rec})*}U^{(2)}(x, y, z_1 + d) = A^{(\text{rec})}U^{(1)*}(x, y, z_1 - d). \qquad (9.25)$$

Equation (9.25) shows that, apart from a simple proportionality factor, the reconstructed field $U^{(2)}$ in the plane $z = z_1 + d$ is the conjugate of the field $U^{(1)}$ in the plane $z = z_1 - d$. The field $U^{(2)}$ is proportional to the conjugate of the object wave $U^*(\mathbf{r})$, it is incoming and is focused into a region located at points obtained by reflecting the object about the plane $z = z_1$ of the hologram, forming in this region a *real pseudoscopic image* of the original object (Fig. 9.2).

9.6. SYSTEM APPROACH TO INVERSE DIFFRACTION

The singularity of the inverse propagator $\mathscr{K}$ can also be interpreted by analyzing the free-space propagation from the framework of the theory of linear systems, extensively used in image formation theory (some aspects of this point of view can be also found in, e.g. Chapter 2 of Ref. 9.6). In order to avoid dealing with unnecessary phase factors, we shall choose the origin of coordinates such that we deal with the reconstruction of the wavefield $U(\mathbf{r})$ in the plane $z = 0$. The propagation integral (Eq. (9.1)) then becomes:

$$U(x_1, y_1, z_1) = \iint_{-\infty}^{\infty} f(x_0, y_0)\mathscr{K}(x_1 - x_0, y_1 - y_0, z_1 - z_0)\, dx_0\, dy_0, \quad (9.26)$$

where we have denoted by $f(x_0, y_0)$ the boundary value:

$$U(x_0, y_0, 0) = f(x_0, y_0), \quad (9.27)$$

then, according to Eq. (9.2), the free-space propagator $\mathscr{K}$ is:

$$\mathscr{K}(x_1 - x_0, y_1 - y_0, z_1 - z_0) = \left(\frac{k}{2\pi}\right)^2 \iint_{-\infty}^{\infty} \exp[ik[p(x_1 - x_0) + q(y_1 - y_0) + mz_1]]\, dp\, dq. \quad (9.28)$$

As was hinted in Section 2.4, Eq. (9.26) is a convolution of f and $\mathscr{K}$, and thus characterizes an *invariant linear system* with *impulse response* $\mathscr{K}$, Eq. (9.28).

By taking the two-dimensional Fourier transform with respect to the x and y variables in both sides of Eq. (9.26), we get:

$$\tilde{U}(p, q, z_1) = \left(\frac{k}{2\pi}\right)^2 a(p, q) \exp(ikmz_1), \quad (9.29)$$

$\tilde{U}(p, q, z_1)$, $a(p, q)$ and $(k/2\pi)^2 \exp(ikmz_1)$ being the two-dimensional Fourier transforms of $U(x, y, z_1)$, $f(x, y)$ and $\mathscr{K}$, respectively:

$$U(x, y, z_1) = \iint_{-\infty}^{\infty} \tilde{U}(p, q, z_1) \exp[ik(px + qy)]\, dp\, dq, \quad (9.30)$$

$$f(x, y) = \iint_{-\infty}^{\infty} a(p, q) \exp[ik(px + qy)]\, dp\, dq, \quad (9.31)$$

$$\mathscr{K}(x, y, z_1) = \left(\frac{k}{2\pi}\right)^2 \iint_{-\infty}^{\infty} \exp[ik(px + qy + mz_1)]\, dp\, dq, \tag{9.32}$$

with p, q, and m satisfying:

$$m = \sqrt{1 - p^2 - q^2}, \qquad p^2 + q^2 \leqslant 1, \tag{9.33a}$$

$$m = i\sqrt{p^2 + q^2 - 1}, \qquad p^2 + q^2 > 1. \tag{9.33b}$$

Note that $\tilde{U}(p, q, z_1)$ physically represents the angular spectrum of $U(\mathbf{r})$, propagated at a distance z_1 from the origin. On the other hand, some aspects of the inverse diffraction problem, which involves the reconstruction of $f(x, y)$ from Eq. (9.26), have already been discussed in Sections 2.3 and 2.4. Emphasizing what was mentioned there, it should be noted that, in accordance with Eq. (9.26), the determination of the angular spectrum $a(p, q)$ from knowledge of $\tilde{U}(p, q, z_1)$ by inverting Eq. (9.29) obviously gives:

$$a(p, q) = \mathbf{H}^{-1}(p, q)\tilde{U}(p, q, z_1). \tag{9.34}$$

In Eq. (9.34), $\mathbf{H}^{-1}(p, q)$ is the inverse of the filter function:

$$\mathbf{H}(p, q) = \left(\frac{k}{2\pi}\right)^2 \exp(ikmz_1), \tag{9.35}$$

i.e., the inverse filter $\mathbf{H}^{-1}(p, q)$ is, according to Eqs. (9.33) and (9.35):

$$\mathbf{H}^{-1}(p, q) = \left(\frac{2\pi}{k}\right)^2 \exp(-ik\sqrt{1 - p^2 - q^2}z_1) \qquad \text{for } p^2 + q^2 \leqslant 1. \tag{9.36a}$$

$$\mathbf{H}^{-1}(p, q) = \left(\frac{2\pi}{k}\right)^2 \exp(k\sqrt{p^2 + q^2 - 1}z_1) \qquad \text{for } p^2 + q^2 > 1. \tag{9.36b}$$

Therefore, this inverse filter diverges as p and q increase beyond the cut-off limit $p^2 + q^2 = 1$ for any distance $z_1 > 0$ at which the field $U(\mathbf{r})$ (and hence, $\tilde{U}(p, q, z_1)$) is being determined.

9.7. DEGREES OF FREEDOM

Equations (9.36) show that the sharp cut-off of the filter $\mathbf{H}(p, q)$ at $p^2 + q^2 = 1$, beyond which it abruptly decreases to zero, is the origin of the divergence of the inverse filter, and thus creates the singularity of the integrals in Eqs. (9.4) and (9.11) for the inverse propagator.

The limitation of the domain of the p, q-space, at which values of the angular spectrum $a(p, q)$ of the propagating field $U(\mathbf{r})$ are accessible, sets a bound to the effective spatial information conveyed by this field. This information is best characterized by the minimum number $\mathcal{N}$ of parameters that are necessary to specify $U(\mathbf{r})$ in an area $\mathbf{W}$ of a certain plane of propagation z = constant.

Following an argument of Von Laue [9.7], further developed by Gabor [9.8]–[9.10], this number $\mathcal{N}$ represents the *spatial number of degrees of freedom* of the wavefield $U(\mathbf{r})$ in the area $\mathbf{W}$.

Let $\Delta\Omega = \cos\theta \sin\theta\Delta\theta\Delta\phi = \Delta p\Delta q$, (cf. Eq. (2.22)), be the element of solid angle in a cone about the z-axis. This cone constitutes the bundle of all possible directions of propagation with directional cosines p, q, and m of the plane wave components of $U(\mathbf{r})$, such that $p^2 + q^2 \leqslant M$, M being a real constant such that $M \leqslant 1$; then, taking into account the angular spectrum representation of $U(\mathbf{r})$, and according to the *Shannon sampling theorem* (see Appendix 4.3), the propagating wavefield $U(\mathbf{r})$ is completely specified at any transversal plane z = constant from its sampling values at points of this plane at the Nyquist frequency, i.e., forming a grid with separation interval $\delta_x = \lambda/\Delta p$ along the x-direction, and $\delta_y = \lambda/\Delta q$ along the y-direction. Hence, in an area $\mathbf{W} = \Delta x\Delta y$ in the plane z = constant, there are $\mathcal{N} = (\Delta x/\delta_x)(\Delta y/\delta_y) = \mathbf{W}/\delta_x\delta_y$ independent values of the wavefunction $U(x, y, z = \text{constant})$, namely:

$$\mathcal{N} = \frac{\mathbf{W}\Delta p\Delta q}{\lambda^2} = \frac{\mathbf{W}\Delta\Omega}{\lambda^2}. \tag{9.37}$$

In fact, according to Gabor, the number $\mathcal{N}$ of spatial degrees of freedom should be multiplied by a factor: $4 = 2 \times 2$. The first factor 2 due to the two possible independent directions of polarization of the electromagnetic signal in the plane z = constant, and the second factor 2 because of the amplitude and the phase of the field. (However, as discussed in Appendix 4.2 and Section 9.10, for bandlimited fields these two quantities are not independent of each other.)

For a bundle of plane waves included within a semiangle with the z-axis equal to $\theta = \arcsin M$, one evidently has from Fig. 9.3:

$$\Delta p\Delta q = \Delta\Omega = \pi M^2. \tag{9.38}$$

Hence, the number of degrees of freedom is:

$$\mathcal{N} = \frac{\pi\mathbf{W}M^2}{\lambda^2}. \tag{9.39}$$

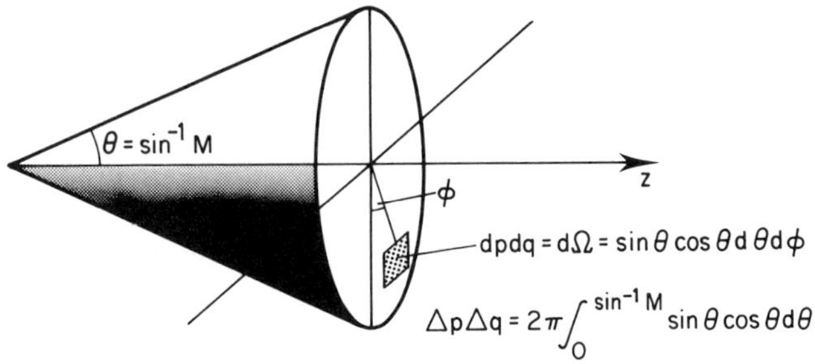

Figure 9.3. Illustration of the solid angle $\Delta\Omega$ of directions of propagation of the plane wave components of a bandlimited wavefield.

Equation (9.39) represents the maximum number of degrees of freedom that can be included within an area $\mathbf{W}$ in a plane $z = \text{constant} > \lambda$.

From the angular spectrum representation for the wavefield $U(\mathbf{r})$ one sees that once we fix the angular divergence M of the bundle of its plane wave components, the number $\mathcal{N}$ of degrees of freedom remains unperturbed at any plane $z = \text{constant}$ as the field propagates in the free half-space $z \geqslant 0$. This invariance of the spatial information content of the wavefield underlines the essence of the Huygens principle [9.10].

In geometrical optics, the constancy of the numerator of Eq. (9.37) represents the well-known *Smith–Helmholtz–Lagrange invariant* (see, e.g. Section 4.4.5 of Ref. 9.11). Observe, however, that if the original boundary value $f(x, y)$ at the plane $z = 0$ contains an infinite number of degrees of freedom (sampling points) within a given area $\mathbf{W}$ of this plane, then, as propagation takes place in the free half-space $z \geqslant 0$, the wavefield $U(\mathbf{r})$ contains at most a number $\mathcal{N}$, given by Eq. (9.39), beyond a certain plane $z = z_1 > \lambda$. This is the irreversible loss of subwavelength information content through propagation due to the evanescent wave cut-off; it was mentioned in Section 4.5 by a different analysis. (Remember the equivalence of the Shannon description and the zero description of the analytic continuation in the complex plane of one-dimensional bandlimited signals.)

In connection with the wavelength resolution limit, it is worth remarking that, in the case of optical waves, this is a practical limitation due to the impossibility of determining the field at distances smaller than a wavelength from the plane of reconstruction. However, at large wavelengths this evanescent wave cut-off limit has been overcome. The authors of Refs. 9.12 and 9.13 report data inversion from the holographic determination of an acoustic wave at a distance $z_1 = 10$ cm, using a wavelength $\lambda = 156$ cm, namely, from knowledge of the field at a distance $z_1 = 0.06\lambda$ from the plane $z = 0$ of reconstruction.

9.8. REPRESENTATION BY EIGENFUNCTIONS

Gabor's expansion theorem [9.8] asserts that the number of degrees of freedom of a propagating wavefield is associated with the number of eigen-solutions of the homogeneous Helmholtz equation. The original work by Gabor did not include the proof of this theorem, which turned out to be extremely difficult. Such a proof has been so far only accomplished by Miyamoto for large values of $\mathcal{N}$ [9.14]. Gabor [9.10] gave, nevertheless, illustrations of expansions in terms of Gaussian beams. This representation, however, is not quite satisfactory on mathematical grounds, because Gaussian beams do not form a complete orthogonal set of functions. From this point of view, a better set is that consisting of the *prolate spheroidal functions* [9.15]–[9.18]. Basically, these are bandlimited eigenfunctions associated with the problem of bandwidth extrapolation. It was shown by Toraldo di Francia [9.19] that they provide a correct description of the transmission of information in imaging optical systems whose impulse response is band-limited to an angular width which is bounded by the effective aperture of the image forming system.

If we write the relationship (Eq. (9.26)) between the boundary value $f(x_0, y_0)$ and the propagating field $U(x_1, y_1, z_1)$ in the operational form:

$$U = \mathbf{K}f, \tag{9.40}$$

$\mathbf{K}$ being the operator involving the convolution (Eq. (9.26)), the eigenfunctions φ_n of the operator $\mathbf{K}$ satisfy, for each eigenvalue λ_n, the following equation:

$$\lambda_n \varphi_n = \mathbf{K}\varphi_n. \tag{9.41}$$

If the eigenfunctions φ_n form a complete orthonormal set, then both U and f can be expanded in terms of these functions, namely:

$$\begin{aligned} U &= \sum_{n=0}^{\infty} g_n \varphi_n, \\ g_n &= \int_{-\infty}^{\infty} U(\mathbf{r})\varphi_n(\mathbf{r})\, d^3r, \end{aligned} \tag{9.42}$$

$$\begin{aligned} f &= \sum_{n=0}^{\infty} f_n \varphi_n, \\ f_n &= \int_{-\infty}^{\infty} f(\mathbf{r})\varphi_n(\mathbf{r})\, d^3r. \end{aligned} \tag{9.43}$$

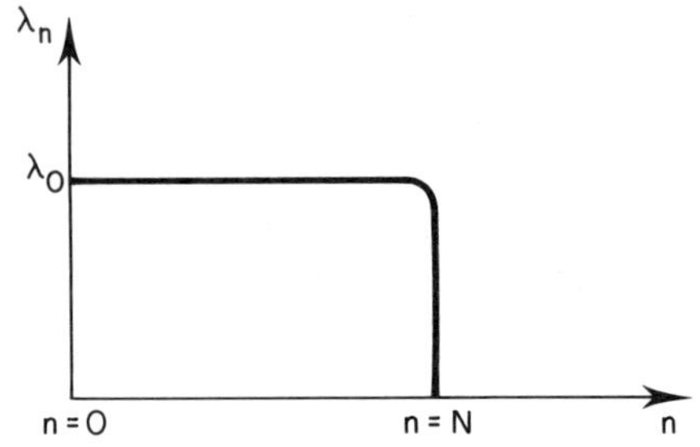

Figure 9.4. Variation of the eigenvalues λ_n with n.

From Eqs. (9.40)–(9.43) we establish that:

$$g_n = f_n \lambda_n. \tag{9.44}$$

All bandlimited operators have, however, the property that the eigenvalues λ_n drop to zero for n larger than a certain value $n = N$, often in a steplike form as shown in Fig. 9.4 [9.19]. Hence, it is seen from Eq. (9.44) that, by associating a degree of freedom with each coefficient g_n of the expansion (Eq. (9.42)) of U, the effective number of degrees of freedom of U is finite, and thus, the sum in Eq. (9.42) actually ends at $n = N$. It can be proven that, in fact, $N = \mathcal{N}$ (see, e.g. Ref. 9.19).

The recovery of f from U by determining the coefficients f_n from those g_n through Eq. (9.44) entails the quotient of g_n by a very small amount λ_n for $n \geqslant \mathcal{N}$. Therefore, due to the unavoidable errors that exist in the determination of U (and, thus, of its coefficients g_n), because of the presence of noise in the detection, the quotient g_n/λ_n will dramatically grow as n increases beyond $n = \mathcal{N}$.

9.9. ILL-POSED NATURE OF INVERSE DIFFRACTION

Since, most generally, a physical wavefield arises from a source distribution of finite extent, and hence, according to Chapter 4, its angular spectrum $a(p, q)$ is an analytic function, some authors (e.g. Wolter [9.20] and Lo [9.21]) propose that it should be possible to continue analytically $a(p, q)$ to frequencies p and q beyond the accessible range $p^2 + q^2 \leqslant 1$. In terms of the expansion into eigenfunctions, this would imply recovery of a number of coefficients (and thus, of degrees of freedom) larger than $n = \mathcal{N}$. As quoted in Section 9.8, this cannot be achieved in practice due to the presence of noise, which makes such an extrapolation unstable. This characteristic is inherent in all inverse optical problems, and, in general, in all inverse problems [9.22]. In mathematical terms, it is said that the data inversion is *ill-posed* or *ill-conditioned* [9.23]. This means that the solution to the inversion does not depend continuously on the data, but rather, small errors in these data produce very large errors in the reconstructed results.

There exists an additional difficulty that appears in all computational calculations involving extrapolation, and which is a consequence of giving values (without noise) of a bandlimited function at a finite number of sampling points (as happens when the wavefield $U(x, y, z_1)$ is known at the Shannon sampling points within a finite aperture of the plane $z = z_1$). It is then possible to show that these data do not specify uniquely the function everywhere (even though no noise is present) [9.24], [9.25]. Hence, extrapolation operations require additional assumptions on the function that should be provided by some appropriate model based on additional *a priori* information [9.25].

Considerable work is currently devoted in optics to the establishment of reconstruction algorithms for the inversion of equations of the form of Eq. (9.40), with the help of additional constraints. (A good account can be found, for example, in Ref. 9.26.)

The modification of the inverse problem in order to obtain an acceptable estimate solution, is generally known as *regularization*. This method, originally proposed by Tikhonov and Arsenin [9.23] (see also Ref. 9.27), attempts to minimize the error from the estimated reconstruction $\tilde{f}$ to Eq. (9.40), namely, one seeks the minimum of:

$$\|U - \mathbf{K}\tilde{f}\| \leqslant \varepsilon, \tag{9.45}$$

where ε is a measure of the noise, or inaccuracy, in the data to be inverted, and $\|(\cdot)\|$ denotes the norm in the space L^2 of square integrable functions, i.e.:

$$\|f\| = \left[\int_{-\infty}^{\infty} |f(x, y)|^2 \, dx \, dy\right]^{1/2} \tag{9.46}$$

The minimization of Eq. (9.45) is attempted, subject to constraints on the estimate $\tilde{f}$. These are represented by:

$$\|\mathbf{L}\tilde{f}\| \leqslant E, \tag{9.47}$$

where $\mathbf{L}$ denotes a certain operation on $\tilde{f}$, and E is a constant. Usually, $\mathbf{L}$

is either some differential operator, or the identity operator. In this last case, Eq. (9.47) restricts the solutions $\tilde{f}$ to having finite energy E.

The regularization method applies Lagrange's multipliers to the functional:

$$\Phi = \|U - \mathbf{K}\tilde{f}\|^2 + \mu\|\mathbf{L}\tilde{f}\|^2. \tag{9.48}$$

In Eq. (9.48) the multiplier μ is an estimate of the inverse of the signal-to-noise ratio, and it is usually taken as: $\mu = (\varepsilon/E)^2$.

Lagrange's method gives, from Eq. (9.48), the following estimate (see Problem 9.4):

$$\tilde{f} = [\mathbf{K}^\dagger\mathbf{K} + \mu\mathbf{L}^\dagger\mathbf{L}]^{-1}\mathbf{K}^\dagger U, \tag{9.49}$$

which is equivalent to a *Wiener filter* [9.28] applied to the wavefunction U. The sign † means the adjoint operator.

This procedure, which is also extensively used in image restoration [9.26], [9.28], leads by means of Eq. (9.49) to the following relationship between the coefficients $\tilde{f}_n$ and g_n of the expansions of $\tilde{f}$ and U into eigenfunctions φ_n (cf. Eqs. (9.42) and (9.43)):

$$\tilde{f}_n = \frac{\lambda_n}{\lambda_n^2 + \mu\beta_n^2}\, g_n; \tag{9.50}$$

where β_n are the eigenvalues of $\mathbf{L}$.

Observe that regularization is a method for increasing slightly the weight of the higher spatial frequency Fourier components (or equivalently, for increasing the higher order eigenvalues) to avoid division by nearly zero quantities. Obviously, if μ is made zero one retrieves the usual relation (Eq. (9.44)) between the coefficients of $\tilde{f}$ and those of U.

Although regularization can improve somewhat the quality of the solution to the inverse problem (usually by using it together with iterative reconstruction algorithms, see, e.g. Chapters 1 and 2 of Ref. 9.26, or Ref. 29), the impossibility of significantly increasing the resolution beyond the evanescent limit cut-off is underlined by the numerical instability of the extrapolation problem. In fact, as a measure of this lack of stability, it can be shown [9.22], [9.30], that for an error ε in the data U as expressed by Eq. (9.45), the resulting error in the estimated solution $\tilde{f}$ to the inverse diffraction problem with respect to the ideal solution f:

$$\delta = \|\tilde{f} - f\|, \tag{9.51}$$

is: $\delta \sim |\ln \varepsilon|^{-1}$. This implies that in order to get an error of $\delta \sim c^{-1}$ (c being a constant greater than zero), a noise level as low as $\varepsilon \sim \exp(-c)$ is required in the data $U(x, y, z_1)$.

A procedure that allows one to estimate the attainable resolution limit in the inverse diffraction problem consists of convolving $f(x, y)$ with a blurring function W_σ. Taking it to be a Gaussian of width σ [9.31]:

$$W_\sigma(x, y) = (\sqrt{2}\pi\sigma)^{-2} \exp[-(x^2 + y^2)/2\sigma^2]. \tag{9.52}$$

The reconstruction of $f_\sigma(x, y)$:

$$f_\sigma(x, y) = f(x, y) * W_\sigma(x, y), \tag{9.53}$$

is then sought. The sign $*$ denotes the convolution product.

If $\tilde{f}_\sigma(x, y)$, given by:

$$\tilde{f}_\sigma(x, y) = \tilde{f}(x, y) * W_\sigma(x, y), \tag{9.54}$$

represents the blurred estimate of the reconstruction; then the error:

$$\rho(\varepsilon, z_1, \sigma) = \| \tilde{f}_\sigma - f_\sigma \|, \tag{9.55}$$

can be interpreted as the error in the restoration of details whose size is of the order of σ.

Figure 9.5 shows the relative error $\delta(\varepsilon, z_1, \eta)$ (Bertero and de Mol [9.31]):

$$\delta(\varepsilon, a, \eta) = \frac{\rho(\varepsilon, a, \eta)}{\rho(\eta)}, \tag{9.56}$$

where $\eta = \sigma/\lambda$, $a = z_1/\lambda$, and $z = z_1$ is the plane at which the propagating wavefield $U(x, y, z_1)$ is determined.

The plot of δ in Fig. 9.5 is made versus η. The solid line corresponds to far-field data ($z_1 = \infty$), and coincides both for $\varepsilon = 10^{-2}$ and $\varepsilon = 10^{-4}$. The two broken lines correspond to data at $z_1 = \lambda$. As seen, the relative error δ does not diminish significantly with the reduction of ε.

9.10. THE PROBLEM OF PHASE RETRIEVAL

So far we have assumed that the propagating wavefield $U(x, y, z)$ is known, both in amplitude and phase, in the plane $z =$ constant. Detectors, however, do not register the phase of signals at optical wavelengths. Only the intensity of the light wave can be directly measured. This adds to the inverse diffraction problem the question of reconstructing the phase of the field from intensity measurements.

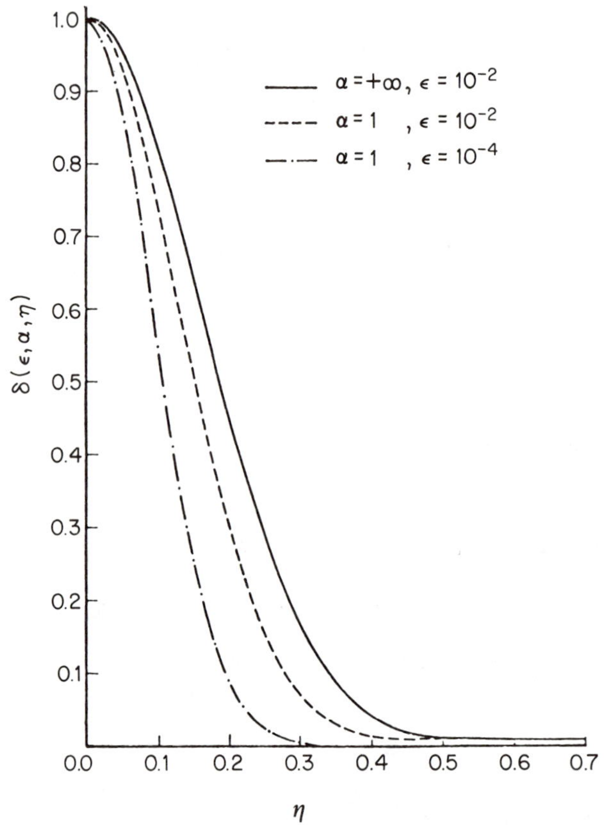

Figure 9.5. Relative error in the restoration of the blurred function $f_\sigma(x, y)$ versus the resolution parameter η for several values of the noise level ε and distance $a = z_1/\lambda$ from the plane $z = 0$. (After M. Bertero and C. de Mol, *I.E.E.E. Trans. Antennas. Propag.* **29**, 368 (1981). © 1981, IEEE).

The phase retrieval problem is inherent in any inverse optical problem. In addition to inverse diffraction, it is common, among others, in X-ray crystallography [9.32], [9.33], astronomy [9.34], microscopy [9.35], image science [9.36] (see also Chapters 6, 7, 8, and 13 of Ref. 9.26), inverse scattering [9.37], and spectroscopy [9.38].

The phase problem for wavefront reconstruction is formulated as the determination of the phase $\phi(u, v)$ of the spectral amplitude $A(u, v)$ of the propagating field from a knowledge of its modulus $|A(u, v)|$. Thus, $A(u, v)$ is expressed in the form:

$$A(u, v) = |A(u, v)| \exp[i\phi(u, v)]. \tag{9.57}$$

This modulus $|A(u, v)|$ is directly provided by the far-field intensity measurement.

As shown by Eq. (2.16), apart from unimportant factors, the relationship between $A(u, v)$ and the solution $f(x, y)$ is then given by the Fourier integral:

$$A(u, v) = \iint_{-\infty}^{\infty} f(x, y) \exp[-i(ux + vy)]\, dx\, dy. \tag{9.58}$$

It should be remarked, however, that a phase retrieval problem may be formulated for the recovery of the phase of the propagating field $U(x, y, z_1)$ at a plane $z = z_1$ from knowledge of its modulus: $M(x, y) = |U(x, y, z_1)|$. In this case the Fourier relationship is established between $U(x, y, z_1)$ and the function $a(p, q) \exp(ikmz_1)$ (cf. Eq. (2.20)).

The most classical and best known reconstruction of the wavefront phase is provided by holography [9.39]–[9.41], in which the phase is registered by interference of the wavefield with a known reference wave. Other procedures of present day investigation involve numerical methods.

There are two main areas in which numerical procedures are under current study: that in which the function $f(x, y)$ is periodic; this includes crystallographic determinations from X-rays, electron microscopy, or neutron diffraction [9.32], [9.33], [9.35], [9.36], and that in which $f(x, y)$ stands for an object of finite extent, and thus, is a function of finite support. In the first case, $f(x, y)$ represents a projection of a three-dimensional distribution of atoms, along the direction of propagation of the incident wave. Direct methods have been established for solving specific structures by using *a priori* information about $f(x, y)$ (reviews are given in, e.g. Refs. 9.35, 9.36, 9.38, and 9.42).

Concerning the second area, a solution was proposed [9.43] for minimal phase functions of one variable by using a dispersion relation of the form of Eq. (A4.14a). Also, based on the properties of the zeros of entire functions of one complex variable, it was observed [9.44], [9.45] that the solution to the phase problem for functions of one variable is not unique, for if the analytic continuation $A(z = u + iu')$ of the one-dimensional Fourier spectrum, Eq. (9.58), has N complex zeros, then its square modulus $|A(z)|^2 = A(z)A^*(z^*)$ has both the N zeros of $A(z)$ and the N complex conjugate zeros of $A^*(z^*)$. Since one-dimensional functions $A(z)$ always factorize in the form of a Hadamard product (cf. Eq. (A4.8a)), there are, then, evidently 2^N possible combinations of those factors, each combination giving a different $A(z)$, but all these functions $A(z)$ having the same square modulus $|A(z)|^2$.

For functions of two or more variables, it was seen in Appendix 4.2 that dispersion relations are not useful. Nevertheless, since factorization is not the rule for these functions (consider that multivariate polynomials do not always factorize, unlike polynomials of one variable), [9.46]–[9.48], the phase problem is not always ambiguous. In fact, although only sufficient conditions for nonfactorizability are known in several variables (see, e.g. Refs. 9.48–9.50), there are empirical indications, from computer simulations of two-dimensional phase reconstruction [9.51], [9.52], that the phase problem for multivariate functions is most likely unique. (However, noise, i.e., uncertainty in the intensity data, can increase the number of solutions compatible with the same data [9.53].)

Although considerable progress has been made in establishing algorithms of phase retrieval [9.51]–[9.58], this problem is still under active study. The brief outline given here does not contain an exhaustive list of references. We refer the reader for greater detail to the more specialized monographs, particularly, Refs. 9.26, 9.34–9.37, 9.42, and 9.45 for more details.

9.11. CONCLUSIONS AND OTHER STUDIES

In this chapter we have discussed the inverse diffraction problem from plane to plane by means of the angular spectrum representation. The limit of resolution of the boundary value $f(x, y)$ at the plane $z = 0$, retrieved from data of the propagating wavefield at $z = \text{constant} > 0$, was discussed in the context of the evanescent wave cut-off. This leads to the concept of degrees of freedom of the wavefunction. We have seen how the attempt to increase the resolution of the reconstruction beyond this cut-off involves singular inverse propagators and numerically unstable extrapolations. In addition, a brief account was presented of the question of phase retrieval, which is associated with inverse optical problems.

Due to their inherent mathematical instabilities, and difficulties in inverting data when there are multiple interactions, the solution of inverse optical problems is in a situation much less satisfactory than that of direct problems. Methods for dealing with reconstructions from spherical surfaces have been discussed in Ref. 9.59. Studies on the determination of the shape of reflectors from far field data have been done, either using geometrical optics [9.60], the Kirchhoff approximation [9.61], or by inverting the Neumann series [9.62]. Also, in electromagnetic theory, both inverse diffraction and its ill-posed nature have been addressed [9.63], [9.64] in connection with the problem of antenna synthesis (i.e., the devising of a particular antenna that produces a desired diffraction pattern).

PROBLEMS

9.1. Prove Eq. (9.9).

9.2. Let $f(x)$ denote the complex amplitude of an object, depending only on the abscissa x. The image formed under coherent illumination by an optical system of numerical aperture W, has a wavefunction $g(x)$ given by:

$$g(y) = \int_{-\infty}^{\infty} f(x) \frac{\sin[kW(y-x)]}{\pi(y-x)} dx.$$

Show that the number of degrees of freedom of this image in a segment of width L is:

$$\mathcal{N} = \frac{kWL}{\pi}.$$

Let $I_g(x)$ and $I_f(x)$ denote the intensity in the image and object, respectively. Under incoherent illumination, one has [9.6]:

$$I_g(y) = \int_{-\infty}^{\infty} I_f(x) \frac{\sin^2[kW(y-x)]}{\pi^2(y-x)^2} dx.$$

Show that the number $\mathcal{N}$ associated with coherent illumination is half the number $\mathcal{N}$ under incoherent illumination (Toraldo di Francia [9.65]). Make a discussion on whether they also contain more structural information.

9.3. Discuss the inversion from plane to plane of the wavefront for a Gaussian beam under free space propagation. Estimate the number of degrees of freedom.

9.4. Prove Eqs. (9.49) and (9.50).

REFERENCES

9.1. J. R. Shewell and E. Wolf, *J. Opt. Soc. Am.* **58**, 1596 (1968).

9.2. G. C. Sherman, *J. Opt. Soc. Am.* **57**, 1490 (1967).

9.3. R. Mittra and R. L. Ramson, in *Modern Optics*, K. Fox, ed., Polytechnic Press, Brooklyn, J. Wiley, New York, 1967; W. Lukosz, *J. Opt. Soc. Am.* **58**, 1084 (1968).

9.4. E. Lalor, *J. Math. Phys.* **9**, 2001 (1968).

9.5. E. Wolf and J. R. Shewell, *J. Math. Phys.* **11**, 2254 (1970).
9.6. J. W. Goodman, *Introduction to Fourier Optics*, McGraw-Hill, New York, 1968.
9.7. M. Von Laue, *Ann. Phys.* **44**, 1197 (1914).
9.8. D. Gabor, in *Proceedings of the Symposium on Astronomical Optics*, Z. Kopal, ed., North-Holland, Amsterdam, 1956, p. 17.
9.9. D. Gabor, in *Progress in Optics*, Vol. 1, E. Wolf, ed., North-Holland, Amsterdam, 1961.
9.10. D. Gabor, *J. Phys. E* **8**, 73 (1975).
9.11. M. Born and E. Wolf, *Principles of Optics*, Pergamon Press, Oxford, 6th edition.
9.12. E. G. Williams and J. D. Maynard, *Phys. Rev. Lett.* **45**, 554 (1980).
9.13. E. G. Williams, J. D. Maynard and E. Skudrzyk, *J. Acoust. Soc. Am.* **68**, 340 (1980).
9.14. K. Miyamoto, *J. Opt. Soc. Am.* **50**, 856 (1960).
9.15. D. Slepian and H. O. Pollack, *Bell Syst. Tech. J.* **40**, 43 (1961).
9.16. H. J. Landau and H. O. Pollack, *Bell Syst. Tech. J.* **40**, 65 (1961).
9.17. D. Slepian, *Bell Syst. Tech. J.* **43**, 3009 (1964).
9.18. B. R. Frieden, in *Progress in Optics*, Vol. 1, E. Wolf, ed., North-Holland, Amsterdam, 1961.
9.19. G. Toraldo di Francia, *J. Opt. Soc. Am.* **59**, 799 (1969).
9.20. H. Wolter, in *Progress in Optics*, Vol. 1, E. Wolf, ed., North-Holland, Amsterdam, 1961.
9.21. Y. T. Lo, *J. Appl. Phys.* **32**, 2052 (1961).
9.22. M. Bertero, C. de Mol and G. A. Viano, in *Inverse Scattering Problems in Optics*, H. P. Baltes, ed., Topics in Current Physics, Vol. 20, Springer-Verlag, Berlin, 1980, Chapter 5.
9.23. A. Tikhonov and V. Y. Arsenin, *Solutions to Ill-Posed Problems*, V. H. Winston, Washington, D.C., 1977.
9.24. R. W. Schafer, R. M. Merserau and M. A. Richards, *Proc. I.E.E.E.* **69**, 432 (1981).
9.25. M. A. Fiddy and T. J. Hall, *J. Opt. Soc. Am.* **71**, 1406 (1981).
9.26. H. Stark, ed., *Image Recovery: Theory and Application*, Academic Press, New York, 1987.
9.27. K. Miller, *S.I.A.M. J. Math. Anal.* **1**, 52 (1970).
9.28. H. C. Andrews and B. R. Hunt, *Digital Image Restoration*, Prentice-Hall, New Jersey, 1977.
9.29. G. Ross, *Opt. Acta* **29**, 1523 (1982).
9.30. M. Bertero, C. de Mol and G. A. Viano, *J. Math. Phys.* **20**, 509 (1979).
9.31. M. Bertero and C. de Mol, *I.E.E.E. Trans. Antenn. Propag.* **29**, 368 (1981).
9.32. M. M. Woolfson, *Direct Methods in Crystallography*, Oxford University Press, Oxford, 1961.
9.33. G. Ramachandran and R. Srinivasan, *Fourier Methods in Crystallography*, J. Wiley, New York, 1970.

9.34. J. C. Dainty, in *Laser Speckle and Related Phenomena*, J. C. Dainty, ed., Topics in Applied Physics, Vol. 9, Springer-Verlag, Berlin, 1984, 2nd edition, p. 255.

9.35. W. O. Saxton, *Computer Techniques for Image Processing in Electron Microscopy*, Academic Press, New York, 1978.

9.36. R. H. T. Bates and M. J. McDonnell, *Image Restoration and Reconstruction*, Clarendon Press, Oxford, 1986.

9.37. G. Ross, M. A. Fiddy and M. Nieto-Vesperinas, in *Inverse Scattering Problems in Optics*, H. P. Baltes, ed., Topics in Current Physics, Vol. 20, Springer-Verlag, Berlin, 1980, Chapter 2.

9.38. G. H. Stout and L. H. Jensen, *X-Ray Structure Determination*, J. Wiley, New York, 1989.

9.39. D. Gabor, *Proc. Roy. Soc. A* **197**, 454 (1949).

9.40. E. N. Leith and J. Upatnieks, *J. Opt. Soc. Am.* **52**, 1123 (1962).

9.41. E. Wolf, *J. Opt. Soc. Am.* **60**, 18 (1970).

9.42. R. P. Millane, *J. Opt. Soc. Am. A* **7**, 394 (1990).

9.43. E. Wolf, *Proc. Phys. Soc.* **80**, 1269 (1962).

9.44. A. Walther, *Opt. Acta* **10**, 41 (1963).

9.45. H. A. Ferwerda, in *Inverse Source Problems in Optics*, H. P. Baltes, ed., Topics in Current Physics, Vol. 8, Springer-Verlag, Berlin, 1978, Chapter 2.

9.46. Yu. M. Bruck and L. G. Sodin, *Opt. Comm.* **30**, 304 (1979).

9.47. M. H. Hayes, *I.E.E.E. Trans. Acoust. Speech Sig. Proc.* **30**, 140 (1982).

9.48. M. Nieto-Vesperinas and J. C. Dainty, *Opt. Comm.* **52**, 94 (1984); *Opt. Comm.* **54**, 333 (1985); *Opt. Comm.* **58**, 83 (1986).

9.49. M. A. Fiddy, B. J. Brames and J. C. Dainty, *Opt. Lett.* **8**, 96 (1983).

9.50. M. Nieto-Vesperinas, *Opt. Acta* **33**, 713 (1986).

9.51. J. R. Fienup, *Opt. Lett.* **3**, 27 (1978); *Appl. Opt.* **21**, 2758 (1982).

9.52. P. J. Napier and R. H. T. Bates, *Astron. Astrophys. Suppl.* **15**, 427 (1974).

9.53. M. J. Perez-Ilzarbe, M. Nieto-Vesperinas and R. Navarro, *J. Opt. Soc. Am. A* **7**, 434 (1990).

9.54. R. H. T. Bates and W. R. Fright, *J. Opt. Soc. Am.* **73**, 358 (1983).

9.55. J. R. Fienup and C. C. Wakerman, *J. Opt. Soc. Am. A* **3**, 1897 (1986).

9.56. T. R. Crimmins, J. R. Fienup and B. J. Thelen, *J. Opt. Soc. Am. A* **7**, 3 (1990).

9.57. A. Levi and H. Stark, *J. Opt. Soc. Am. A* **1**, 932 (1984).

9.58. M. Nieto-Vesperinas, R. Navarro and F. J. Fuentes, *J. Opt. Soc. Am. A* **5**, 30 (1988); R. Navarro, F. J. Fuentes and M. Nieto-Vesperinas, *Astron. Astrophys.* **208**, 374 (1989).

9.59. B. J. Hoenders, in *Inverse Source Problems in Optics*, H. P. Baltes, ed., Topics in Current Physics, Vol. 8, Springer-Verlag, Berlin, 1978, Chapter 3.

9.60. J. B. Keller, *I.R.E. Trans.* **4**, 312 (1956).

9.61. R. M. Lewis, *I.E.E.E. Trans. Antenn. Propag.* **17**, 308 (1969).

9.62. R. T. Prosser, *J. Math. Phys.* **10**, 1819 (1969).

9.63. G. S. Deschamps and H. S. Cabayan, *I.E.E.E. Trans. Antenn. Propag.* **20**, 268 (1972).

9.64. H. S. Cabayan, R. C. Murphy and T. J. F. Pavlasek, *I.E.E.E. Trans. Antenn. Propag.* **21**, 346 (1973).

9.65. G. Toraldo di Francia, *J. Opt. Soc. Am.* **45**, 497 (1955).

10

Inverse Source and Scattering Problems in Optics

In Chapters 3 and 4 we studied radiated and scattered fields and some mathematical properties of their complex amplitudes. In this chapter we shall address the determination of the structural characteristics of media that produce those fields, either by radiation or scattering, from knowledge of their complex wavefunctions. We shall consider the time dependence of the fields given by a factor: $\exp(-i\omega t)$.

The question of extracting information about the medium that gives rise to a certain field is generally referred to as an *inverse source problem* when that medium is a *radiating source*, and as an *inverse scattering problem* when the medium is a *scatterer*.

10.1. INTRODUCTION

The inversion of data in *inverse source* and *scattering* problems share mathematical difficulties with inverse diffraction; specifically as regards the finite number of degrees of freedom of information conveyed by the field data, and thus, the resolution limit present in the object reconstruction. This is underlined by the ill-posed nature which is common to these problems. Besides, there are two additional difficulties: first, even if the phase of the field is retrieved, the structure to be determined is often three-dimensional, whereas the collection of radiated, or scattered, complex amplitude data in one experiment corresponds to a function of only two independent variables; this entails the need for collecting data from a large number of mutually complementary experiments which are often difficult to perform. Second, as regards inverse scattering, the reconstruction of the potential is straightforward analytically (and thus well established), only within the *first Born approximation*, or related approaches like the *Rytov approximation.* Methods for more complicated situations that involve multiple scattering, are usually devised in each case for the specific system under study.

As seen in Chapter 3, in optical scattering the structure of the medium (assumed to be nonmagnetic) is often characterized by the refractive index $n(\mathbf{r})$ (or equivalently, by the dielectric susceptibility $\chi(\mathbf{r})$) as a function of the point $\mathbf{r}$ of the material. The *scattering potential* is given in terms of $\chi(\mathbf{r})$, or $n(\mathbf{r})$, as described in Sections 3.8 and 3.10. In inverse source problems, on the other hand, the source may be characterized by e.g. the electric current density generating electromagnetic radiation; in the case of acoustic waves, the source may be induced and given in terms of the amplitude of some insonifying wave.

In radiology, straight-through propagation of rays, described by geometrical optics, leads to successful reconstructions of objects by means of *computed tomography* (see, e.g., the reviews of Refs. 10.1–10.3). Nevertheless, at larger wavelengths, at which scattering has to be included, the theory of data inversion is considerably more complicated.

We shall consider in this chapter deterministic sources only; this involves dealing directly with the complex amplitudes of the fields. Accounts on the reconstruction of random sources from field correlations can be found in works by Devaney [10.4], Hoenders and Baltes [10.5], and Baltes et al. [10.6] (see also the references listed in Chapter 1 of the book of Ref. 10.6, and those quoted in Ref. 10.7).

We shall first address the inverse source and scattering problems based on the angular spectrum formalism discussed in Chapter 3. This follows a pioneering work by Wolf [10.8], and subsequently developed further by Friberg and Wolf [10.9], Devaney and Wolf [10.10], Carter and Ho [10.11]–[10.13], and the authors of Ref. 10.14. We shall then discuss the imaging integral equation for sources and scatterers as obtained by Porter [10.15], [10.16] in connection with generalized holography, and a related equation then independently derived by Bleistein and Bojarski [10.17] (see also Bleistein and Cohen [10.18] and Bojarski [10.19]–[10.21]).

The problem of the lack of uniqueness due to the three-dimensionality of the structure to be determined will also be analyzed according to the procedure of Devaney [10.22] based on Wolf's model [10.8], and as discussed by Bleistein and Cohen [10.18] (see also the review of Ref. 10.16) for the imaging equation. Other methods can be found in the work by Hoenders [10.23], and references cited therein. The subject of *diffraction tomography* (Devancy [10.24], [10.25]), which permits the reconstruction of weak scatterers by a generalization of conventional computed tomography, will also be discussed.

In general, inverse source and scattering problems in optics constitute a current active subject of research in which methods, ranking from acoustics and electromagnetic theory to quantum mechanics [10.26], are incorporated (e.g. the distorted-wave Born and Rytov approximations, or the Gelfand–

Levitan–Marchenko method). In addition to the references quoted above, the reader can find useful information about the state of the art in the vast number of references contained in the material edited by: Sabatier [10.27], Boerner et al. [10.28], Devaney [10.29], Boerner [10.30], and Sabatier [10.31].

10.2. FORMULATION OF INVERSE SOURCE AND SCATTERING PROBLEMS

In Section 3.2 we discussed the following expression for the electric vector of a monochromatic field emitted by a system of charges and currents, contained in a volume V situated within a strip limited by two half-spaces $\mathscr{R}^+$ and $\mathscr{R}^-$ (cf. Fig. 3.1):

$$\mathbf{E}(\mathbf{r}_>) = \iint_{-\infty}^{\infty} \mathbf{e}^{(\pm)}(\mathbf{K}) \exp[i(\mathbf{K}\cdot\mathbf{R} \pm k_z z)]\, d^2K, \tag{10.1}$$

where $\mathbf{K} = k(p, q)$, $k_z = km$ and $k = \omega/c$, p, q, and m being the directional cosines of the plane wave components in the angular spectrum representation in Eq. (10.1), which satisfy Eqs. (2.9). $\mathbf{r}_> = (\mathbf{R}_>, z_>)$ is the position vector of the field point, considered either in $\mathscr{R}^+$ or $\mathscr{R}^-$ according to whether the sign "plus" or "minus" holds in the integrand of Eq. (10.1).

On the other hand, we recall that the angular spectrum $\mathbf{e}^{(\pm)}(\mathbf{K})$ is given by (cf. Eqs. (3.6) and (3.9)):

$$\mathbf{e}^{(\pm)}(\mathbf{K}) = -\mathbf{k}^{(\pm)} \times \left[\mathbf{k}^{(\pm)} \times \left(-\frac{(2\pi)^2}{kck_z}\tilde{\mathbf{j}}(\mathbf{k}^{\pm})\right)\right], \tag{10.2}$$

where $\mathbf{k}^{\pm} = (\mathbf{K}, \pm k_z)$, and $\tilde{\mathbf{j}}(\mathbf{k}^{\pm})$ is the three-dimensional Fourier transform of the electric current density $\mathbf{j}(\mathbf{r})$:

$$\tilde{\mathbf{j}}(\mathbf{k}) = \frac{1}{(2\pi)^3}\int_{-\infty}^{\infty} \mathbf{j}(\mathbf{r}) \exp(-i\mathbf{k}\cdot\mathbf{r})\, d^3r, \tag{10.3}$$

evaluated at values $\mathbf{k} = \mathbf{k}^{\pm}$ of its argument. Since $\mathbf{j}(\mathbf{r})$ is zero outside the volume V the integral in Eq. (10.3) is extended to V only. An equation similar to Eq. (10.1) holds for the magnetic vector (cf. Section 3.2).

In the case of a field $(\mathbf{E}^{(i)}, \mathbf{H}^{(i)})$ incident on a scatterer, the expression for the field scattered into the half-spaces $\mathscr{R}^+$ and $\mathscr{R}^-$ is given, as shown in

Section 3.6, by equations similar to Eqs. (10.1)–(10.3) for the electric vector (and the corresponding equations for the magnetic vector), namely by Eqs. (3.7)–(3.10), (3.59a), and (3.59b), where, instead of the source function $\mathbf{j}(\mathbf{r})$, the induced polarization $\mathbf{P}(\mathbf{r})$ appears (assuming the scatterer to be nonmagnetic, the induced magnetization $\mathbf{M}(\mathbf{r})$ in those equations is zero; otherwise it would have to be included).

In the scalar theory, as shown in Section 3.4, the wavefield emitted by the time-harmonic source density distribution $\rho(\mathbf{r})$ is (cf. Section 3.4):

$$U(\mathbf{r}_>) = \iint_{-\infty}^{\infty} A^{(\pm)}(\mathbf{K}) \exp[i(\mathbf{K}\cdot\mathbf{R}_> \pm k_z z_>)]\, d^2K, \tag{10.4}$$

where the angular spectrum $A^{(\pm)}(\mathbf{K})$ is:

$$A^{(\pm)}(\mathbf{K}) = \frac{i}{2\pi k_z} \int_V \rho(\mathbf{r}') \exp[-i(\mathbf{K}\cdot\mathbf{R}' \pm k_z z')]\, d^3r'. \tag{10.5}$$

It should be equally recalled that for the case of a scattered field, Eqs. (10.4) and (10.5) hold by replacing $\rho(\mathbf{r})$ with $(-1/4\pi)F(\mathbf{r})U(\mathbf{r})$, $F(\mathbf{r})$ being the potential and $U(\mathbf{r})$ the field wavefunction inside the scatterer (Section 3.10).

The *inverse source problem* consists of determining $\mathbf{j}(\mathbf{r})$ (or $\rho(\mathbf{r})$ in the scalar theory) from knowledge of the emitted field, namely, from $\mathbf{E}(\mathbf{r}_>)$ or $\mathbf{H}(\mathbf{r}_>)$ (or from $U(\mathbf{r}_>)$ for scalar wavefields).

On the other hand, the *inverse scattering problem* in optics pertains to determining the dielectric susceptibility $\chi(\mathbf{r})$ (and the magnetic susceptibility $\eta(\mathbf{r})$ if the medium is magnetic) from the scattered field. As shown in Sections 3.7 and 3.8, the determination of $\chi(\mathbf{r})$ and $\eta(\mathbf{r})$ (and ultimately, the refractive index of the medium) from the "equivalent" source distributions, represented by the induced polarization $\mathbf{P}(\mathbf{r})$ and induced magnetization $\mathbf{M}(\mathbf{r})$, is generally quite involved. Since $\mathbf{P}(\mathbf{r}) = \chi(\mathbf{r})\mathbf{E}(\mathbf{r})$ and $\mathbf{M}(\mathbf{r}) = \eta(\mathbf{r})\mathbf{H}(\mathbf{r})$, this determination requires specification of the fields inside the medium. (Analogously, in the scalar theory, since $\rho(\mathbf{r}) = -(1/4\pi)F(\mathbf{r})U(\mathbf{r})$, the determination of $F(\mathbf{r})$ entails knowledge of the interior wavefunction $U(\mathbf{r})$.) Hence, the solution first involves devising a model that solves the *direct* scattering problem by addressing the multiple interactions between field and scatterer, and then either determining the field inside or inverting the scheme to find the corresponding potential. As shown in Sections 3.8 and 3.10, only within the domain of validity of the first Born approximation (or related solutions like the Rytov approximation), the "equivalent" sources and potentials are simply related to the incident field $\mathbf{E}^{(i)}$ and $\mathbf{H}^{(i)}$ (or $U^{(i)}$ in the scalar theory), namely:

$$\mathbf{P}(\mathbf{r}) = \chi(\mathbf{r})\mathbf{E}^{(i)}(\mathbf{r}); \tag{10.6}$$

$$\mathbf{M}(\mathbf{r}) = \eta(\mathbf{r})\mathbf{H}^{(i)}(\mathbf{r}); \tag{10.7}$$

$$\rho(\mathbf{r}) = -\frac{1}{4\pi} F(\mathbf{r})U^{(i)}(\mathbf{r}). \tag{10.8}$$

In this case, we recall that, assuming the incident field to be a plane wave of wavevector $\mathbf{k}_i$, and assuming, for instance, the medium to be nonmagnetic, the electric vector of the scattered field is given by Eq. (10.1), with the angular spectrum $\mathbf{e}_B^{(\pm)}(\mathbf{K})$ given by (cf. Eq. (3.86a)):

$$\mathbf{e}_B^{(\pm)}(\mathbf{K}) = -i\frac{(2\pi)^2}{k_z}[\mathbf{k}^{(\pm)} \times (\mathbf{k}^{(\pm)} \times \mathbf{e}_i)]\tilde{\chi}(\mathbf{k}^{(\pm)} - \mathbf{k}_i), \tag{10.9}$$

$\mathbf{e}_i$ being the constant amplitude of the incident electric vector, and $\tilde{\chi}(\mathbf{r})$ being the three-dimensional Fourier transform of $\chi(\mathbf{r})$.

Analogously, we recall that in the scalar theory, the scattered wavefield is given by Eq. (10.4) with the angular spectrum $A_B^{(\pm)}(\mathbf{K})$ given by (cf. Eq. (3.119)):

$$A_B^{(\pm)}(\mathbf{K}) = -i\frac{\pi}{k_z}\tilde{F}(\mathbf{k}^{(\pm)} - \mathbf{k}_i), \tag{10.10}$$

$\tilde{F}(\mathbf{r})$ being the three-dimensional Fourier transform of $F(\mathbf{r})$.

10.3. INFORMATION CONTENT OF ONE EXPERIMENT

The determination of $\mathbf{e}^{(\pm)}(\mathbf{K})$ (or $A^{(\pm)}(\mathbf{K})$) from the scattered field in either $\mathscr{R}^+$ or $\mathscr{R}^-$ implies knowledge of both the modulus and the phase of this field. Thus, in optics this field has to be determined either by holographic means (see, e.g. Ref. 10.8), or by some of the computational methods outlined in Section 9.10. The measurements may be performed either in the far zone (then, as seen in Section 3.3, one obtains directly the angular spectrum up to the evanescent wave cut-off, namely, for values $|\mathbf{K}| \leqslant k$), or in a plane $z = z_1$, situated either in $\mathscr{R}^+$ or $\mathscr{R}^-$, at a finite distance from the scattering volume (cf. Fig. 3.1). By a procedure analogous to the one used in Section 2.3, the angular spectrum $\mathbf{e}^{(\pm)}(\mathbf{K})$ is obtained from Eq. (10.1) and shown to be:

$$\mathbf{e}^{(\pm)}(\mathbf{K}) = \frac{1}{(2\pi)^2}\exp(\mp ik_z z_1)\int_{-\infty}^{\infty} \mathbf{E}(\mathbf{R}, z_1)\exp(-i\mathbf{K}\cdot\mathbf{R})\, d^2R, \tag{10.11}$$

in the electromagnetic theory; or by:

$$A^{(\pm)}(\mathbf{K}) = \frac{1}{(2\pi)^2} \exp(\mp ik_z z_1) \int_{-\infty}^{\infty} U(\mathbf{R}, z_1) \exp(-i\mathbf{K}\cdot\mathbf{R})\, d^2R, \quad (10.12)$$

in the scalar case. From one experiment, namely, from the angular spectrum of the radiated field (or scattered field, determined for a given incident wave of wavevector $\mathbf{k}_i$) one has to invert Eq. (10.2) or (10.5) for the inverse source problem, or Eq. (10.9) or (10.10) for the inverse scattering problem within the first Born approximation. (For the Rytov approximation, the conclusions derived next equally apply.) The angular spectrum is, however, the limiting value of the three-dimensional Fourier transform of the source distribution, evaluated at the argument: $\mathbf{k} = \mathbf{k}^{(\pm)}$ for the inverse source problem, and the limiting value of the three-dimensional Fourier transform of the potential, evaluated at: $\mathbf{K}^{(\pm)} = \mathbf{k}^{(\pm)} - \mathbf{k}_i$ for the inverse scattering problem in the first Born approximation.

Since, however, $|\mathbf{k}^{(\pm)}|^2 = |\mathbf{K}|^2 + k_z^2 = \omega^2/c^2$, the Fourier transform of the source distribution is known only over the surface of a sphere (this is *the Ewald sphere*, analogous to that of X-ray diffraction theory [10.32]), whose center is at the origin of the **k**-space, and whose radius is $k = \omega/c$. Hence, the inversion of this Fourier transform is impossible.

Analogously, for the inverse scattering problem, since $|\mathbf{k}^{(\pm)}|^2 = \omega^2/c^2 = |\mathbf{k}_i|^2$, then the three-dimensional Fourier transform of either $\chi(\mathbf{r})$ or $F(\mathbf{r})$ is known only over the surface of the sphere:

$$|\mathbf{K}^{(\pm)} + \mathbf{k}_i|^2 = |\mathbf{k}^{(\pm)}|^2 = \omega^2/c^2 = |\mathbf{k}_i|^2, \quad (10.13)$$

whose analytic equation in the **K**-space may be explicitly written in terms of the Cartesian components of the wavevectors as:

$$(\mathbf{K}_x^{(\pm)} + kp_i)^2 + (\mathbf{K}_y^{(\pm)} + kq_i)^2 + (\mathbf{K}_z^{(\pm)} + km_i)^2 = k^2, \quad (10.14)$$

where, as indicated above, $\mathbf{K}^{(\pm)} = (\mathbf{K}_x^{(\pm)}, \mathbf{K}_y^{(\pm)}, \mathbf{K}_z^{(\pm)})$ is the *wavevector transfer* $\mathbf{k}^{(\pm)} - \mathbf{k}_i$, and $\mathbf{k}_i = k(p_i, q_i, m_i)$, p_i, q_i, and m_i being the cosine directors of $\mathbf{k}_i$.

Equation (10.14) shows that, in the first Born approximation, the three-dimensional Fourier transform of the potential is known only over the Ewald sphere centered at the point $\mathbf{K} = -\mathbf{k}_i$ and whose radius is ω/c.

The above facts imply that the field emitted, or scattered, outside a localized volume carries only two-dimensional information. Therefore, the data given from one experiment as described above, are insufficient to reconstruct the three-dimensional source or potential.

In scattering experiments, and within the first Born (or Rytov) approximation, the Ewald sphere can be varied by changing either the magnitude ω/c or the direction $\mathbf{k}_i$ of the incident wavevector. The former procedure corresponds to a wide frequency band and a narrow range of propagation directions of the incident field, whereas the latter configuration implies a single frequency and a wide range of angles of incidence. Either method yields values of the three-dimensional Fourier transform $\tilde{F}(\mathbf{K}^{\pm})$ of the potential at points contained in a volume of the $\mathbf{K}$-space. For instance, varying the direction of propagation of the incident field, gives $\tilde{F}(\mathbf{K}^{\pm})$ inside the so-called *limiting Ewald sphere*. This limiting sphere is the envelope of all the Ewald spheres whose center $\mathbf{K} = -\mathbf{k}_i$ varies with the direction of $\mathbf{k}_i$ (see Fig. 10.1). This permits the inversion of the three-dimensional potential Fourier transform. Note, however, that since the volume enclosed by this limiting sphere is finite, there is a bound to the accessible $\mathbf{K}$-space, namely, only $\mathbf{K}$-vectors such that:

$$|\mathbf{K}^{(\pm)}| \leqslant 2\frac{\omega}{c}, \tag{10.15}$$

are accessible. This, in turn, implies by means of arguments similar to those

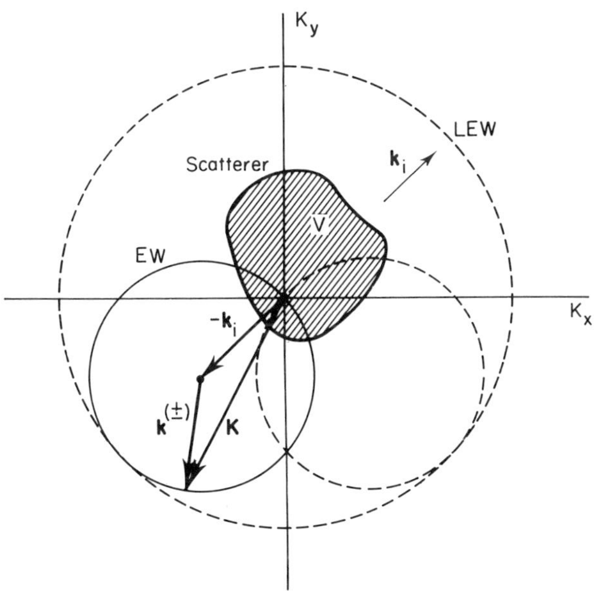

Figure 10.1. The Ewald sphere of accessible data (*EW*), and the limiting Ewald sphere (*LEW*).

of Section 2.4, that one can retrieve the scattering potential only with a resolution of linear size L satisfying (Wolf [10.8]):

$$\frac{1}{L^2} \leqslant \left(\frac{2}{\lambda}\right)^2,$$

namely:

$$L \geqslant \frac{\lambda}{2}, \tag{10.16}$$

λ being the wavelength.

Note that, however, it is obvious that this procedure for scanning the **k**-space cannot be used in inverse source problems. Nevertheless, these problems are usually of a statistical nature, that is, the source function is random. Thus, interferometric measurements that determine field autocorrelations are done in order to retrieve source statistical information, such as source correlation functions or probability densities [10.4], [10.33].

Moreover, note the additional difficulty in retrieving the vector functions $\mathbf{j}(\mathbf{r})$ or $\mathbf{P}(\mathbf{r})$ since only the transversal part of these functions is related to the angular spectrum of the field (cf. Eqs. (10.2) and (10.9)).

When multiple scattering takes place, another complication appears because the "equivalent source distribution" $\mathbf{P}(\mathbf{r})$ (or $\rho(\mathbf{r}) = (-1/4\pi)F(\mathbf{r})U(\mathbf{r})$ in the scalar theory) changes as the incident wavevector $\mathbf{k}_i$ varies, thus being different for each Ewald sphere.

For the above reasons, solutions to three-dimensional inverse source and scattering problems have been rigorously obtained so far only for some statistical sources [10.4], and for scatterers within the first Born approximation (or related approaches like Rytov or distorted wave Born) (see, e.g. Refs. 10.24 and 10.25), or in specific cases with additional *a priori* information about the object (see works in Refs. 10.26–10.31).

When the body is a perfect reflector, the structure of its surface can be determined from one experiment (see articles in Refs. 10.28–10.30). In this case, one should also mention the approach known as the "physical optics far field inverse scattering identity," or POFFIS, established by Bojarski [10.34], relating the far field under the Kirchhoff approximation with the shape function of the scatterer.

10.4. INTEGRAL EQUATIONS FOR FIELDS OVER ARBITRARY SURFACES

In this section we shall discuss the inverse source and scattering problems in terms of some integral equations directly derived from the theorems addressed in Section 1.6. They lead to a real image of the source or scattering potential from the field recorded (by some holographic means) on an arbitrary close surface that surrounds that source volume V, independently of the particular shape of the recording surface (Porter [10.15], [10.16]). We shall also see its relationship with an equation derived by Bleistein and Cohen [10.18] in connection with the uniqueness of the inverse source problem both in acoustics and electromagnetics (see also Refs. 10.17, 10.19, 10.35, and 10.36). A subsequent analysis was also performed by Porter and Devaney [10.37], [10.38] for the source in either homogeneous or inhomogeneous media. Generalizations to sources embedded in inhomogeneous attenuative media have been done in Refs. 10.39 and 10.40. Also, the information content of these equations and their relationship with the method of data inversion from the holographic determination of this field at a plane [10.8] has been analyzed by Wolf and Porter [10.41].

10.4.1. Vector Theory

Let us consider a source distribution $\mathbf{F}_e(\mathbf{r})$, localized inside a volume V, and producing a field outside V whose electric vector is $\mathbf{E}(\mathbf{r})$. According to Section 1.2, $\mathbf{E}(\mathbf{r})$ satisfies the equation:

$$\nabla \times \nabla \times \mathbf{E}(\mathbf{r}) - k^2\mathbf{E}(\mathbf{r}) = \mathbf{F}_e(\mathbf{r}), \tag{10.17}$$

together with the vector form, Eq. (1.32), of Sommerfeld's radiation condition for outgoing fields.

$\mathbf{F}_e(\mathbf{r})$ in (10.17) can represent either a radiating source distribution or an "equivalent" source distribution induced by an incident field on a scatterer occupying the volume V. In the latter case $\mathbf{F}_e(\mathbf{r})$ is related to the induced polarization $\mathbf{P}(\mathbf{r})$ and induced magnetization $\mathbf{M}(\mathbf{r})$ by (cf. Eq. (3.56a)):

$$\mathbf{F}_e(\mathbf{r}) = 4\pi k[k\mathbf{P}(\mathbf{r}) + i\nabla \times \mathbf{M}(\mathbf{r})]. \tag{10.18}$$

Let $\mathscr{G}(\mathbf{r}, \mathbf{r}')$ be the dyadic Green function, Eq. (1.27), then the application of the vector form, Eq. (1.33), of Green's theorem to the volume V' that includes V and is bounded by a surface S (Fig. 10.2) gives at any point $\mathbf{r}$ of V (cf. Eq. (1.34), we change now slightly the notation):

$$\mathbf{E}(\mathbf{r}) = \frac{1}{4\pi}\int_V \mathbf{F}_e(\mathbf{r}')\mathscr{G}(\mathbf{r}, \mathbf{r}')\, d^3r' - \frac{1}{4\pi}\Sigma_{[\mathscr{G}]}^{(-)}(\mathbf{r}), \tag{10.19}$$

where, according to Eq. (1.39):

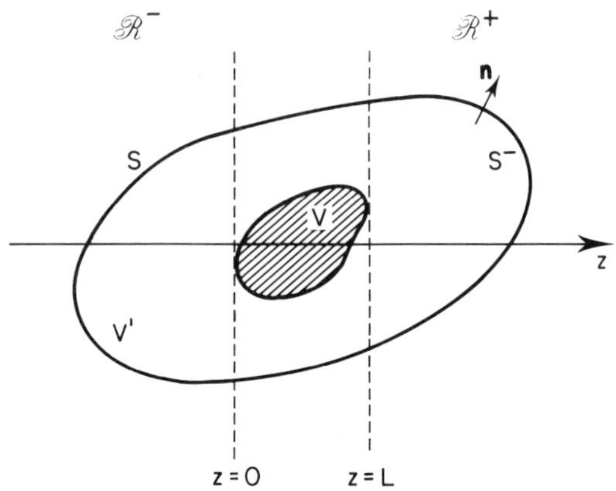

Figure 10.2. Illustration of the notation.

$$\sum_{[\mathscr{G}]}^{(-)}(\mathbf{r}) = \int_{S^-} [(\mathbf{n} \times \nabla \times \mathbf{E}(\mathbf{r}'))\cdot\mathscr{G}(\mathbf{r}, \mathbf{r}') + (\mathbf{n} \times \mathbf{E}(\mathbf{r}'))\cdot(\nabla \times \mathscr{G}(\mathbf{r}, \mathbf{r}'))]\, dS. \tag{10.20}$$

The symbol S^- in Eq. (10.20) indicates that the limiting values in the integrand are taken by approaching the surface S from inside the volume V'; $\mathbf{n}$ represents the unit outward normal.

By writing an equation analogous to Eq. (10.20), using the complex conjugate dyadic: $\mathscr{G}^*(\mathbf{r}, \mathbf{r}')$, recalling that $G(\mathbf{r}, \mathbf{r}') - G^*(\mathbf{r}, \mathbf{r}') = 2ikj_0(kr)$ (j_0 being the spherical Bessel function of the first kind and zero order, and $G^*(\mathbf{r}, \mathbf{r}')$ being the complex conjugate of the free-space Green function, Eq. (1.28), which is a convergent spherical wave), and subtracting it from Eq. (10.20), we obtain:

$$\int_V \mathbf{F}_e(\mathbf{r}')\cdot\mathbf{J}(\mathbf{r}, \mathbf{r}')\, d^3r' = \sum_{[\mathbf{J}]}^{(-)}(\mathbf{r}), \tag{10.21}$$

where:

$$\mathbf{J}(\mathbf{r}, \mathbf{r}') = \left(\mathscr{I} + \frac{1}{k^2}\nabla\nabla\right) j_0(k|\mathbf{r} - \mathbf{r}'|), \tag{10.22}$$

and:

$$\sum_{[\mathbf{J}]}^{(-)}(\mathbf{r}) = \int_{S^-} [(\mathbf{n} \times \nabla \times \mathbf{E}(\mathbf{r}'))\cdot\mathbf{J}(\mathbf{r}, \mathbf{r}') + (\mathbf{n} \times \mathbf{E}(\mathbf{r}'))\cdot(\nabla \times \mathbf{J}(\mathbf{r}, \mathbf{r}'))]\, dS. \tag{10.23}$$

Equation (10.21) is the vector form of the imaging equation (Porter [10.16], [10.42]; Bleistein and Cohen [10.18]).

We are next going to discuss the scalar form of this equation as well as its physical meaning.

10.4.2. Scalar Theory

Let $\rho(\mathbf{r})$ be a source density, localized in a volume V; $\rho(\mathbf{r})$ may represent either a primary source or an "equivalent source distribution" induced in a scatterer, as quoted in Section 10.2. As is well known, the field in the presence of this source satisfies the inhomogeneous Helmholtz equation (3.16). (The wavefunctions in this section correspond to either radiated or scattered fields only; namely, contrary to Section 1.6.3 no incident field $U^i(\mathbf{r})$ is included in $U(\mathbf{r})$.)

The application of Green's theorem to the volume V' that contains V then leads to (cf. Eq. (1.56)):

$$U(\mathbf{r}) = \int_V \rho(\mathbf{r}')G(\mathbf{r}, \mathbf{r}')\, d^3r' - \frac{1}{4\pi}\sum(\mathbf{r}), \tag{10.24}$$

where:

$$\sum(\mathbf{r}) = \int_S \left(U\frac{\partial G}{\partial \mathbf{n}} - G\frac{\partial U}{\partial \mathbf{n}}\right) dS. \tag{10.25}$$

By writing an equation similar to Eq. (10.24) with the incoming Green's function $G^*(\mathbf{r}, \mathbf{r}')$, and subtracting it from Eq. (10.24), we are led to:

$$\int_V \rho(\mathbf{r}')j_0(k|\mathbf{r} - \mathbf{r}'|)\, d^3r' = \frac{1}{4\pi}\sum_{[j]}(\mathbf{r}), \tag{10.26}$$

where:

$$\sum_{[j]}(\mathbf{r}) = \int_S \left(U\frac{\partial j_0}{\partial \mathbf{n}} - j_0\frac{\partial U}{\partial \mathbf{n}}\right) dS. \tag{10.27}$$

Equation (10.26) is the imaging equation as given by Bleistein and Cohen [10.18] (see also Refs. 10.17, 10.19–10.21).

It is possible to express Eq. (10.26) in the form given by Porter [10.15] (see also Porter and Devaney [10.37]) for the inverse source problem. To do that, let us consider Eq. (1.51), which represents the extinction theorem, and take into account that in this case:

$$\sum(\mathbf{r}) = 0, \tag{10.28}$$

for points $\mathbf{r}$ inside V'.

From Eqs. (10.28) and (10.27) we obtain:

$$\sum_{[U]}(\mathbf{r}) = -\frac{1}{2ik}\sum_{[*]}(\mathbf{r}), \tag{10.29}$$

where:

$$\sum_{[*]}(\mathbf{r}) = \int_S \left(U\frac{\partial G^*}{\partial \mathbf{n}} - G^*\frac{\partial U}{\partial \mathbf{n}} \right) dS. \tag{10.30}$$

By comparing the complex conjugate of Eqs. (10.26) and (10.29), one gets directly the aforementioned alternative form of the imaging equation for the inverse source problem (Porter [10.15]):

$$8\pi ki \int_V \rho^*(\mathbf{r}')j_0(k|\mathbf{r}-\mathbf{r}'|)\, d^3r' = \Gamma(\mathbf{r}), \tag{10.31}$$

where $\Gamma(\mathbf{r})$ is the complex conjugate of $\Sigma_{[*]}(\mathbf{r})$, namely:

$$\Gamma(\mathbf{r}) = \int_S \left(U^*\frac{\partial G}{\partial \mathbf{n}} - G\frac{\partial U^*}{\partial \mathbf{n}} \right) dS. \tag{10.32}$$

The left-hand side of Eq. (10.31) contains an image of the complex conjugate of the source distribution $\rho(\mathbf{r})$. As will be seen in the next section, the propagator $j_0(k|\mathbf{r}-\mathbf{r}'|)$ corresponds to propagation, from a point at the surface S at which U^* is recorded, back into V'. We shall refer to the form, Eq. (10.31), of the imaging equation as the *backpropagation equation.*

10.5. INFORMATION CONTAINED IN THE IMAGING EQUATION. PHYSICAL MEANING OF THE BACKPROPAGATION EQUATION

Let us consider Eq. (10.31). The discussion that follows can be equally applied to the corresponding equation in the vector theory.

Since the spherical Bessel function $j_0(kr)$ can be expressed as: $j_0(kr) = (1/2ik)[G(r) - G^*(r)]$, the left-hand side of Eq. (10.31) is equal to the difference between an outgoing and an incoming field; therefore (see Problem 2.11), this term represents a source-free field that satisfies the homogeneous Helmholtz equation everywhere and has plane wave components propagating in all directions. Namely, it admits an angular spectrum representation as

Eq. (2.30), with the spectral amplitude $F(\theta, \phi)$ being:

$$F(\theta, \phi) = 8\pi i \int_V \rho^*(\mathbf{r}') \exp(-ik\mathbf{s}\cdot\mathbf{r}')\, d^3r'. \tag{10.33}$$

Wolf and Porter [10.41] derived an equation similar to Eq. (10.33) by introducing into the left-hand side of Eq. (10.31) the plane wave expansion of the spherical Bessel function (cf. also Eq. (2.30)):

$$j_0(k|\mathbf{r} - \mathbf{r}'|) = \frac{1}{4\pi} \int_{-\pi}^{\pi} d\phi \int_0^{\pi} d\theta \sin\theta \exp(ik\mathbf{s}\cdot|\mathbf{r} - \mathbf{r}'|), \tag{10.34}$$

where $\mathbf{s} = (\sin\theta\cos\phi, \sin\theta\sin\phi, \cos\theta)$. In addition, with reference to Fig. 10.2 and following the result established at the end of Section 2.7.2, on comparing Eq. (10.33) with Eq. (10.5) and introducing Eq. (10.34) into the left-hand side of Eq. (10.31), we obtain:

$$8\pi k i \int_V \rho^*(\mathbf{r}') j_0(k|\mathbf{r} - \mathbf{r}'|)\, d^3r' = -[U_h^{(+)*}(\mathbf{r}) + U_h^{(-)*}(\mathbf{r})], \tag{10.35}$$

throughout the whole space.

In Eq. (10.35), $U_h^{(+)}(\mathbf{r})$ and $U_h^{(-)}(\mathbf{r})$ are source-free fields that represent the waves emitted by the source distribution $\rho(\mathbf{r})$ into $\mathscr{R}^+$ and $\mathscr{R}^-$, respectively, having excluded from them the evanescent components. Moreover, as expressed by the asymptotic forms, Eqs. (2.40), the analytic continuation of $U_h^{(+)}(\mathbf{r})$ into $\mathscr{R}^-$ is a convergent field in $\mathscr{R}^-$; similarly, the analytic continuation of $U_h^{(-)}(\mathbf{r})$ into $\mathscr{R}^+$ is a convergent field in $\mathscr{R}^+$. Therefore, Eq. (10.35) shows that the image of $\rho^*(\mathbf{r})$ is obtained by conjugating these fields and backpropagating them into V'. The holographic process that achieves this is characterized by the field $\Gamma(\mathbf{r})$ of Eq. (10.32).

In fact, within a scalar formulation, $\Gamma(\mathbf{r})$ is the field obtained from a *generalized hologram* recorded on the surface S; this field is source-free and backpropagates from S into V' producing a real pseudoscopic image of the source distribution $\rho(\mathbf{r})$. To see this, let us assume that a hologram is formed in the surface S by recording in this surface the intensity given by the interference of the field $U(\mathbf{r})$ radiated (or scattered) from the source distribution $\rho(\mathbf{r})$ according to Eq. (3.17), with a reference field of known phase and unit amplitude. The field amplitude transmittance of the hologram is given by an expression similar to Eq. (9.19). When the surface hologram S is illuminated from outside by the conjugate of the reference field, then, as seen in the third term of the right-hand side of Eq. (9.20) and as discussed in Section 9.5, one obtains in V' a real pseudoscopic image of the object that created the emitted field $U(\mathbf{r})$, namely, of the source distribution $\rho(\mathbf{r})$. The field that forms this image is given in terms of the boundary value of $U^*(\mathbf{r})$ on S by the Helmholtz–Kirchhoff integral representation, Eq. (10.32). In addition, when the expression (Eq. (3.17)) for $U(\mathbf{r})$ is introduced into the surface integral of Eq. (10.32), one obtains the left-hand side of Eq. (10.31) which is clearly a real image of $\rho^*(\mathbf{r})$ (the pseudoscopic image of $\rho(\mathbf{r})$), whose imaging kernel is the backpropagator $j_0(k|\mathbf{r} - \mathbf{r}'|)$ (observe that this backpropagator is independent of the recording surface S of the hologram (Porter [10.15])).

In connection with the representation (Eqs. (10.4) and (10.5)) of the field, either radiated or scattered from V into the half-spaces $\mathscr{R}^+$ and $\mathscr{R}^-$, if the closed surface S is formed by two planes Σ^+ and Σ^- in $\mathscr{R}^+$ and $\mathscr{R}^-$, respectively, and normal to the z-axis, and the two faces parallel to the z-axis that join these two planes at infinity, then, when the integral of Eq. (10.32) is applied to this field and this surface S, one straightforwardly obtains (Wolf and Porter [10.41]):

$$\int_{\Sigma^\pm} U^*(\mathbf{r}') \frac{\partial G(\mathbf{r}, \mathbf{r}')}{\partial \mathbf{n}} \, d\Sigma^\pm = -\tfrac{1}{2} U_h^{(\pm)*}(\mathbf{r}) - \tfrac{1}{2} U_e^{(\pm)}(\mathbf{r}), \qquad (10.36a)$$

$$\int_{\Sigma^\pm} G(\mathbf{r}, \mathbf{r}') \frac{\partial U^*(\mathbf{r}')}{\partial \mathbf{n}} \, d\Sigma^\pm = \tfrac{1}{2} U_h^{(\pm)*}(\mathbf{r}) - \tfrac{1}{2} U_e^{(\pm)}(\mathbf{r}). \qquad (10.36b)$$

In Eqs. (10.36), $U_h^{(+)}(\mathbf{r})$ and $U_h^{(-)}(\mathbf{r})$ are the homogeneous parts of the fields emitted by the source (see also Eq. (10.35)), and $U_e^{(\pm)}(\mathbf{r})$ are the contributions of the evanescent components to these fields. Observe that the right-hand side of Eq. (10.32) given by the subtraction of Eqs. (10.36a) and (10.36b) cancels out the evanescent parts $U_e^{(\pm)}(\mathbf{r})$, and through Eq. (10.31) it leads again to Eq. (10.35).

Also, by inspecting Eq. (10.33) it is easy to see that the three-dimensional Fourier transform $\tilde{\rho}(\mathbf{k})$ of the source density $\rho(\mathbf{r})$ is given from one experiment at values of the argument $\mathbf{k}$ that satisfy:

$$|\mathbf{k}|^2 = \frac{\omega^2}{c^2}. \qquad (10.37)$$

The procedure to retrieve $\tilde{\rho}(\mathbf{k})$ inside the limiting Ewald sphere is thus identical to the one described in Section 10.3. The inversion of data from values of the field at a certain plane either in $\mathscr{R}^+$ or $\mathscr{R}^-$, yields the same information as the inversion through the backpropagation equation formalism. In the former case, evanescent waves are lost unless the field is retrieved on a plane at a distance of the order of a wavelength from the source. In the latter case, however, the addition of diverging and converging fields (or what is equivalent, the addition of boundary values of both the field and its normal derivative) automatically cancels out the evanescent wave contribution.

10.6. AMBIGUITY IN INVERSE SOURCE AND SCATTERING PROBLEMS. NONRADIATING SOURCES AND NONSCATTERING SCATTERERS

The lack of information necessary to retrieve the three-dimensional structure of a source distribution or potential from one experiment poses the question of uniqueness of the sources compatible with the given data. Namely, whether there can be different sources, or potentials, giving rise in one experiment to the same radiated, or scattered, field outside the volume V. One can even ask whether there may be sources, or potentials, producing in one experiment a null angular spectrum of the emitted, or scattered, field outside V. Due to the properties of fields discussed in Chapter 4, a vanishing angular spectrum would imply a zero field everywhere outside V. Such sources, or potentials, could be added to any other radiating source or potential without changing the exterior field.

Since the angular spectrum is the three-dimensional Fourier transform of the source, or potential, distribution, and assuming V to be finite, then this angular spectrum can be extended as an entire function of exponential type in the complex $\mathbf{z}$-space ($\mathbf{z} = (z_1, z_2, z_3)$, $z_1 = K_1 + iK'_1$, $z_2 = K_2 + iK'_2$, $z_3 = K_3 + iK'_3$) (cf. Appendix 4.1); as such it may have zero lines in this complex $\mathbf{z}$-space [10.43], [10.44] (namely, four-dimensional surfaces in this six-dimensional $\mathbf{z}$-space at which this function vanishes). Therefore, one can conceive one experiment for a certain source distribution, or potential, such that its three-dimensional Fourier transform has one of its zero lines at those values of $\mathbf{z}$ being exactly equal to $\mathbf{k}^{(\pm)}$, i.e., at: $|\mathbf{k}^{(\pm)}|^2 = \omega^2/c^2$ (or satisfying Eq. (10.13) for scattering in the first Born approximation) [10.45]. This argument proves that the vanishing of the angular spectrum, namely, of the three-dimensional Fourier transform of the source, or potential, at a certain hypersurface of values of its argument, is sufficient for this to be nonradiating or nonscattering. This is, however, also a necessary condition, for if the field, radiated or scattered outside V, is zero, then the far field is also zero; this, as seen in Sections 3.3 and 3.6, implies that the angular spectrum is zero in the domain of propagating plane wave components. However, according to Section 4.4, this would imply also that the angular spectrum identically vanishes.

Also, from Eq. (3.2a) it can be easily seen (Devaney and Wolf [10.10]) that any nonradiating current distribution, contained in a volume V, can be represented by the current density:

$$\mathbf{j}_{NR}(\mathbf{r}) = \frac{c}{4\pi ik}[\nabla \times (\nabla \times \mathbf{E}(\mathbf{r})) - k^2\mathbf{E}(\mathbf{r})], \tag{10.38}$$

$\mathbf{E}(\mathbf{r})$ being the electric vector of a field that vanishes outside V. In addition, this field is unique, being the electric field generated by the current distribution (see Problem 10.2).

Expressions similar to Eq. (10.38) can be found from Eqs. (3.56) for the induced polarization $\mathbf{P}(\mathbf{r})$ or magnetization $\mathbf{M}(\mathbf{r})$ in a nonscattering scatterer (Problem 10.3).

Analogously, in the scalar theory, a nonradiating source distribution is found from Eq. (3.16) to be:

$$\rho_{NR}(\mathbf{r}) = -\frac{1}{4\pi}[\nabla^2 U(\mathbf{r}) + k^2 U(\mathbf{r})], \tag{10.39}$$

where $U(\mathbf{r})$ is an arbitrary field that vanishes outside V, and corresponds to the field produced by the distribution of Eq. (10.39).

From Eq. (10.13) it is straightforward to see (Devaney [10.22]) that, within the first Born approximation, a class of nonscattering potentials is formed by those whose three-dimensional Fourier transform is:

$$\tilde{F}(\mathbf{K}) = (|\mathbf{K}|^2 + 2\mathbf{k}_i \cdot \mathbf{K})\tilde{\Lambda}(\mathbf{K}), \tag{10.40}$$

where $\tilde{\Lambda}(\mathbf{K})$ is an entire function of exponential type (Problem 10.4).

Nonradiating sources and nonscattering scatterers are subjects of active current research (see, e.g. Refs. 10.10, 10.18, 10.22, and 10.23). Earlier work was done in investigations on the electromagnetic self-force and radiation reaction (Sommerfeld [10.46], Herglotz [10.47], Hertz [10.48], Schott [10.49], Bohm and Weinstein [10.50], and Goedecke [10.51]).

In previous chapters we have already seen examples of nonradiating distributions. For instance, Problem 3.7 discusses an example due to Schott [10.49], [10.51] (see also Problem 10.5). In addition, Problem 2.7 shows that a source distribution with spherical symmetry $\rho(\mathbf{r}) = \rho(r)$, localized in a sphere of radius a, radiates a field $U(\mathbf{r})$ in $r > a$ given by:

$$U(\mathbf{r}) = A(k)\frac{\exp(ikr)}{r}, \tag{10.41a}$$

$$A(k) = 4\pi \int_0^a \rho(r)\sin(kr)r\,dr. \tag{10.41b}$$

Therefore, even if we know that the source has spherical symmetry, we cannot infer its radial dependence $\rho(r)$, since at a given wavenumber k, we only know the complex constant $A(k)$ from the exterior field.

Also, if we attempt to recover $\rho(r)$ from the imaging equation (10.26) by considering the recording surface S to be a sphere of radius $b > a$, we obtain on introducing Eq. (10.41b) into Eq. (10.27) (Devaney and Sherman [10.52]):

$$\sum_{[j]}(\mathbf{r}) = \frac{A(k)}{k}\frac{\sin kr}{r}. \tag{10.42}$$

On the other hand, the left-hand side of Eq. (10.26) yields:

$$\int_V \rho(r')j_0(k|\mathbf{r} - \mathbf{r}'|)\, d^3r' = \frac{4\pi}{k}\frac{\sin kr}{r}\int_0^a \rho(r')\sin(kr')r'\, dr'. \tag{10.43}$$

Equations (10.42) and (10.43) show that the imaging equation (10.26) reduces again to Eq. (10.41b).

It is worth noting that if $\rho(r)$ is zero for $r > a$, then according to Eq. (10.41b) $A(k)$ can be regarded as the Fourier transform of $r\rho(r)$, $\rho(r)$ being an even function in $-\infty \leqslant r \leqslant \infty$. Consequently, due to the Paley–Wiener theorem (cf. Appendix 4.1) $A(k)$ given by Eq. (10.41b) is the limit for real k of an entire function of exponential type in the complex $k + ik'$-plane. Hence, it will have a numerable infinity of zeros at certain values $k_n + ik'_n$ ($n = 0, 1, 2, \ldots$). Therefore, if the Fourier transform of $\rho(r)r$ has real zeros k_m ($m = 0, 1, \ldots, M$, M may be either a finite or infinite integer), then $A(k)$ will be zero at those values k_m of k. Thus: *Any spherically symmetric source distribution $\rho(r)$, localized to a sphere of radius a and such that the Fourier transform of $\rho(r)r$ ($\rho(r)$ being even in $-\infty \leqslant r \leqslant \infty$) possesses real zeros k_m ($m = 0, 1, \ldots, M$, M being an integer), is nonradiating at the values of $k = k_m$.*

The examples of Problems 3.7 and 10.5 are just particular instances of the above statement. In the case of spherical symmetry the sphere $|\mathbf{k}^{(\pm)}|^2 = \omega^2/c^2$ acquires the form $k^2 = \omega^2/c^2$ and thus the lack of uniqueness of the inversion at fixed k is a consequence of the discussion of Section 10.3. The existence of values of k at which the source could be nonradiating is thus linked to that of functions whose Fourier transform has real zeros. (The interested reader can find a discussion and references of entire functions of exponential type with real zeros, e.g. in Sections 8.2 and 8.3 of Ref. 10.53 and in Ref. 10.54.)

As another example, it is well known from potential theory (see, e.g. Stratton [10.55]) that one can replace a source distribution in a volume by a single layer–double layer distribution over any surface enclosing that volume such that each produces the same exterior field. The difference between these distributions then constitutes a nonradiating source (Bleistein and Cohen [10.18]). (See Problems 10.6 and 10.7 for additional examples.)

10.6.1. Eigenfunction Analysis

The question of radiating and nonradiating sources can also be analyzed in the context of eigenfunctions of the imaging operator contained in the left-hand side of Eq. (10.26). These eigenfunctions are orthonormal and satisfy for each eigenvalue λ_n the equation (Bleistein and Cohen [10.18]):

$$\int_V \varphi_n(\mathbf{r}') j_0(k|\mathbf{r} - \mathbf{r}'|)\, d^3r' = \lambda_n \varphi_n(\mathbf{r}). \tag{10.44}$$

By expanding the imaging field, Eq. (10.27), and the source $\rho(\mathbf{r})$ in terms of these eigenfunctions:

$$\sum_{[j]} (\mathbf{r}) = \sum_{n=0}^{\infty} b_n \varphi_n(\mathbf{r}), \tag{10.45}$$

$$\rho(\mathbf{r}) = \sum_{n=0}^{\infty} a_n \varphi_n(\mathbf{r}), \tag{10.46}$$

we obtain from Eqs. (10.26), (10.27), (10.44)–(10.46):

$$b_n = a_n \lambda_n. \tag{10.47}$$

Since, as discussed in Section 9.8, the eigenvalues λ_n decrease sharply to zero for n larger than a certain $n = \mathcal{N}$, the corresponding coefficients b_n are zero and, therefore, for the part $\rho_{NR}(\mathbf{r})$ of $\rho(\mathbf{r})$ represented by the expansion (Eq. (10.46)) with $n > \mathcal{N}$ the following equation is satisfied:

$$\int_V \rho_{NR}(\mathbf{r}') j_0(k|\mathbf{r} - \mathbf{r}'|)\, d^3r' = 0. \tag{10.48}$$

Namely, nonradiating source distributions $\rho_{NR}(\mathbf{r})$ are associated with null eigenvalues λ_n $(n > \mathcal{N})$, and their corresponding eigenfunctions of the imaging operator are expressed in Eq. (10.44).

On the other hand, denoting as $\rho_R(\mathbf{r})$ the part of $\rho(\mathbf{r})$ given by the expansion (Eq. (10.46)) with $n \leqslant \mathcal{N}$, one evidently has that $\rho_R(\mathbf{r})$ is radiating and it is associated with nonzero eigenvalues. As a consequence, we have decomposed $\rho(\mathbf{r})$ as:

$$\rho(\mathbf{r}) = \rho_R(\mathbf{r}) + \rho_{NR}(\mathbf{r}). \tag{10.49}$$

Also, from Eq. (10.48), or its equivalent for $n > \mathcal{N}$ in Eq. (10.44), we conclude that nothing can be known about $\rho_{NR}(\mathbf{r})$ from one experiment.

Porter and Devaney [10.38] have shown that the decomposition, Eq. (10.49), leads to a unique solution of the inverse source problem when one knows that the source distribution $\rho(\mathbf{r})$ has minimum energy. In fact, the energy E of the source is:

$$E = \int_V |\rho(\mathbf{r})|^2 \, d^3r. \tag{10.50}$$

On introducing Eq. (10.49) into (10.50) we obtain:

$$E = \int_V [|\rho(\mathbf{r})|^2 + |\rho_{NR}(\mathbf{r})|^2 + \rho_{NR}(\mathbf{r})\rho_R^*(\mathbf{r}) + \rho_{NR}^*(\mathbf{r})\rho_R(\mathbf{r})] \, d^3r. \tag{10.51}$$

However, since the eigenfunctions φ_n are orthogonal, we have:

$$\int_V \rho_{NR}^*(\mathbf{r})\rho_R(\mathbf{r}) \, d^3r = 0. \tag{10.52}$$

Therefore:

$$\int_V |\rho_R(\mathbf{r})|^2 \, d^3r < \int_V [|\rho_{NR}(\mathbf{r})|^2 + |\rho_R(\mathbf{r})|^2] \, d^3r = E. \tag{10.53}$$

Equation (10.53) shows that $\rho_R(\mathbf{r})$ is the minimum energy solution to the inverse source problem.

Since the backpropagator $j_0(k|\mathbf{r} - \mathbf{r}'|)$ satisfies the homogeneous Helmholtz equation, it is evident from Eq. (10.44) that the eigenfunctions $\varphi_n(\mathbf{r})$ also satisfy:

$$\lambda_n(\nabla^2 + k^2)\varphi_n(\mathbf{r}) = 0. \tag{10.54}$$

Therefore, for $\lambda_n \neq 0$ these eigenfunctions also satisfy the Helmholtz equation. Since, as mentioned above, $\rho_R(\mathbf{r})$ is represented by the eigenfunctions $\varphi_n(\mathbf{r})$ associated with nonvanishing eigenvalues λ_n, i.e., with $n < \mathcal{N}$ (cf. Eq. (10.46)), the radiating distributions also satisfy the homogeneous Helmholtz equation, namely:

$$(\nabla^2 + k^2)\rho_R(\mathbf{r}) = 0. \tag{10.55}$$

Observe that since an image is formed in a source-free region, Eq. (10.55) is also a necessary condition for the imaging field $\sum_{[j]}(\mathbf{r})$ of Eq. (10.27) to yield a perfect image of the source in the absence of noise (Porter and Devaney [10.38]). Hence, these sources ρ_R are determined by solving Eq. (10.44) for $\varphi_n(\mathbf{r})$ and λ_n; then, knowing also the imaging field $\sum_{[j]}(\mathbf{r})$, the coefficients of the expansion (Eq. (10.45)) are found from the projections of $\sum_{[j]}(\mathbf{r})$ on to the wavefunctions $\varphi_n(\mathbf{r})$:

$$b_n = \int_V \sum_{[j]} (\mathbf{r}) \varphi_n(\mathbf{r}) \, d^3r. \tag{10.56}$$

Finally, the coefficients a_n that characterize the source through Eq. (10.46) are determined from those b_n and the eigenvalues λ_n by means of Eq. (10.47).

10.7. RECONSTRUCTION USING FOURIER SERIES

As shown in previous sections, the reconstruction of the three-dimensional structure of a source distribution, or scattering potential, is generally not possible from one experiment. Only in the case in which the source is a perfect reflector, can a single experiment permit the reconstruction of its boundary surface $D(x, y, z) = 0$ [10.56]. Demonstrations of perfectly conducting rough surface and grating reconstruction from one set of scattering data have been given in Refs. 10.57–10.59; however, the calculation difficulties have made this possible so far only within the Kirchhoff approximation or the Rayleigh method.

Under the first Born and Rytov approximations there are computational procedures based on the model originally proposed by Wolf [10.8]. The field is given at a certain plane $z = z_0$, z_0 being a constant; the angular spectrum is then obtained from Eqs. (10.11) or (10.12), thus rendering the three-dimensional Fourier transform of the potential at the Ewald sphere, Eq. (10.13).

Assuming that the potential $F(\mathbf{r})$ is included within a sphere of radius R, it is expanded into a Fourier series:

$$F(\mathbf{r}) = \sum_{l,m,n=0}^{\infty} \tilde{F}_{l,m,n} \exp\left[i \frac{\pi}{R} (lx + my + nz) \right], \tag{10.57}$$

where the Fourier coefficients are:

$$\tilde{F}_{l,m,n} = \tilde{F}\left(\frac{l\pi}{R}, \frac{m\pi}{R}, \frac{n\pi}{R} \right), \tag{10.58}$$

$\tilde{F}(\mathbf{k})$ being the three-dimensional Fourier transform of $F(\mathbf{r})$, and l, m, and n being integers; namely, from Eqs. (10.10) and (10.58) we have for the wavevector transfer:

$$\mathbf{K}^{(\pm)} \equiv \mathbf{k}^{(\pm)} - \mathbf{k}_i = \left(\frac{l\pi}{R}, \frac{m\pi}{R}, \frac{n\pi}{R}\right). \tag{10.59}$$

However, $\mathbf{K}^{(\pm)}$ is in the limiting Ewald sphere (Fig. 10.1) (see, e.g. Ref. 10.60) and, according to Eq. (10.14) as $k = \omega/c$ varies, one has [10.33], [10.61]:

$$|\mathbf{K}^{(\pm)}| = 2k, \tag{10.60}$$

namely, from Eq. (10.59), the integers l, m, and n satisfy:

$$\frac{\pi^2}{R^2}(l^2 + m^2 + n^2) \leqslant 4k^2. \tag{10.61}$$

For instance, making $\mathbf{K}^{(\pm)}$ parallel to the x-axis one can determine the Fourier coefficient $F(l\pi/R, 0, 0)$ for l taking values in the interval $(-2kR/\pi, 2kR/\pi)$ (Devaney [10.33]). This is done by performing a series of $(4kR/\pi) + 1$ experiments, varying the magnitude of the $\mathbf{K}$-vector in discrete steps from $-2k$ to $2k$, with a sampling interval equal to π/R. By changing the orientation of $\mathbf{K}^{(\pm)}$ one can obtain the other Fourier coefficients $F_{l,m,n}$. This renders a Fourier series (Eq. (10.57)) truncated to the cubic grid each of whose sides is included in the interval $(-2kR/\pi, 2kR/\pi)$.

Reconstructions of simple potential functions using this procedure have been done by Carter and Ho [10.11]–[10.13], and by Fercher *et al.* [10.14]. However, as pointed out by these researchers, this method involves great experimental problems in accurately collecting the data from different sets of measurements, then making necessary extrapolations of these data, an operation to which this reconstruction algorithm is very sensitive. Other related methods proposed [10.62] have not been actually implemented in experiments.

10.8. DIFFRACTION TOMOGRAPHY

In this section we address the method for reconstructing three-dimensional weak scattering potentials known as *diffraction tomography* (Devaney [10.24], [10.25], [10.63], [10.64]). This procedure applies under the first Born and Rytov approximations.

This technique is based on the *filtered backpropagation algorithm*. As will be seen, an interesting feature of this method is that it represents a generalization to weak scatterers of the well-known *filtered backprojection algorithm* used under the geometrical optics approximation in *computed tomography* [10.1]–[10.3]. Moreover, it constitutes an improvement over previous procedures [10.65]–[10.68] that use the Fourier series approach outlined in Section 10.7. In this connection, there are also current techniques of three-dimensional structure reconstruction from data on the Ewald sphere in X-ray diffraction [10.69], [10.70], or electron microscopy [10.71]. Filtered backprojection algorithms are also extensively used in image recovery [10.72].

In order to establish the basic concepts used in diffraction tomography, it is convenient to review first the essentials of the *central slice theorem* (see, e.g. Ref. 10.3) and computed tomography.

10.8.1. The Central Slice Theorem

Theorem. *Let $\tilde{F}(K_x, K_y, K_z)$ be the three-dimensional Fourier transform of a function $F(x, y, z)$, then the two-dimensional slice $\tilde{F}(K_x, 0, K_z)$ is the two-dimensional Fourier transform in the x and z variables of the projection of $(1/2\pi)F(x, y, z)$ along the y-axis, namely, of $\int (1/2\pi)F(x, y, z)\, dy$.* ■

The proof of this theorem is straightforward. Let us write $\tilde{F}(K_x, K_y, K_z)$ as:

$$\tilde{F}(K_x, K_y, K_z) = \frac{1}{(2\pi)^3}\int_{-\infty}^{\infty} F(\mathbf{r}) \exp(-i\mathbf{k}\cdot\mathbf{r})\, d^3r, \tag{10.62}$$

where $\mathbf{r} = (x, y, z)$ and $\mathbf{k} = (k_x, k_y, k_z)$.

Making $K_y = 0$ in Eq. (10.62) we obtain:

$$\tilde{F}(K_x, 0, K_z) = \frac{1}{(2\pi)^2}\iint_{-\infty}^{\infty} dx\, dz \exp[-i(K_x x + K_z z)]\left[\frac{1}{2\pi}\int_{-\infty}^{\infty} dy F(\mathbf{r})\right]. \tag{10.63}$$

Equation (10.63) constitutes the proof of the theorem.

This result can be generalized for any other arbitrary two-dimensional projection of $F(\mathbf{r})$. Let us rotate the system of coordinates (x, y, z) to another Cartesian system (ξ, ζ, η), so that the η-axis is characterized in the (x, y, z) system by the polar and azimuthal angles θ and ϕ, respectively (Fig. 10.3(a)); then, the projection:

$$P_{\theta,\phi}(\xi, \zeta) = \frac{1}{2\pi}\int_{-\infty}^{\infty} F(x, y, z)\, d\eta \tag{10.64}$$

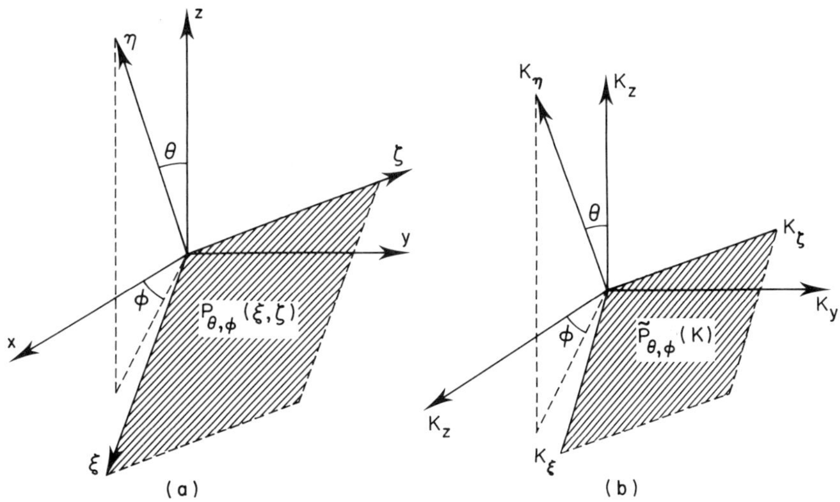

Figure 10.3. (a) Geometry for defining the projections $P_{\theta,\phi}(\xi, \zeta)$ of the three-dimensional function $F(x, y, z)$ along an arbitrary direction η. The Cartesian system of coordinates ξ, ζ, η is rotated with respect to the x, y, z system with η characterized in the x, y, z coordinates by the polar angle θ and the azimuthal angle ϕ. (b) Slice $\tilde{P}_{\theta,\phi}(K)$ given by the two-dimensional Fourier transform of $P_{\theta,\phi}(\xi, \zeta)$. This slice is contained in a plane normal to the direction characterized by the angles θ and ϕ in the K_x, K_y, K_z system of coordinates.

has a two-dimensional Fourier transform given by the slice in the (K_x, K_y, K_z)-plane (Fig. 10.3(b)):

$$\tilde{P}_{\theta,\phi}(|K|) \equiv \tilde{P}_{\theta,\phi}(K_\xi, K_\zeta) = \frac{1}{(2\pi)^2} \iint_{-\infty}^{\infty} P_{\theta,\phi}(\xi, \zeta) \exp[-i(K_\xi \xi + K_\zeta \zeta)]\, d\xi\, d\zeta, \tag{10.65}$$

where $|K| = \sqrt{K_\xi^2 + K_\zeta^2}$.

By successively rotating the system of coordinates (ξ, ζ, η), the projections $P_{\theta,\phi}(\xi, \zeta)$ yield by Fourier transformation a set of slices $\tilde{P}_{\theta,\phi}(|K|)$ of the three-dimensional Fourier transform $\tilde{F}(K_x, K_y, K_z)$. The complete three-dimensional structure of $F(\mathbf{r})$ can then be obtained from this set of slices by means of the inverse three-dimensional Fourier transform, which in spherical coordinates reads:

$$F(x, y, z) = \int_0^{2\pi} d\phi \int_0^{\pi} d\theta \sin\theta \int_0^{M} d|K||K|^2 \tilde{P}_{\theta,\phi}(|K|) \times \exp[i|K|(x \sin\theta \cos\phi + y \sin\theta \sin\phi + z \cos\theta)]. \tag{10.66}$$

In the representation (Eq. (10.66)) the $|K|$-integration is restricted to the maximum bound M of the accessible Fourier space.

10.8.2. Principles of Computed Tomography

In conventional transmission computed tomography [10.1]–[10.3] the field propagates rectilinearly across the object, following the laws of geometrical optics. Usually a plane parallel beam impinges upon the sample so that a section of the object in the beam plane is illuminated. Typically, an absorption coefficient $f(x, y, z)$ acting upon the passing radiation is to be determined. In this way, if the plane $z = a$ contains the incident beam, and hence the illuminated section of the object, the function to be determined is $F(x, y) = f(x, y, a)$. By continuously varying the value of a one obtains a set of sections that constitutes a sampling of the three-dimensional function $f(x, y, z)$.

If the beam propagates parallel to the ζ-axis in a Cartesian system of coordinates (ξ, ζ), which is rotated at an angle α with respect to the (x, y)-system (Fig. 10.4), one records the projections:

$$F_\alpha(\xi) = \int_{-\infty}^{\infty} F(x, y)\, d\zeta. \tag{10.67}$$

The one-dimensional Fourier transform $\tilde{F}_\alpha(K)$ of $F_\alpha(\xi)$:

$$\tilde{F}_\alpha(K) = \frac{1}{2\pi} \int_{-\infty}^{\infty} F_\alpha(\xi) \exp(-iK\xi)\, d\xi \tag{10.68}$$

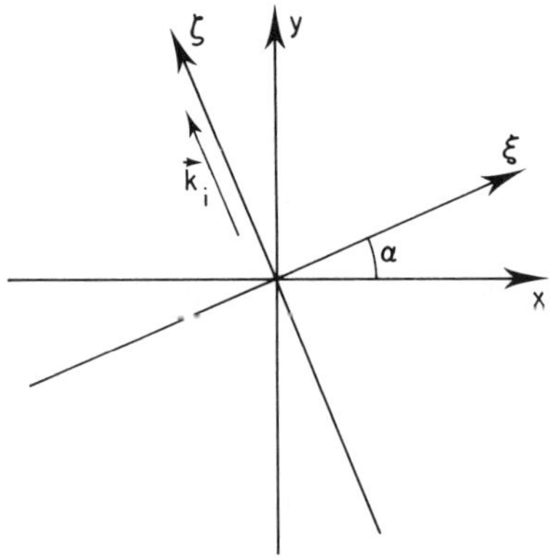

Figure 10.4. Rotation of the x, y-axes under an arbitrary angle α.

is a one-dimensional slice in the (K_x, K_y)-plane of the two-dimensional Fourier transform $\tilde{F}(K_x, K_y)$ of $F(x, y)$ in a direction at an angle α with the K_x-axis.

By varying α one obtains a set of projections $F_\alpha(\xi)$. To each projection there corresponds a slice $\tilde{F}_\alpha(K)$ of the two-dimensional Fourier transform $\tilde{F}(K_x, K_y)$. The function $F(x, y)$ is then reconstructed from this set of slices by inverse Fourier transformation, which in polar coordinates reads:

$$F(x, y) = \int_0^\pi d\alpha \int_{-M}^{M} dK|K|\tilde{F}_\alpha(K) \exp[iK(x \cos \alpha + y \sin \alpha)], \quad (10.69)$$

where K is defined in the accessible interval of data $(-M, M)$ in the Fourier space.

By defining the filter:

$$g(t) = \int_{-M}^{M} dK|K| \exp(-iKt), \quad (10.70)$$

Equation (10.69) may be expressed as:

$$F(x, y) = \int_0^\pi d\alpha[g(x \cos \alpha + y \sin \alpha) * F_\alpha(x \cos \alpha + y \sin \alpha)], \quad (10.71)$$

where the convolution $g(t) * F_\alpha(t)$ is defined as follows:

$$T_\alpha(t) = g(t) * F_\alpha(t) = \int_{-\infty}^{\infty} g(t - t')F_\alpha(t')\, dt'. \quad (10.72)$$

The functions $T_\alpha(x \cos \alpha + y \sin \alpha)$ are known as the *filtered projections* of $F(x, y)$. In terms of these functions, $F(x, y)$, given by Eq. (10.71), may be represented as:

$$F(x, y) = \int_0^\pi T_\alpha(x \cos \alpha + y \sin \alpha)\, d\alpha. \quad (10.73)$$

Equations (10.72) and (10.73) underlie the *filtered backprojection algorithm* for reconstructing a two-dimensional function $F(x, y)$ in computed tomography. In the next sections we shall discuss a generalization of this procedure to weak scatterers known as the *filtered backpropagation algorithm* (Devaney [10.63]), which constitutes the main tool of diffraction tomography, and we shall show that in the geometrical optics limit it reduces to the filtered backprojection method.

10.8.3. The Basic Equation of Diffraction Tomography

We shall consider scalar fields to illustrate the approach. Let $U^{(i)}(\mathbf{r})$ be a plane wave propagating along the z-axis with wavevector $\mathbf{k}_i = (0, 0, k)$ and constant amplitude A, incident on the scatterer, i.e.:

$$U^{(i)}(\mathbf{r}) = A \exp(ikz). \tag{10.74}$$

We assume that the potential $F(\mathbf{r})$ to be reconstructed occupies a volume V (Fig. 10.2), and the scattered field is determined by some recording means at a plane $z = z_0$ of $\mathscr{R}^+$, z_0 being constant.

According to Section 3.10.1, in the first Born approximation the scattered field U_B is related to the three-dimensional Fourier transform $\tilde{F}(\mathbf{k})$ of $F(\mathbf{r})$ by:

$$U_B(\mathbf{R}, z_0) = -i\pi A \iint_{-\infty}^{\infty} \tilde{F}(\mathbf{K}, k_z - k) \exp(i\mathbf{K}\cdot\mathbf{R}) \exp(ik_z z_0) \frac{d^2K}{k_z}. \tag{10.75}$$

By a two-dimensional Fourier transformation in $\mathbf{R}$ one obtains from Eq. (10.75) the basic equation of diffraction tomography:

$$\frac{ik_z}{4\pi^3 A} \exp(-ik_z z_0) \iint_{-\infty}^{\infty} U_B(\mathbf{R}, z_0) \exp(-i\mathbf{K}\cdot\mathbf{R})\, d^2R = \tilde{F}(\mathbf{K}, k_z - k). \tag{10.76}$$

On the other hand, according to Eq. (10.74) and Eqs. (3.133) and (3.134), the scattered field is expressed under the Rytov approximation by:

$$U_R(\mathbf{r}) = A \exp(ik_i z) \exp[A \exp(-ik_i z) U_B(\mathbf{r})]; \tag{10.77}$$

where $U_B(\mathbf{r})$ is given by Eq. (10.75). Thus, since the Born field U_B determines the Rytov field U_R through the complex phase of Eq. (10.77), the procedure to obtain the Fourier transform $\tilde{F}(\mathbf{K}, k_z - k)$ of the scattering potential within the Rytov approximation is straightforward by expressing U_B in terms of the complex phase of U_R. We shall therefore concentrate on Eq. (10.76) only; this is the basic expression for the reconstruction of $F(\mathbf{r})$. For instance, by particularizing to $K_y = 0$ we obtain:

$$\frac{ik_z}{4\pi^3 A} \exp(-ik_z z_0) \int_{-\infty}^{\infty} dy \int_{-\infty}^{\infty} dx U_B(x, y, z_0) \exp(-iK_x x) = \tilde{F}(K_x, 0, k_z - k). \tag{10.78}$$

The right-hand side of Eq. (10.78) gives the three-dimensional Fourier transform $\tilde{F}(K_x, K_y, K_z)$ of $F(x, y, z)$ evaluated on the slice $K_y = 0$ and for values of $K_z = k_z - k$. If this Fourier transform were known for all values of K_z, then, according to the central slice theorem, the function $\tilde{F}(K_x, 0, K_z)$ would provide by two-dimensional inverse Fourier transformation the projection $(1/2\pi)\int_{-\infty}^{\infty} F(\mathbf{r})\,dy$. By successively rotating the direction of the y-axis, one would then obtain a set of slices of $\tilde{F}(K_x, K_y, K_z)$ as shown in Section 10.8.1, which would allow the determination of $F(x, y, z)$ by means of Eqs. (10.64)–(10.66).

The problem faced in diffraction tomography is, therefore, the reconstruction of $\tilde{F}(K_x, 0, K_z)$ (or equivalently, of the projection $(1/2\pi)\int_{-\infty}^{\infty} F(\mathbf{r})\,dy$) from the one-dimensional slices $\tilde{F}(K_x, 0, k_z - k)$ at $K_z = k_z - k$, namely, on the circle $|\mathbf{K}|^2 = K^2 + (k_z - k)^2$, $\mathbf{K}$ being the wavevector transfer (cf. Eq. (10.13)). This is achieved by successively rotating the wavevector of the incident field.

10.8.4. The Filtered Backpropagation Equation

Let us denote by $G(x, z)$ the projection $(1/2\pi)\int_{-\infty}^{\infty} F(\mathbf{r})\,dy$, and by $\tilde{G}(K_x, K_z)$ its two-dimensional Fourier transform. In other words, $\tilde{G}(K_x, K_z) = \tilde{F}(K_x, 0, K_z)$. We then write Eq. (10.78) in terms of these functions as:

$$\tilde{G}(K_x, k_z - k) = \frac{\exp[-i(k_z - k)z_0]}{2\pi}\int_{-\infty}^{\infty} M(x)\exp(-iK_x x)\,dx, \qquad (10.79)$$

where we have introduced the function:

$$M(x) = \frac{ik_z}{\pi A}\exp(-ikz_0)U_B^P(x, z_0), \qquad (10.80)$$

$U_B^P(x, z_0)$ being the projection:

$$U_B^P(x, z_0) = \frac{1}{2\pi}\int_{-\infty}^{\infty} U_B(x, y, z_0)\,dy. \qquad (10.81)$$

On introducing the Fourier representation of $\tilde{G}(K_x, K_z)$ in terms of $G(x, z)$ into the left-hand side of Eq. (10.79) we obtain:

$$\frac{1}{(2\pi)^2}\iint_{-\infty}^{\infty} G(x, z)\exp[-i(K_x x + (k_z - k)z)]\,dx\,dz$$
$$= \frac{\exp[-i(k_z - k)z_0]}{2\pi}\int_{-\infty}^{\infty} M(x)\exp(-iK_x x)\,dx. \qquad (10.82)$$

Equation (10.82) can be generalized for any arbitrary direction of propagation of the incident field. Choosing this direction along the η-axis in a Cartesian system of coordinates (ξ, η) rotated at an angle α with respect to the system (x, z), Eq. (10.82) generalizes to:

$$\frac{1}{(2\pi)^2}\iint_{-\infty}^{\infty} G(\xi, \eta)\exp[-i(K\xi + (\sqrt{k^2 - K^2} - k)\eta)]\, d\xi\, d\eta$$
$$= \frac{\exp[-i(\sqrt{k^2 - K^2} - k)\eta_0]}{2\pi}\int_{-\infty}^{\infty} M_\alpha(\xi)\exp(-iK\xi)\, d\xi, \quad (10.83)$$

where now the projection $M_\alpha(\xi)$ is:

$$M_\alpha(\xi) = \frac{i\sqrt{k^2 - K^2}}{\pi A}\exp(-ik\eta_0)\int_{-\infty}^{\infty} U_B^\alpha(\xi, y, \eta_0)\, dy, \quad (10.84)$$

$U_B^\alpha(\xi, y, \eta_0)$ now being the scattered field $U_B(x, y, z)$ expressed in terms of the (ξ, y, η)-coordinates and evaluated at $\eta = \eta_0$.

Let us denote by $\tilde{M}_\alpha(K)$ the one-dimensional Fourier transform of the projection $M_\alpha(\xi)$ that appears in the integrand of the right-hand side of Eq. (10.83). By varying α, the functions $\tilde{M}_\alpha(K)$ create a set of slices of the two-dimensional Fourier transform $\tilde{G}(K_x, K_z)$ of $G(x, y)$. Thus, this set gives $G(x, y)$ through inverse Fourier transformation:

$$G(x, z) = \iint_{-\infty}^{\infty} \tilde{M}_\alpha(K)\exp[-i(\sqrt{k^2 - K^2} - k)\eta_0]$$
$$\times \exp[i(\mathsf{K}_x x + \mathsf{K}_z z)]\, d\mathbf{K}_x\, d\mathbf{K}_z, \quad (10.85)$$

K_x and K_z being the components of the wavevector transfer $\mathbf{K} = \mathbf{k} - \mathbf{k}_i$, which may be expressed in terms of K and α as (Fig. 10.4):

$$\mathsf{K}_x = K + k\sin\alpha, \quad (10.86a)$$

$$\mathsf{K}_z = \sqrt{k^2 - K^2} - k\cos\alpha. \quad (10.86b)$$

From Eqs. (10.86) we obtain:

$$d\mathsf{K}_x\, d\mathsf{K}_z = \left(k\sin\alpha + \frac{k|K|}{\sqrt{k^2 - K^2}}\cos\alpha\right) dK\, d\alpha. \quad (10.87)$$

On introducing Eqs. (10.86) and (10.87) into Eq. (10.85) we obtain:

$$G(x, z) = \frac{1}{2}\int_{-\pi}^{\pi} d\alpha \int_{0}^{k} dK\left(k \sin \alpha + \frac{k}{\sqrt{k^2 - K^2}}|K| \cos \alpha\right)\tilde{M}_\alpha(K) \times \exp[i[(K + k \sin \alpha)x + (\sqrt{k^2 - K^2} - k \cos \alpha)(z - \eta_0)]]. \tag{10.88}$$

Observe that in Eq. (10.88) k is the maximum value of K since in the reconstruction process we are considering only source-free fields. On the other hand, the representation (Eq. (10.88)) is also valid by expressing the integrand in the system of coordinates (ξ, η) instead of (x, y). In the system (ξ, η) the wavevector $\mathbf{k}_i$ of the incident wave is along the η-axis and $k \cos \alpha = k, k \sin \alpha = 0$. It is convenient to integrate in K, establishing a lower limit $-k$ instead of zero, so that the K-integration leads to the filtering operator of Eq. (10.70); this requires one to divide the integral by 2, since then there will be double coverage of the integration domain (Devaney [10.63]). Therefore, we can express Eq. (10.88) as:

$$G(x, z) = \frac{k}{2}\int_{-\pi}^{\pi} d\alpha \int_{-k}^{k} dK \frac{|K|}{\sqrt{k^2 - K^2}} \tilde{M}_\alpha(K) \times \exp[i(K\xi + (\sqrt{k^2 - K^2} - k)(\eta - \eta_0))]. \tag{10.89}$$

Equation (10.89) is the *filtered backpropagation equation*; it constitutes a generalization for diffraction tomography of Eq. (10.69).

Recalling Eq. (10.70) and using the well-known theorem for Fourier transforms of convolutions, as well as the associative property of this operation, Eq. (10.89) can also be expressed as:

$$G(x, z) = \int_{-\pi}^{\pi} d\alpha \left[\left[g(\xi) * \mathscr{F}^{-1}\left[\frac{k}{2}\frac{\tilde{M}_\alpha(K)}{\sqrt{k^2 - K^2}}\right]\right] * \mathscr{F}^{-1}[\exp[i(\sqrt{k^2 - K^2} - k)(\eta - \eta_0)]]\right]. \tag{10.90}$$

The asterisk in Eq. (10.90) denotes convolution in the ξ variable; $g(\xi)$ is the filter of Eq. (10.70). In addition:

$$\mathscr{F}^{-1}\left[\frac{k}{2}\frac{\tilde{M}_\alpha(K)}{\sqrt{k^2 - K^2}}\right] = \frac{k}{2}\int_{-\infty}^{\infty} \frac{\tilde{M}_\alpha(K)}{\sqrt{k^2 - K^2}} \exp(iK\xi)\, dK, \tag{10.91}$$

and:

$$H(\xi, \eta - \eta_0) \equiv \mathscr{F}^{-1}[\exp[i(\sqrt{k^2 - K^2} - k)(\eta - \eta_0)]]$$
$$= \int_{-\infty}^{\infty} \exp[i[K\xi + (\sqrt{k^2 - K^2} - k)(\eta - \eta_0)]]\, dK. \quad (10.92)$$

The quantities:

$$N_\alpha(\xi) = g(\xi) * \mathscr{F}^{-1}\left[\frac{k}{2}\frac{\tilde{M}_\alpha(K)}{\sqrt{k^2 - K^2}}\right] \quad (10.93)$$

are called the *filtered projections* of $G(x, y)$ (Devaney [10.63]), in analogy with Eq. (10.72).

On the other hand, the quantity:

$$T_\alpha(\xi, \eta) = N_\alpha(\xi) * H(\xi, \eta - \eta_0) = \int_{-\infty}^{\infty} H(\xi - \xi', \eta - \eta_0) N_\alpha(\xi')\, d\xi' \quad (10.94)$$

is the *filtered backpropagated field* at each angle of incidence α. In fact, it is clear upon comparison of Eq. (10.92) with (9.4), that the kernel $H(\xi, \eta - \eta_0)$ yields in Eq. (10.94) a field backpropagated into the half-space $\eta < \eta_0$, generated from the filtered projections $N_\alpha(\xi)$. Observe that the filter $g(\xi)$ prevents the existence of evanescent components in this field in analogy with the discussion of Section 10.5.

By introducing Eqs. (10.93) and (10.94) into Eq. (10.90), and using the expressions for the transformation of coordinates under a rotation, we finally obtain the function $G(x, y)$ by angularly superimposing the filtered backpropagated fields $T_\alpha(\xi, \eta)$; namely:

$$G(x, z) = \int_{-\pi}^{\pi} T_\alpha(x \sin\alpha - y\cos\alpha, x\cos\alpha + y\sin\alpha)\, d\alpha. \quad (10.95)$$

In practice, Eq. (10.95) is approximated by a finite sum over the available range of angles of incidence α. This limits the resolution of the reconstruction.

Diffraction tomography has found use in ultrasonic imaging (Devaney [10.25], [10.64]) and seismic profiling (Devaney [10.73]). The filtered backpropagation algorithm has been further generalized [10.74] to recordings of the scattered field over arbitrary surfaces. An alternative procedure known as *holographic tomography*, which is of use for arbitrary receiver geometries, has been developed by Porter [10.75] (see also Ref. 10.16). In this technique the scattered field and its normal derivative are recorded over an arbitrary surface surrounding the object upon illumination under several angles, then the reconstruction process is performed on the basis of Eqs. (10.31) and (10.32). For detection on planar screens, the difference between this method and the diffraction tomography discussed above stems from the different filter used, which in the case of holographic tomography arises from a mapping of the circles of recording into Cartesian coordinates. Other backpropagation methods have been applied to seismic data [10.76].

10.8.5. The Geometrical Optics Limit

In the limit of geometrical optics one has $k_z \to k$, and therefore, after generalizing Eq. (10.78) to any direction of propagation of the incident wave by rotating the x, z axes at an angle α, Eq. (10.78) becomes in this limit:

$$\frac{ik}{\pi A} \exp(-ik\eta_0) \int_{-\infty}^{\infty} d\xi \exp(-iK_x\xi) \int_{-\infty}^{\infty} dy U_B^{\alpha}(\xi, y, \eta_0)$$
$$= \iint_{-\infty}^{\infty} F_y(\xi, \eta) \exp(-iK_x\xi)\, d\xi\, d\eta. \quad (10.96)$$

Equation (10.96) may be expressed in the form:

$$\int_{-\infty}^{\infty} M_\alpha(\xi) \exp(-iK_x\xi)\, d\xi = \iint_{-\infty}^{\infty} F_y(\xi, \eta) \exp(-iK_x\eta)\, d\xi\, d\eta. \quad (10.97)$$

In Eq. (10.97), $M_\alpha(\xi)$ is given by an expression identical to Eq. (10.84) with $\sqrt{k^2 - K^2}$ replaced by k. Equation (10.97) also shows that $M_\alpha(\xi)$ is equal to the projection of $F_y(\xi, \eta)$ along η, i.e.:

$$M_\alpha(\xi) = \int_{-\infty}^{\infty} F_y(\xi, \eta)\, d\eta, \quad (10.98)$$

then $F_y(\xi, \eta)$ can be reconstructed by using the central slice theorem. Namely, one gets from Eq. (10.98):

$$F_y(\xi, \eta) = \int_0^{\pi} d\alpha \int_{-W}^{W} \tilde{M}_\alpha(K) \exp[iK(\xi \cos\alpha + \eta \sin\alpha)]|K|\, dK, \quad (10.99)$$

where:

$$\tilde{M}_\alpha(K) = \int_{-\infty}^{\infty} M_\alpha(\xi) \exp(-iK\xi)\, d\xi. \quad (10.100)$$

The interval of integration $(-W, W)$, $W < k$, in Eq. (10.99) is the finite bandwidth available in practice.

In fact by introducing the filter function $g(t)$, as in Eq. (10.70), with the K-integration limited to the interval $(-W, W)$, one can show as discussed in Section 10.8.2 that:

$$F_y(\xi, \eta) = \frac{1}{2\pi} \int_0^{\pi} N_\alpha(\xi \cos \alpha + \eta \sin \alpha)\, d\alpha, \tag{10.101}$$

where the filtered projection $N_\alpha(t)$ is:

$$N_\alpha(t) = \int_{-\infty}^{\infty} M_\alpha(\xi) g(t - \xi)\, d\xi, \tag{10.102}$$

$M_\alpha(\xi)$ being the projection of $F_y(\xi, \eta)$ along η, given by Eq. (10.84), with $\sqrt{k^2 - K^2}$ replaced by k.

Taking into account Eqs. (10.77) and (3.151), Eq. (10.101) leads to the method of conventional computed tomography [10.1]–[10.3] (cf. Eq. (10.73)). When the potential is a function of two variables, one has: $F_y(\xi, \eta) = F(\xi, \eta)$ and one does not need further operations.

10.8.6. An Example

As an illustration of the filtered backpropagation algorithm, the following figures due to Devaney [10.24] show reconstructions of a two-dimensional object, consisting of four overlapping disks, from several view angles.

Figure 10.5 corresponds to one, five, nine, and twenty-one angles. The wavelength employed is one unit, and the radius of the smallest circle is fifteen units; as seen, the quality of the reconstruction increases with the number of view angles. Figure 10.6 shows reconstructions of the same object from fifty-one view angles and wavelengths of one, three, five, and ten units; as the wavelength increases, the higher frequency loss involved in the filtering operation of the scattered field spectrum is more noticeable, and results in a decrease in resolution of the reconstruction.

A comparison of the filtered backpropagation algorithm, versus the filtered backprojection algorithm of conventional computed tomography, is shown in Fig. 10.7 for fifty-one view angles and wavelengths of one, three, five, and ten units. As expected, the geometrical optics approximation yields a poorer performance as the wavelength increases.

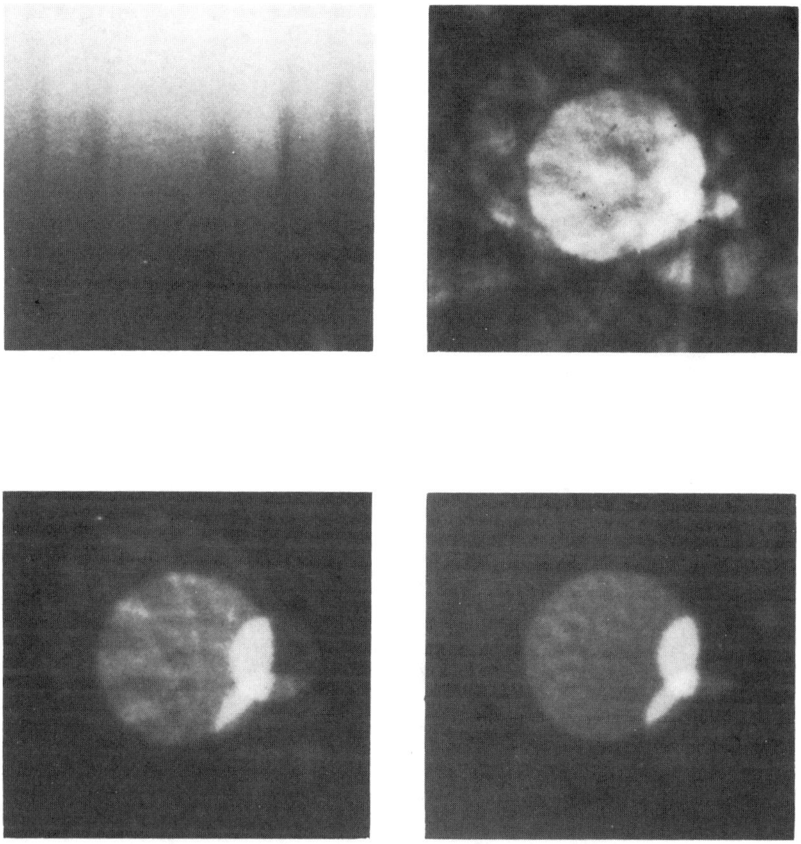

Figure 10.5. From left to right, top to bottom, reconstruction by filtered backpropagation for one, five, nine, and twenty-one view angles. The wavelength is one unit. (After A. J. Devaney, *Proc. S.P.I.E.* **358**, 10 (1982).)

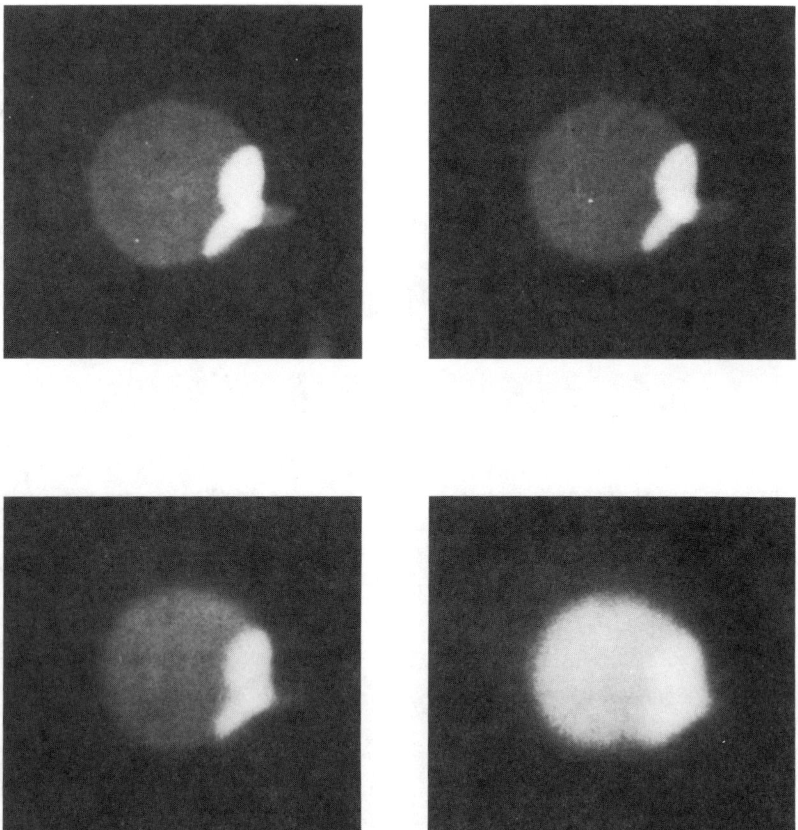

Figure 10.6. From left to right, top to bottom, reconstruction by filtered backpropagation for fifty-one view angles and wavelengths of one, three, five, and ten units. (After A. J. Devaney, *Proc. S.P.I.E.* **358**, 10 (1982).)

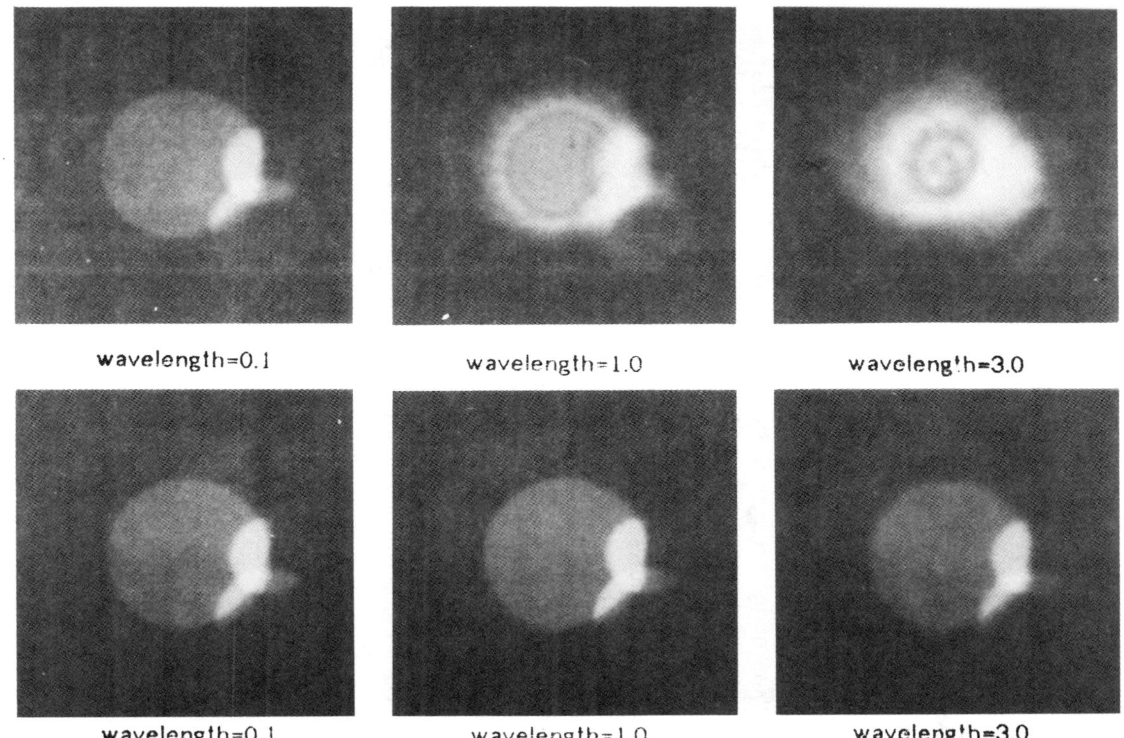

Figure 10.7. Filtered backpropagation (bottom) versus filtered backprojection (top) for fifty-one view angles and wavelengths of one-tenth, one, and three units. (After A. J. Devaney, *Proc. S.P.I.E.* **358**, 10 (1982).)

10.9. OTHER INVERSE SCATTERING METHODS. MULTIPLE SCATTERING APPROACHES

Optical reconstruction methods that make use of more general scattering theories than those discussed in this chapter involve multiple scattering models. These approaches are largely adapted from potential theory in quantum mechanics. A few of them will be quoted in this section; the reader can find a comprehensive bibliography, both for optics and electromagnetics, in Refs. 10.7, 10.27–10.31, and 10.77.

A method has been proposed, using the Shannon sampling theorem for the Fourier transform of the potential (Devaney and Wolf [10.78]); iterations are then carried out by means of the Neumann series. This iterative scheme is based on the inverse scattering procedure put forward by Moses [10.79] and Prosser [10.80] for inverting the Neumann–Born series of the direct scattering problem.

The inversion under stronger scattering interactions has been addressed [10.81] by means of the distorted-wave Born approximation [10.82], in which the scattering potential is decomposed into the sum of a known strong potential and an unknown weak perturbation.

Optimization procedures have been applied to refractive index reconstructions formulated as the minimization of a nonlinear functional with constraints [10.83].

The inverse scattering method of Gelfand–Levitan–Marchenko [10.84], [10.85] for the one-dimensional Schrödinger equation has been applied to the determination of the profile of the permittivity of a slab of dielectric stratified medium [10.86]. Refractive index profiles have also been reconstructed by means of other nonlinear methods [10.87], [10.88]. Techniques to overcome the nonlinearity of data inversion are under current investigation [10.89]–[10.91].

The question of reconstruction, and its uniqueness, for random sources was considered by La Haie [10.92] by extending the methods of generalized holography to field correlation functions.

It can be concluded that, in general, inverse scattering methods require first an appropriate solution technique to the corresponding *direct* problem. At present, while procedures for weak scatterers are established, allowing reconstructions in a rather systematic way, there are as yet no general theories for dealing with multiple scattering situations, the solution largely being connected with the nature of the specific problem. The quality of the reconstruction, therefore, will depend on how well the scattering potential satisfies the assumptions involved in the approximate model in use (see for instance [10.93]).

PROBLEMS

10.1. Derive from Eqs. (10.21) and (10.22) a vector form of the back-propagation equation, analogous in the electromagnetic formulation to the scalar equation (10.31). Use the formalism of Sections 10.4.1 and 1.6.2.

Write an equivalent equation by means of the Stratton–Chu integral of Problem 1.3.

10.2. Prove that the electric field associated with the nonradiating current distribution of Eq. (10.38) is unique.

10.3. Using Eqs. (3.56) write expressions for the induced polarization $\mathbf{P}(\mathbf{r})$ and the induced magnetization $\mathbf{M}(\mathbf{r})$ of a nonscattering scatterer, analogous to Eq. (10.38).

10.4. Prove that, under the first Born approximation, a class of nonscattering potentials is given by the expression:

$$F(\mathbf{r}) = -[\nabla^2 + 2ik\mathbf{n}_i \cdot \nabla]\Lambda(\mathbf{r}),$$

where $\Lambda(\mathbf{r})$ is a continuous function localized within a volume V, and $\mathbf{n}_i$ is a unit vector denoting the direction of propagation of an incident plane wave. (Devaney [10.22].)

10.5. Consider the spherically symmetric source distribution, orbiting with period T:

$$\rho(\mathbf{r}) = \frac{A}{r}\cos\omega_\rho r, \qquad \text{when } r \leqslant R,$$

$$\rho(\mathbf{r}) = 0, \qquad \text{when } r > R,$$

where A is a constant and $\omega_\rho = 2\pi/T$.

Show that the field radiated by this distribution is zero if $ka = 2\pi n$ and $\omega_\rho R = 2\pi l$, $l \neq n$, l, and n, being positive integers (Goedecke [10.51]).

10.6. Use Eqs. (10.41a) and (10.41b) to prove that the source distribution:

$$\rho_{NR}(\mathbf{r}) = 2\left[\frac{1}{r}j_0(kr) + \frac{dj_0(kr)}{dr}\right]\delta(r-a) - j_0(kr)\frac{d}{dr}\delta(r-a),$$

is nonradiating for any k (Porter [10.16]).

10.7. Use the spherical harmonic expansion of the Green function $G(\mathbf{r}, \mathbf{r}')$ of Problem 2.8 and the expression for the field radiated by a source distribution $\rho(\mathbf{r})$, localized in a volume V:

$$U(\mathbf{r}) = \int_V \rho(\mathbf{r}')G(\mathbf{r}, \mathbf{r}')\, d^3r',$$

then, expand $\rho(\mathbf{r})$ as:

$$\rho(\mathbf{r}) = \sum_{l=0}^{\infty} \sum_{m=-l}^{l} \rho_l^m(r) Y_l^m(\theta, \phi),$$

with:

$$\rho_l^m(r) = \int_{-\pi}^{\pi} d\phi \int_0^{\pi} d\theta \sin\theta \rho(\mathbf{r}) Y_l^{m*}(\theta, \phi).$$

Show that the source localized to the sphere $r \leqslant a$:

$$\rho_{NR}(\mathbf{r}) = \delta(\mathbf{r}) - \frac{j_0(kr)}{\int_0^a j_0^2(kr) r^2\, dr},$$

is nonradiating for any k (Bleistein and Cohen [10.18]).

REFERENCES

10.1. G. T. Herman, ed., *Image Reconstruction from Projections*, Topics in Applied Physics, Vol. 32, Springer-Verlag, Berlin, 1979.

10.2. A. C. Kak, *Proc. I.E.E.E.* **67**, 1245 (1979).

10.3. H. H. Barrett and W. Swindell, *Radiological Imaging: The Theory of Image Formation, Detection and Processing*, Academic Press, New York, 1981.

10.4. A. J. Devaney, *J. Math. Phys.* **20**, 1687 (1979).

10.5. B. J. Hoenders and H. P. Baltes, *J. Phys. A* **13**, 995 (1980).

10.6. H. P. Baltes, J. Geist and A. Walther, in *Inverse Source Problems in Optics*, H. P. Baltes, ed., Topics in Current Physics, Vol. 9, Springer-Verlag, Berlin, 1978, Chapter 5.

10.7. H. P. Baltes, ed., *Inverse Scattering Problems in Optics*, Topics in Current Physics, Vol. 20, Springer-Verlag, Berlin, 1980.

10.8. E. Wolf, *Opt. Comm.* **1**, 153 (1969).

10.9. A. T. Friberg and E. Wolf, *J. Opt. Soc. Am.* **73**, 26 (1983).

10.10. A. J. Devaney and E. Wolf, *Phys. Rev. D* **8**, 1044 (1973).

10.11. W. H. Carter, *J. Opt. Soc. Am.* **60**, 306 (1970).

10.12. W. H. Carter and P. C. Ho, *Appl. Opt.* **13**, 162 (1974).

10.13. P. C. Ho and W. H. Carter, *Appl. Opt.* **15**, 313 (1976).

10.14. A. F. Fercher, H. Bartelt, H. Becker and E. Wiltschko, *Appl. Opt.* **18**, 2427 (1979).

10.15. R. P. Porter, *Phys. Lett. A* **29**, 193 (1969); *J. Opt. Soc. Am.* **60**, 1051 (1970).

10.16. R. P. Porter, in *Progress in Optics*, Vol. 27, E. Wolf, ed., North-Holland, Amsterdam, 1989.

10.17. N. Bleistein and N. N. Bojarski, *Office of Naval Research NR 041-434/1-6-72* (available as: AD-A003 588, Natl. Tech. Inf. Serv., Springfield, VA, 1974).

10.18. N. Bleistein and J. K. Cohen, *J. Math. Phys.* **18**, 194 (1977).

10.19. N. N. Bojarski, Contract No. N00019-72-C-0462 (Department of the Navy, Naval Airsystems Command), 1972.

10.20. N. N. Bojarski, *J. Math. Phys.* **22**, 1647 (1981).

10.21. N. N. Bojarski, *I.E.E.E. Trans. Antenn. Propag.* **30**, 975 (1982).

10.22. A. J. Devaney, *J. Math. Phys.* **19**, 1526 (1978).

10.23. B. J. Hoenders, in *Inverse Source Problems in Optics*, H. P. Baltes, ed., Topics in Current Physics, Vol. 9, Springer-Verlag, Berlin, 1978, Chapter 2.

10.24. A. J. Devaney, *Proc. S.P.I.E.* **358**, 10 (1982).

10.25. A. J. Devaney, *I.E.E.E. Trans. Sonics Ultrasonics* **30**, 355 (1983).

10.26. K. Chadan and P. C. Sabatier, *Inverse Problems in Quantum Scattering Theory*, Springer-Verlag, Berlin, 1989, 2nd edition.

10.27. P. C. Sabatier, ed., *Applied Inverse Problems*, Lecture Notes in Physics, Vol. 85, Springer-Verlag, Berlin, 1978.

10.28. W. M. Boerner, A. K. Jordan and I. W. Kay, eds., *Inverse Methods in Electromagnetics, I.E.E.E. Trans. Antenn. Propag.* **29** (2) (1981).

10.29. A. J. Devaney, ed., *Inverse Problems in Propagation and Scattering, J. Opt. Soc. Am. A* **2** (11) (1985).

10.30. W. M. Boerner, ed., *Inverse Methods in Electromagnetic Imaging*, N.A.T.O. A.S.I. Series, Serie C, Vol. 143, Reidel, Dordrecht, 1985.

10.31. P. C. Sabatier, ed., *Some Topics on Inverse Problems*, World Scientific, Singapore, 1988.

10.32. R. Hosemann and S. H. Baghi, *Direct Analysis of Diffraction of Matter*, North-Holland, Amsterdam, 1962.

10.33. A. J. Devaney, in *Optics in Four Dimensions*, M. A. Machado and L. M. Narducci, eds., American Institute of Physics Conference Proceedings, Vol. 65, New York, 1981.

10.34. N. N. Bojarski, *Syracuse University Research Corporation*, SURC N SPL 67-3, AD-845 126, New York, 1967.

10.35. N. N. Bojarski, *Syracuse University Research Corporation*, SURC SPL TR 68-70, New York, 1968.

10.36. W. R. Stone, *Radio Sci.* **16**, 1029 (1981).

10.37. R. P. Porter and A. J. Devaney, *J. Opt. Soc. Am.* **72**, 327 (1982).

10.38. R. P. Porter and A. J. Devaney, *J. Opt. Soc. Am.* **72**, 1707 (1982).

10.39. A. J. Devaney and R. P. Porter, *J. Opt. Soc. Am. A* **2**, 2006 (1985).

10.40. L. Tsang, A. Ishimaru, R. P. Porter and D. Rouseff, *J. Opt. Soc. Am. A* **4**, 1783 (1987).

10.41. E. Wolf and R. P. Porter, *Radio Sci.* **21**, 627 (1986).

10.42. R. P. Porter, *J. Opt. Soc. Am.* **61**, 789 (1971).

10.43. B. A. Fuks, *Introduction to the Theory of Analytic Functions of Several Complex Variables*, Am. Math. Soc. Trans. Math. Monogr., Vol. 8, Providence, Rhode Island, 1963.

10.44. I. Manolitsakis, *J. Math. Phys.* **23**, 2291 (1982).

10.45. M. Nieto-Vesperinas, *J. Math. Phys.* **25**, 2109 (1984).

10.46. A. Sommerfeld, *Nachr. Akad. Wiss. Göttingen, Math. Phys. Kl. IIa Math.-Phys. Chem.* **1904**, 99 and 363; **1905**, 201.

10.47. G. Herglotz, *Math. Ann.* **65**, 87 (1908).

10.48. P. Hertz, *Math. Ann.* **65**, 1 (1908).

10.49. G. A. Schott, *Phil. Mag. Suppl.* 7 **15**, 752 (1933).

10.50. D. Bohm and M. Weinstein, *Phys. Rev.* **74**, 1789 (1948).

10.51. G. H. Goedecke, *Phys. Rev.* **135**, B281 (1964).

10.52. A. J. Devaney and G. C. Sherman, *I.E.E.E. Trans. Antenn. Propag.* **30**, 1034 (1982); *I.E.E.E. Trans. Antenn. Propag.* **30**, 1038 and 1041 (1982).

10.53. R. P. Boas, *Entire Functions*, Academic Press, New York, 1954.

10.54. L. de Brange, *Hilbert Spaces of Entire Functions*, Prentice-Hall, New York, 1968.

10.55. J. A. Stratton, *Electromagnetic Theory*, McGraw-Hill, New York, 1941.

10.56. A. G. Ramm, *Phys. Rev. Lett.* **52**, 13 (1984).

10.57. K. H. Rieder, N. Garcia and V. Celli, *Surf. Sci.* **108**, 169 (1981).

10.58. D. S. Kaufman, R. James and T. Engel, *Surf. Sci.* **148**, 72 (1984).

10.59. A. M. J. Huiser, A. Quattropani and H. P. Baltes, *Opt. Comm.* **41**, 149 (1981).

10.60. R. Dandliker and K. Weiss, *Opt. Comm.* **1**, 323 (1970).

10.61. A. J. Devaney, *Opt. Lett.* **7**, 11 (1982).

10.62. M. Nieto-Vesperinas, M. Gea, G. Ross and M. A. Fiddy, in *Optics in Four Dimensions*, M. A. Machado and L. M. Narducci, eds., American Institute of Physics Conference Proceedings, Vol. 65, New York, 1981, p. 679.

10.63. A. J. Devaney, *Ultrasonic Imaging* **4**, 336 (1982).

10.64. A. J. Devaney, *I.E.E.E. Trans. Biomed. Eng.* **30**, 377 (1983).

10.65. R. K. Mueller, M. Kaveh and G. Wade, *Proc. I.E.E.E.* **67**, 567 (1979).

10.66. F. Stenger and S. A. Johnson, *Appl. Math. Notes* **4**, 102 (1979).

10.67. K. Iwata and N. Nagata, *Japan J. Appl. Phys. 14*, Suppl. 14-1, 379 (1974).

10.68. S. J. Norton and M. Linzer, *I.E.E.E. Trans. Biomed. Eng.* **28**, 202 (1981).

10.69. H. Lipson and W. Cochran, *The Determination of Crystal Structures*, Cornell University Press, Ithaca, New York, 1966.

10.70. B. K. Vainhstein, *Diffraction of X-Rays by Chain Molecules*, J. Wiley, New York, 1974.

10.71. P. W. Hawkes, ed., *Computer Processing of Electron Microscopy Images*, Topics in Current Physics, Vol. 13, Springer-Verlag, Berlin, 1978.

10.72. H. Stark, ed., *Image Recovery: Theory and Application*, Academic Press, New York, 1987.

10.73. A. J. Devaney, *I.E.E.E. Trans. Geosci. and Remote Sensing* **22**, 3 (1984).

10.74. A. J. Devaney and G. Beylkin, *Ultrasonic Imaging* **6**, 181 (1984).

10.75. R. P. Porter, *J. Acoust. Soc. Am.* **80**, 1220 (1986).

10.76. C. Esmersoy and B. C. Levy, *Proc. I.E.E.E.* **74**, 466 (1986).

10.77. J. A. De Santo, A. W. Saenz and W. W. Zachary, *Mathematical Methods and Applications of Scattering Theory*, Lecture Notes in Physics, Vol. 130, Springer-Verlag, Berlin, 1980.

10.78. A. J. Devaney and E. Wolf, *Phys. Lett. A* **89**, 269 (1982).

10.79. H. E. Moses, *Phys. Rev.* **102**, 559 (1956).

10.80. R. T. Prosser, *J. Math. Phys.* **10**, 1819 (1969); **17**, 1775 (1976); **21**, 2648 (1980).

10.81. A. J. Devaney and M. L. Oristiglio, *Phys. Rev. Lett.* **51**, 237 (1983).

10.82. J. R. Taylor, *Scattering Theory*, J. Wiley, New York, 1972.

10.83. W. H. Weston, *J. Math. Phys.* **20**, 53 (1979); **21**, 758 (1980); **22**, 2523 (1981).

10.84. I. M. Gelfand and B. M. Levitan, *Am. Math. Soc. Trans. Ser. 2* **1**, 253 (1955).

10.85. V. A. Marchenko, *Math. Rev.* **26**, 797 (1978).

10.86. E. Vogelzang, H. A. Ferwerda and D. Yevick, *Opt. Acta* **31**, 541 (1984).

10.87. D. L. Jaggard and K. E. Olson, *J. Opt. Soc. Am. A* **2**, 1932 (1985).

10.88. D. L. Jaggard, Y. Kim, K. I. Schultz and P. V. Frangos, *J. Opt. Soc. Am. A* **4**, 1228 (1987).

10.89. M. Cheney and J. H. Rose, *Inverse Problems* **4**, 435 (1988).

10.90. M. Cheney and G. Kristensson, *Inverse Problems* **4**, 625 (1988).

10.91. W. C. Chew, in *Future Directions of Electromagnetic Research*, J. A. Kong, ed., National Science Foundation Workshop Report, Boston, MA, 1989, p. 327.

10.92. I. J. La Haie, *J. Opt. Soc. Am. A* **2**, 35 (1985); *A* **3**, 1073 (1986).

10.93. M. A. Fiddy, in *Future Directions of Electromagnetic Research*, J. A. Kong, ed., National Science Foundation Workshop Report, Boston, MA, 1989, p. 346.

Author Index

Subject Index